Coffee, Tea, Chocolate, and the Brain

NUTRITION, BRAIN, AND BEHAVIOR

Series Editor: Chandan Prasad, PhD
Professor and Vice Chairman (Research)
Department of Medicine
LSU Health Sciences Center
New Orleans, LA, USA

Series Editorial Advisory Board

Janina R. Galler, MD
Director and Professor of Psychiatry
and Public Health
Center for Behavioral Development
and Mental Retardation
Boston University School of Medicine
Boston, MA, USA

R.C.A. Guedes, MD, PhD
Departmento de Nutrição
Centro de Ciencias Da Saúde
Universidade Federal de Pernambuco
Recife/PE, BRASIL

Gerald Huether, PhD
Department of Psychiatric Medicine
Georg-August-Universitat Gottingen
D-37075 Göttingen, Germany

Abba J. Kastin, MD, DSc
Editor-in-chief, PEPTIDES
Endocrinology Section,
Medical Service, V.A. Medical Center,
601 Perdido Street,
New Orleans, LA, USA

H.R. Lieberman, PhD
Military Performance and Neuroscience
Division,
USARIEM
Natick, MA, USA

Published Titles:

DHEA and the Brain
Edited by Robert Morfin
ISBN 0-415-27585-7

Coffee, Tea, Chocolate, and the Brain

Edited by

Astrid Nehlig

INSERM
Strasbourg, France

CRC PRESS

Boca Raton   London   New York   Washington, D.C.

Library of Congress Cataloging-in-Publication Data

Coffee, tea, chocolate, and the brain / edited by Astrid Nehlig.
     p. ; cm. — (Nutrition, brain, and behavior ; v. 2)
Includes bibliographical references and index.
     ISBN 0-415-30691-4 (hardback : alk. paper)
    1. Caffeine—Physiological effect. 2. Coffee—Physiological effect. 3. Tea—Physiological effect.
4. Chocolate—Physiological effect. 5. Neurochemistry. 6. Brain—Effect of drugs on.
     [DNLM: 1. Brain—drug effects. 2. Coffee—Physiology. 3. Cacao—physiology.
4. Caffeine—pharmacology. 5. Cognition—drug effects. 6. Tea—physiology. WB 438 C674 2004]
I. Nehlig, Astrid. II. Series.
     QP801.C24 C64 2004
     612.8'2—dc21
                                                                                                                                                                   2003011477

This book contains information obtained from authentic and highly regarded sources. Reprinted material is quoted with permission, and sources are indicated. A wide variety of references are listed. Reasonable efforts have been made to publish reliable data and information, but the author and the publisher cannot assume responsibility for the validity of all materials or for the consequences of their use.

Neither this book nor any part may be reproduced or transmitted in any form or by any means, electronic or mechanical, including photocopying, microfilming, and recording, or by any information storage or retrieval system, without prior permission in writing from the publisher.

All rights reserved. Authorization to photocopy items for internal or personal use, or the personal or internal use of specific clients, may be granted by CRC Press LLC, provided that $1.50 per page photocopied is paid directly to Copyright clearance Center, 222 Rosewood Drive, Danvers, MA 01923 USA. The fee code for users of the Transactional Reporting Service is ISBN 0-415-30691-4/04/$0.00+$1.50. The fee is subject to change without notice. For organizations that have been granted a photocopy license by the CCC, a separate system of payment has been arranged.

The consent of CRC Press LLC does not extend to copying for general distribution, for promotion, for creating new works, or for resale. Specific permission must be obtained in writing from CRC Press LLC for such copying.

Direct all inquiries to CRC Press LLC, 2000 N.W. Corporate Blvd., Boca Raton, Florida 33431.

Trademark Notice: Product or corporate names may be trademarks or registered trademarks, and are used only for identification and explanation, without intent to infringe.

Visit the CRC Press Web site at www.crcpress.com

© 2004 by CRC Press LLC

No claim to original U.S. Government works
International Standard Book Number 0-415-30691-4
Library of Congress Card Number 2003011477
Printed in the United States of America 1 2 3 4 5 6 7 8 9 0
Printed on acid-free paper

Preface

This book is the second in the series "Nutrition, Brain and Behavior." The purpose of this series is to provide a forum whereby basic and clinical scientists can share their knowledge and perspectives regarding the role of nutrition in brain function and behavior. The breadth and diversity of the topics covered in this book make it of great interest to specialists working on coffee/caffeine/tea/chocolate research, to nutritionists and physicians, and to anyone interested in obtaining objective information on the consequences of the consumption of coffee, tea, and chocolate on the brain.

Coffee is a very popular beverage, the second most frequently consumed after water. Likewise, tea is a fundamental part of the diet of Asian countries and the U.K. and is becoming progressively more popular in Western countries. Chocolate is also widely consumed all over the world. The pleasure derived from the consumption of coffee, tea, and chocolate is accompanied by a whole range of effects on the brain, which may explain their attractiveness and side effects. Coffee, tea, and chocolate all contain methylxanthines, mainly caffeine, and a large part of their effects on the brain are the result of the presence of these substances.

As part of this series on nutrition, the brain, and behavior, the present book brings new information to the long-debated issue of the beneficial and possible negative effects on the brain from the consumption of coffee, tea, or chocolate. Most of the book is devoted to the effects of coffee or caffeine, which constitute the majority of the literature and research on these topics. Much less is known about the other constituents in roasted coffee or about the effects of tea or chocolate on the brain.

In this book, we have selected world specialists to update our knowledge on the effects of these three methylxanthine-containing substances. Together with a collection of the data on the effects of coffee and caffeine on sleep, cognition, memory and performance, and mood, this book contains specific information on new avenues of research, such as the effect of caffeine on Parkinson's disease, ischemia, and seizures, and on the mostly unknown effects of the chlorogenic acids found in coffee. The effects of caffeine on the stress axis and development of the brain are also updated. Finally, the potential for addiction to coffee, caffeine, and chocolate is debated, as well as both the possible headache-inducing effect of chocolate consumption and the alleviating effect of caffeine on various types of headaches.

Altogether, these updates and new findings are reassuring and rather positive, showing again that moderate coffee, tea, or chocolate consumption has mostly beneficial effects and can contribute to a balanced and healthy diet.

We would like to take this opportunity to thank all the authors for their excellent contributions and cooperation in the preparation of this book.

Astrid Nehlig, Ph.D.
Strasbourg, France
Editor

Chandan Prasad, Ph.D.
New Orleans, Louisiana, USA
Series Editor

Editor

Astrid Nehlig, Ph.D., earned a master's degree in physiology and two Ph.D. degrees in physiology and functional neurochemistry from the scientific University of Nancy, France. She is a research director at the French Medical Research Institute, INSERM, in Strasbourg. Her main research interests are brain metabolism, brain development, the effects of coffee and caffeine on the brain, and temporal lobe epilepsy. She has authored or co-authored approximately 200 articles, books, and book chapters and has been invited to deliver more than 50 lectures at international meetings and research centers. She has received several grants for her work, mainly from the Medical Research Foundation, NATO, and private companies, and a 2002 award from the American Epilepsy Society.

Dr. Nehlig has spent two years in the United States working in a highly recognized neuroimaging laboratory at the National Institute for Mental Health in Bethesda, Maryland. She has led an INSERM research team of 10 to 15 persons for 20 years, resulting in the education of more than 15 Ph.D. students and several postdoctoral fellows. She is on the editorial board of the international journal *Epilepsia* and is a member of the commission of neurobiology of the International League Against Epilepsy and of the French Society of Cerebral Blood Flow and Metabolism. She is also the scientific advisor of PEC (Physiological Effects of Coffee), the European Scientific Association of the Coffee Industry. She acts as an expert for numerous scientific journals and international societies, such as NATO, the British Wellcome Trust, and the Australian Medical Research Institute.

Contributors

Mustafa al'Absi
University of Minnesota School of Medicine
Duluth, Minnesota

Alberto Ascherio
Harvard School of Public Health
Boston, Massachusetts

David Benton
University of Wales
Swansea, Wales

Miguel Casas
Hospital Universitari Vall d'Hebron
Barcelona, Spain

John W. Daly
National Institute of Health
Laboratory of Bioorganic Chemistry
Bethesda, Maryland

Tomas de Paulis
Institute for Coffee Studies
Nashville, Tennessee

Bertil B. Fredholm
Karolinska Institutet
Stockholm, Sweden

Heather Jones
University of Maryland
College Park, Maryland

Monicque M. Lorist
Univeristy of Groningen
Groningen, the Netherlands

William R. Lovallo
University of Oklahoma Health Sciences
  Center
and
VA Medical Center Behavioral Sciences
  Laboratories
Oklahoma City, Oklahoma

Mark Mann
University of Maryland
College Park, Maryland

Peter R. Martin
Institute for Coffee Studies
Nashville, Tennessee

Tetsuo Nakamoto
Louisiana State University Health Sciences
  Center
New Orleans, Louisiana

Astrid Nehlig
INSERM
Strasbourg, France

Amanda Osborne
University of Maryland
College Park, Maryland

Gemma Prat
Hospital Universitari Vall d'Hebron
Barcelona, Spain

Adil Qureshi
Hospital Universitari Vall d'Hebron
Barcelona, Spain

Josep Antoni Ramos-Quiroga
Hospital Universitari Vall d'Hebron
Barcelona, Spain

Lidia Savi
Primary Headache Center
Torino, Italy

Michael A. Schwarzschild
Harvard School of Public Health
Boston, Massachusetts

Jeroen A. J. Schmitt
Universiteit Maastricht
Maastricht, the Netherlands

Barry D. Smith
University of Maryland
College Park, Maryland

Jan Snel
University of Amsterdam
Amsterdam, the Netherlands

Zoë Tieges
University of Amsterdam
Amsterdam, the Netherlands

Martin P. J. van Boxtel
Universiteit Maastricht
Maastricht, the Netherlands

Thom White
University of Maryland
College Park, Maryland

Contents

Chapter 1 Mechanisms of Action of Caffeine on the Nervous System 1
John W. Daly and Bertil B. Fredholm

Chapter 2 Effects of Caffeine on Sleep and Wakefulness: An Update 13
Jan Snel, Zoë Tieges, and Monicque M. Lorist

Chapter 3 Arousal and Behavior: Biopsychological Effects of Caffeine................................35
*Barry D. Smith, Amanda Osborne, Mark Mann,
Heather Jones, and Thom White*

Chapter 4 Coffee, Caffeine, and Cognitive Performance 53
Jan Snel, Monicque M. Lorist, and Zoë Tieges

Chapter 5 Effects of Coffee and Caffeine on Mood and Mood Disorders................................73
Miguel Casas, Josep Antoni Ramos-Quiroga, Gemma Prat, and Adil Qureshi

Chapter 6 Age-Related Changes in the Effects of Coffee on Memory and Cognitive
Performance 85
Martin P. J. van Boxtel and Jeroen A. J. Schmitt

Chapter 7 Neurodevelopmental Consequences of Coffee/Caffeine Exposure 97
Tetsuo Nakamoto

Chapter 8 Caffeine's Effects on the Human Stress Axis..113
Mustafa al'Absi and William R. Lovallo

Chapter 9 Dependence upon Coffee and Caffeine: An Update...133
Astrid Nehlig

Chapter 10 Caffeine and Parkinson's Disease 147
Michael A. Schwarzschild and Alberto Ascherio

Chapter 11 Caffeine in Ischemia and Seizures: Paradoxical Effects of Long-Term
Exposure 165
Astrid Nehlig and Bertil B. Fredholm

Chapter 12 Caffeine and Headache: Relationship with the Effects of Caffeine on Cerebral Blood Flow 175

Astrid Nehlig

Chapter 13 Cerebral Effects of Noncaffeine Constituents in Roasted Coffee 187

Tomas de Paulis and Peter R. Martin

Chapter 14 Can Tea Consumption Protect against Stroke? 197

Astrid Nehlig

Chapter 15 The Biology and Psychology of Chocolate Craving 205

David Benton

Chapter 16 Is There a Relationship between Chocolate Consumption and Headache? 219

Lidia Savi

Index 227

1 Mechanisms of Action of Caffeine on the Nervous System

John W. Daly and Bertil B. Fredholm

CONTENTS

Introduction 1
Potential Sites of Action 2
    Adenosine Receptors: Blockade by Caffeine 2
    Inhibition of Phosphodiesterases by Caffeine 4
    Ion Channels: I. Effects of Caffeine on Calcium 4
    Ion Channels: II. Effects of Caffeine on $GABA_A$ and Glycine Receptors 5
    Other Effects of Caffeine 6
Conclusions 6
References 6

INTRODUCTION

Because of its presence in popular drinks, caffeine is doubtlessly the most widely consumed of all behaviorally active drugs (Serafin, 1996; Fredholm et al., 1999). Although caffeine is the major pharmacologically active methylxanthine in coffee and tea, cocoa and chocolate contain severalfold higher levels of theobromine than caffeine, along with trace amounts of theophylline. Paraxanthine is a major metabolite of caffeine in humans, while theophylline is a minor metabolite. Thus, not only caffeine, but also the other natural methylxanthines are relevant to effects in humans. In animal models, caffeine, theophylline, and paraxanthine are all behavioral stimulants, whereas the effects of theobromine are weak (Daly et al., 1981). Caffeine, theophylline, and theobromine have been or are used as adjuncts or agents in medicinal formulations. Methylxanthines have been used to treat bronchial asthma (Serafin, 1996), apnea of infants (Bairam et al., 1987; Serafin 1996), as cardiac stimulants (Ahmad and Watson, 1990), as diuretics (Eddy and Downes, 1928), as adjuncts with analgesics (Sawynok and Yaksh, 1993; Zhang, 2001), in electroconvulsive therapy (Coffey et al., 1990), and in combination with ergotamine for treatment of migraine (Diener et al., 2002). An herbal dietary supplement containing ephedrine and caffeine is used as an anorectic (Haller et al., 2002). Other potential therapeutic targets for caffeine include diabetes (Islam et al., 1998; Islam, 2002), Parkinsonism (Schwarzschild et al., 2002), and even cancer (Lu et al., 2002). Caffeine has been used as a diagnostic tool for malignant hyperthermia (Larach, 1989). Clinical uses of caffeine have been reviewed (Sawynok, 1995). In the following chapter, we will focus on the actions of caffeine on the nervous system.

POTENTIAL SITES OF ACTION

Three major mechanisms must be considered with respect to the actions of caffeine on the peripheral and central nervous system: (1) blockade of adenosine receptors, in particular A_1- and A_{2A}-adenosine receptors; (2) blockade of phosphodiesterases, regulating levels of cyclic nucleotides; and (3) action on ion channels, in particular those regulating intracellular levels of calcium and those regulated by the inhibitory neurotransmitters γ-aminobutyric acid (GABA) and glycine (Fredholm, 1980; Daly, 1993; Nehlig and Debry, 1994; Fredholm et al., 1997, 1999; Daly and Fredholm, 1998).

Caffeine's effects are biphasic. The stimulatory behavioral effects in humans (and rodents) become manifest with plasma levels of 5 to 20 μM, whereas higher doses are depressant. The only sites of action where caffeine would be expected to have a major pharmacological effect at levels of 5 to 20 μM are the A_1- and the A_{2A}-adenosine receptors, where caffeine is a competitive antagonist (Daly and Fredholm, 1998). Major effects at other sites of action, such as phosphodiesterases (inhibition), GABA and glycine receptors (blockade), and intracellular calcium-release channels (sensitization to activation by calcium) would be expected to require at least tenfold higher *in vivo* levels of caffeine. At such levels, toxic effects of caffeine, often referred to at nonlethal levels as "caffeinism" in humans, become manifest. Convulsions and death can occur at levels above 300 μM. However, it cannot be excluded that subtle effects of 5 to 20 μM caffeine at sites of action other than adenosine receptors might have some relevance to both acute and chronic effects of caffeine. Extensive *in vitro* studies of the actions of caffeine at such sites are usually performed at concentrations of caffeine of 1 mM or more, clearly levels that *in vivo* are lethal.

ADENOSINE RECEPTORS: BLOCKADE BY CAFFEINE

Four adenosine receptors have been cloned and pharmacologically characterized: A_1-, A_{2A}-, A_{2B}-, and A_3-adenosine receptors (Fredholm et al., 2000, 2001a). Of these the A_3-adenosine receptor in rodent species has very low sensitivity to blockade by theophylline, with K_i values of 100 μM or more (Ji et al., 1994). Human A_3-adenosine receptors are somewhat more sensitive to xanthines, but at *in vivo* levels of 5 to 20 μM caffeine will have virtually no effect even on the human A_3 receptors. By contrast, results from rodents and humans show that caffeine binds to A_1, A_{2A}, or A_{2B} receptors with K_d values in the range of 2 to 20 μM (see Fredholm et al., 1999, 2001b). Thus, caffeine at the levels reached during normal human consumption could exert its actions at A_1, A_{2A}, or A_{2B} receptors, but not by blocking A_3 receptors.

If caffeine is to exert its actions by blocking adenosine receptors, a prerequisite is that there be a significant ongoing (tonic) activation of A_1, A_{2A}, or A_{2B} receptors. All the evidence suggests that at these receptors, adenosine is the important endogenous agonist (Fredholm et al., 1999, 2000, 2001b). Only at A_3 receptors does inosine seem to be a potential agonist candidate (Jin et al., 1997; Fredholm et al., 2001b). In his original proposal of P1 (adenosine) and P2 (ATP) receptors, Burnstock (1978) included the provision that the adenosine receptors would be blocked by theophylline, while the ATP receptors would be insensitive to theophylline. However, there have also been reports of ATP responses that are inhibited by theophylline (Silinksy and Ginsberg, 1983; Shinozuka et al., 1988; Ikeuchi et al., 1996; Mendoza-Fernandez et al., 2000). Such effects have been suggested to indicate novel receptors or to be caused by heteromeric association of A_1-adenosine and P2Y receptors (Yoshioka et al., 2001). However, the most parsimonious explanation is that the effects are due to rapid breakdown of ATP to adenosine and actions on classical adenosine receptors (Masino et al., 2002). Therefore, caffeine (as well as theophylline and paraxanthine) should act by antagonizing the actions of endogenous adenosine at A_1, A_{2A}, or A_{2B} receptors. This requires that the endogenous levels be sufficiently high to ensure an ongoing tonic activation. In the case of A_1 and A_{2A} receptors, this requirement is fulfilled, at least at those locations where the receptors are abundantly expressed (Fredholm et al., 1999, 2001a,b). By contrast, A_{2B} receptors may not be expressed at sufficiently high abundance to ensure tonic activation by endogenous adenosine during physiological conditions. It must, however, be remembered that the potency of an agonist is not a fixed value but depends on factors such as receptor number and also the effect studied (Kenakin, 1995). It is therefore interesting to note that when activation of mitogen-activated protein kinases is studied, adenosine is as potent on A_{2B} as on A_1 and A_{2A} receptors (Schulte and Fredholm, 2000). Hence, the idea that A_{2B} receptors are "low-affinity" receptors activated only at supraphysiological levels of adenosine may not be absolutely true. Nevertheless, the available evidence suggests that most of the effects of caffeine are best explained by blockade of tonic adenosine activation of A_1 and A_{2A} receptors.

In chapters to follow, the relative roles of the different adenosine receptor subtypes in mediating *in vivo* effects of caffeine will be discussed. Here it will suffice to point out that blockade of A_1 receptors by caffeine could remove either a G_i input to adenylyl cyclase or tonic effects mediated through $G_{\beta,\gamma}$ on calcium release, potassium channels, and voltage-sensitive calcium channels. Conversely, blockade of A_{2A}-adenosine receptors could remove stimulatory input to adenylyl cyclase. In the complex neuronal circuitry of the central nervous system, the ultimate effects will depend on the site and nature of physiological input by endogenous adenosine. Hints about the biological roles of adenosine are also provided by the distribution of the receptors.

Adenosine A_1 receptors are found all over the brain and spinal cord (Fastbom et al., 1986; Jarvis et al., 1987; Weaver, 1996; Svenningsson et al., 1997a; Dunwiddie and Masino, 2001). In the adult rodent and human brain, levels are particularly high in the hippocampus, cortex, and cerebellum. By contrast, A_{2A} receptors have a much more restricted distribution, being present in high amounts only in the dopamine-rich regions of the brain, including the nucleus caudatus, putamen, nucleus accumbens, and tuberculum olfactorium (Jarvis et al., 1989; Parkinson and Fredholm, 1990; Svenningsson et al., 1997b, 1998, 1999a; Rosin et al., 1998). They are virtually restricted to the GABAergic output neurons that compose the so-called indirect pathway and that also are characterized by expressing enkephalin and dopamine D_2 receptors. There is, indeed, very strong evidence for a close functional relationship between A_{2A} and D_2 receptors (Svenningsson et al., 1999a).

The adenosine A_1 receptors appear to play two major roles: (1) activation of potassium channels leading to hyperpolarization and to decreased rates of neuronal firing and (2) inhibition of calcium channels leading to decreased neurotransmitter release. This will lead to inhibition of excitatory neurotransmission, and there is good evidence for interactions between A_1 and NMDA receptors (Harvey and Lacey, 1997; de Mendonça and Ribeiro, 1993). Adenosine A_{2A} receptors regulate the function of GABAergic neurons of the basal ganglia. The effects are opposite those of dopamine acting at D_2 receptors. It is now clear that these receptors are predominantly involved in the stimulant effects of caffeine (Svenningsson et al., 1995; El Yacoubi et al., 2000).

The two caffeine metabolites, theophylline and paraxanthine, are even more potent inhibitors of adenosine receptors than the parent compound (Svenningsson et al., 1999a; Fredholm et al., 2001b). Therefore, the weighted sum of all of them must be considered when evaluating the effective concentration of antagonist at the adenosine receptors.

Investigation of roles of adenosine receptors has been greatly facilitated by the development of a wide variety of potent and/or selective antagonists. Some are xanthines, deriving from caffeine and theophylline as lead compounds, while others are based on other compounds containing instead of a purine other heterocyclic ring systems (Hess, 2001). In addition, the development of receptor knock-out mice has been instrumental in our current understanding. Thus, experiments using A_{2A} knock-out mice have conclusively shown that blockade of striatal A_{2A} receptors is the reason why caffeine can induce its behaviorally stimulant effects (Ledent et al., 1997; El Yacoubi et al., 2000) and the mechanisms involved have been clarified in considerable molecular detail (Svenningsson et al., 1999b; Lindskog et al., 2002). In addition, A_{2A} knock-out mice showed increased aggressiveness and anxiety (Ledent et al., 1997), a characteristic shared by A_1 knock-out mice (Johansson et al., 2001). The fact that elimination of either receptor leads to anxiety could provide the basis for the well-known fact that anxiety is produced by high doses of caffeine in humans (Fredholm et al., 1999); whereas A_{2A} knock-out mice showed hypoalgesia, A_1 knock-out mice showed hyperalgesia. Finally, using A_1 and A_{2A} knock-out mice it was shown that at least part of the behaviorally depressant effect of higher doses of caffeine depends on a mechanism other than adenosine receptor blockade (Halldner-Henriksson et al., 2002).

INHIBITION OF PHOSPHODIESTERASES BY CAFFEINE

The potentiation of a hormonal response by caffeine or theophylline (Butcher and Sutherland, 1962) was considered for years as a criterion for involvement of cyclic AMP in the response, and such xanthines became the prototypic phosphodiesterase inhibitors. Both caffeine and theophylline now are considered rather weak and nonselective phosphodiesterase inhibitors, requiring concentrations far above 5 to 20 μM for significant inhibition of such enzymes (Choi et al., 1988). In 1970, it was demonstrated that caffeine/theophylline blocked adenosine-mediated cyclic AMP formation (Sattin and Rall, 1970), and attention shifted to the importance of adenosine receptor blockade in the effects of alkylxanthines. Agents have been sought that would be selective either towards phosphodiesterases or towards adenosine receptors (Daly, 2000). It has been proposed that the behavioral depressant effects of xanthines are due to inhibition of phosphodiesterases, while the behavioral stimulation by caffeine and other xanthines is due to blockade of adenosine receptors (Choi et al., 1988; Daly, 1993). Indeed, many nonxanthine phosphodiesterase inhibitors are behavioral depressants (Beer et al., 1972). The depressant effects of high concentrations of caffeine will depend, as with any centrally active agent, on the specific neuronal pathways that are affected. The central pathways where there might be a further elevation of cyclic AMP, due to inhibition of phosphodiesterase by caffeine, have not been defined. A limited number of xanthines and other agents that are selective towards different subtypes of phosphodiesterases are available (Daly, 2000). Unfortunately, many have other activities, such as blockade of adenosine receptors, that decrease their utility as research tools.

ION CHANNELS: I. EFFECTS OF CAFFEINE ON CALCIUM

Caffeine at high concentrations has been reported to have a multitude of effects on calcium channels, transporters, and modulatory sites (Daly, 2000). Caffeine has been known for more than four decades to cause muscle contracture due to release of intracellular calcium. It is now known that caffeine enhances the calcium-sensitivity of a cyclic ADP-ribose-sensitive calcium release channel, the so-called ryanodine-sensitive channel, thereby causing release of intracellular calcium from storage sites in the sarcoplasmic reticulum of muscle and the endoplasmic reticulum of muscle and other cells, including neuronal cells (McPherson et al., 1991; Galione, 1994). Caffeine has been extensively used as a research tool to investigate *in vitro* the role of release of calcium stores through what is now called the ryanodine-sensitive receptor. In pancreatic β-cells, caffeine-induced calcium release appears to depend on elevated cAMP (Islam et al., 1998). In most cases, significant release of calcium from storage sites in cells or in isolated sarcoplasmic reticulum has required concentrations of caffeine of 1 mM or higher. However, it is uncertain whether slight acute or chronic effects of low concentrations of caffeine on intracellular calcium might have a significant functional impact on the central nervous system. Caffeine targets not only the ryanodine-sensitive calcium-release channel, but has also been reported to have effects on several other entities that are involved in calcium homeostasis (Daly, 2000). These include inhibition of IP_3-induced release of calcium from intracellular storage sites (Parker and Ivorra, 1991; Brown et al., 1992; Missiaen et al., 1992, 1994; Bezprozvanny et al., 1994; Ehrlich et al., 1994; Hague et al., 2000; Sei et al., 2001; however, see Teraoka et al., 1997) and/or inhibition of receptor-mediated IP_3 formation (Toescu et al., 1992; Seo et al., 1999). Both require millimolar concentrations of caffeine. Caffeine at high millimolar concentrations appears to elicit influx of calcium in several cell types (Avidor et al., 1994; Guerrero et al., 1994; Ufret-Vincenty et al., 1995; Sei et al., 2001; Cordero and Romero, 2002); the nature of the channels is unknown. A functional coupling of the caffeine-sensitive calcium-release channels and the voltage-sensitive L-type calcium channels has been reported in neurons (Chavis et al., 1996). Caffeine at millimolar concentrations has been reported to inhibit L-type calcium channels (Kramer et al., 1994; Yoshino et al., 1996). Evidence suggesting both activation and inhibition of L-type calcium channels by caffeine has been reported for pancreatic β-cells, and the former was attributed to inhibition of K_{ATP} channels (Islam et al., 1995). Caffeine at high concentrations reduces uptake of calcium into cardiac mitochondria (Sardão et al., 2002).

As yet, no xanthines have been developed with high potency/selectivity for the ryanodine-sensitive calcium release channels for use as tools to probe possible significance of the inhibition of this channel by caffeine (see Daly, 2000; Shi et al., 2003 and references therein). Ryanodine, 4-chloro-*m*-cresol, and eudistomins represent other compounds that activate ryanodine receptors, but ryanodine and the cresol are too toxic for *in vivo* studies, while eudistomins have poor solubility and hence availability for *in vivo* studies.

ION CHANNELS: II. EFFECTS OF CAFFEINE ON GABA$_A$ AND GLYCINE RECEPTORS

Caffeine has been known for two decades to interact with GABA$_A$ receptors, based primarily on the inhibition by caffeine and theophylline of binding of benzodiazepine agonists to that receptor in brain membranes (Marangos et al., 1979). The binding of a benzodiazepine antagonist, RO15-1788, also is inhibited (Davies et al., 1984). However, the IC_{50} values for caffeine were about 350 μM at such benzodiazepine sites. A variety of evidence suggests that blockade of GABA$_A$ receptors is responsible for the convulsant activity of high doses of caffeine (Amabeoku, 1999; also see Daly, 1993) but is not involved in behavioral stimulation observed at low dosages of caffeine. There are other reported effects of caffeine and/or theophylline on binding of ligands to the GABA$_A$ receptor, including reversal of the inhibitory effect of GABA on binding of a convulsant, (+/−)-*t*-butylcyclo-phosphothionate (TBPS) (Squires and Saederup, 1987), a slight stimulatory effect on binding of TBPS (Shi et al., 2003), and an inhibition of binding of GABA (Ticku and Birch, 1980) or of the GABA antagonist SR-95531 (Shi et al., 2003) to the GABA site. It appears likely that caffeine at high concentrations affects GABA$_A$ receptors in a complex, allosteric manner. Functionally, caffeine at 50 μM was reported to inhibit the chloride flux elicited in synaptoneurosomes by a GABA agonist, muscimol (Lopez et al., 1989). At a higher 100 μM concentration, caffeine had no effect, suggestive of a bell-shaped dose-response curve. In the same study with mice, relatively low doses of caffeine (20 mg/kg) appeared to reduce GABA$_A$ receptor-mediated responses, measured *ex vivo* with muscimol in synaptoneurosomes. Functional inhibition of GABA$_A$ receptors, in such studies, might involve inhibition of the GABA receptor by elevated calcium, resulting from caffeine-induced release from intracellular calcium stores (Desaulles et al., 1991; Kardos and Blandl, 1994). In hippocampal neurons inhibition of GABA receptor-elicited chloride currents by millimolar concentrations of caffeine did not appear to involve elevation of calcium (Uneyama et al., 1993). Caffeine was almost tenfold more potent in inhibiting glycine-elicited chloride currents with an IC_{50} of 500 μM. Further studies on inhibition of glycine responses do not seem to have been forthcoming. *In toto*, the low potency of caffeine at GABA$_A$ receptors makes it unlikely that such effects contribute to the behavioral stimulant effects of caffeine. However, it is possible that subtle blocking effects at GABA receptors could contribute to both acute and chronic effects by affecting the role of inhibitory GABA- and glycine-neuronal pathways. Apparent alterations in GABAergic activities have been reported after chronic caffeine intake in rodents (Mukhopadhyay and Poddar, 1998, 2000). Chronic caffeine intake does result in changes in receptors for several neurotransmitters, including GABA$_A$ receptors (Shi et al., 1993), but whether such alterations are the result of direct effects or are "downstream" of effects at adenosine receptors is unknown. No xanthines selective for GABA$_A$ receptors have been forthcoming, and other agents that interact with the GABA$_A$ receptor channel complex do not appear suitable as research tools to investigate the unique functional significance of complex interactions of caffeine with GABA receptors.

OTHER EFFECTS OF CAFFEINE

There are a wide range of other effects of caffeine on ion channels (Reisser et al., 1996; Schroder et al., 2000; Teramoto et al., 2000; Kotsias and Venosa, 2001), enzymes (see Daly, 1993), including lipid and protein kinases (Foukas et al., 2002) and cell cycles (Jiang et al., 2000; Qi et al., 2002), but virtually all require high concentrations of caffeine (see Daly, 1993, 2000 and references therein). Such effects are probably not relevant to the behavioral stimulant properties of caffeine that occur at plasma levels of 5 to 20 μM.

There are peripheral effects of caffeine, some perhaps mediated through adenosine receptors and others through inhibition of phosphodiesterase, that could indirectly affect the function of the central nervous system. Conversely, certain peripheral effects of caffeine may be centrally mediated. The elevation of plasma levels of epinephrine by moderate doses of caffeine in humans was noted as early as the 1960s (see Robertson et al., 1978). The released epinephrine appears likely to be responsible for the caffeine-elicited reduction in insulin sensitivity in humans (Keijzers et al., 2002; Thong and Graham, 2002). The mechanism by which caffeine elicits release of epinephrine from adrenal gland appears likely to be due to increases in sympathetic input, since direct effects of caffeine on release of catecholamines from adrenal chromaffin cells requires millimolar concentrations (Ohta et al., 2002). Thus, direct effects on release of epinephrine from the adrenal gland seem unlikely in human studies. Caffeine also increases free fatty acids (Kogure et al., 2002; Thong and Graham, 2002), presumably in part through blockade of A_1-adenosine receptors on adipocytes. Theophylline has been proposed to induce histone deacetylase activity, thereby reducing gene transcription and, for instance, cytokine-mediated inflammatory responses, apparently by mechanisms not involving adenosine receptors or inhibition of phosphodiesterases (Ito et al., 2002). *In vivo* effects of caffeine on expression of nitric oxide synthetase and Na^+/K^+ ATPase in rat kidney have been reported (Lee et al., 2002). Whether there are similar effects in the central nervous system is unknown. Caffeine, in addition to increasing plasma epinephrine, increases corticosterone and renin (Robertson et al., 1978; Uhde et al., 1984), an effect often associated with stress (see Henry and Stephens, 1980).

CONCLUSIONS

Caffeine and other methylxanthines are potentially able to affect a large number of molecular targets. Nevertheless, the current best evidence indicates that the only effect in the central nervous system that is relevant at lower doses of caffeine is blockade of A_1 and A_{2A} receptors. Higher doses that are related to toxicity and depressant effects appear to exert their effects, at least in part, by mechanisms other than adenosine receptor blockade.

REFERENCES

Ahmad, R.A. and Watson, R.D. (1990) Treatment of postural hypotension. A review. *Drugs*, 39, 74–85.
Amabeoku, G.J. (1999) Gamma-aminobutyric acid and glutamic acid receptors may mediate theophylline-induced seizures in mice. *General Pharmacology*, 32, 365–372.
Avidor, T., Clementi, E., Schwartz, L. and Atlas, D. (1994) Caffeine-induced transmitter release is mediated via ryanodine-sensitive channel. *Neuroscience Letters*, 165, 133–136.
Bairam, A., Boutroy, M.J., Badonnel, Y. and Vert, P. (1987) Theophylline versus caffeine: comparative effects in treatment of idiopathic apnea in the preterm infant. *Journal of Pediatrics*, 110, 636–639.
Beer, B., Chasin, M., Clody, D.E., Vogel, J.R. and Horovitz, Z.P. (1972) Cyclic adenosine monophosphate phosphodiesterase in brain: effect on anxiety. *Science*, 176, 428–430.
Bezprozvanny, I., Bezprozvannaya, S. and Ehrlich B.E. (1994) Caffeine-induced inhibition of inositol (1,4,5)-trisphosphate-gated calcium channels from cerebellum. *Molecular Biology of the Cell*, 5, 97–103.

Brown, G.R., Sayers, L.G., Kirk, C.J., Michell, R.H. and Michelangeli, F. (1992) The opening of the inositol 1,4,5-triphosphate-sensitive Ca channel in rat cerebellum is inhibited by caffeine. *Biochemical Journal*, 282, 309–312.

Burnstock, G. (1978) A basis for distinguishing two types of purinergic receptors, in *Cell Membrane Receptors for Drugs and Hormones: A Multidisciplinary Approach*, Straub, R.W. and Bolis, I.L., Eds., Raven Press, New York, pp. 107–118.

Butcher, R.W. and Sutherland, E.W. (1962) Adenosine 3', 5'-phosphate in biological materials. I. Purification and properties of cyclic 3', 5'-nucleotide phosphodiesterase and use of this enzyme to characterize adenosine 3', 5'-phosphate in human urine. *Journal of Biological Chemistry*, 237, 1244–1250.

Chavis, P., Fagni, L., Lansman, J.B. and Bockaert, J. (1996) Functional coupling between ryanodine receptors and L-type calcium channels in neurons. *Nature*, 382, 719–722.

Choi, O.H., Shamim, M.T., Padgett, W.L. and Daly, J.W. (1988) Caffeine and theophylline analogues: correlation of behavioral effects with activity as adenosine receptor antagonists and as phosphodiesterase inhibitors. *Life Sciences*, 43, 387–398.

Coffey, C.E., Figiel, S. and Weiner, R.D. (1990) Caffeine augmentation of ECT. *American Journal of Psychiatry*, 147, 579–585.

Cordero, J.F. and Romero, P.J. (2002) Caffeine activates a mechanosensitive Ca^{2+} channel in human red cells. *Cell Calcium*, 31, 189–200.

Daly, J.W. (1993) Mechanism of action of caffeine, in *Caffeine, Coffee and Health*, Garattini, S., Ed., Raven Press, New York, pp. 97–150.

Daly, J.W. (2000) Alkylxanthines as research tools. *Journal of the Autonomic Nervous System*, 81, 44–52.

Daly, J.W. and Fredholm, B.B. (1998) Caffeine: an atypical drug of dependence. *Drug and Alcohol Dependence*, 51, 199–206.

Daly, J.W., Bruns, R.F. and Snyder, S.H. (1981) Adenosine receptors in the central nervous system: relationship to the central actions of methylxanthines. *Life Sciences*, 28, 2083–2097.

Davies, L.P., Chow, C.E. and Johnston, G.A.R. (1984) Interaction of purines and related compounds with photo affinity-labelled benzodiazepine receptors at rat brain membranes. *European Journal of Pharmacology*, 97, 325–329.

de Mendonça, A. and Ribeiro, J.A. (1993) Adenosine inhibits the NMDA receptor-mediated excitatory postsynaptic potential in the hippocampus. *Brain Research*, 606, 351–356.

Desaulles, E., Boux, O. and Feltz, P. (1991) Caffeine-induced Ca^{2+} release inhibits $GABA_A$ responsiveness in rat identified native primary afferents. *European Journal of Pharmacology*, 203, 11137–11140.

Diener, H.-C., Jansen, J.-P., Reches, A., Pascual, J., Pitei, D. and Steiner, T.J. (2002) Efficacy, tolerability and safety of oral eletriptan and ergotamine plus caffeine (cafergot) in the acute treatment of migraine: a multicentre, randomized, double-blind, placebo-controlled comparison. *European Neurology*, 47, 99–107.

Dunwiddie, T.V. and Masino, S.A. (2001) The role and regulation of adenosine in the central nervous system. *Annual Review of Neuroscience*, 24, 31–55.

Eddy, N.B. and Downes, A.W. (1928) Tolerance and cross-tolerance in the human subject to the diuretic effect of caffeine, theobromine and theophylline. *Journal of Pharmacology and Experimental Therapeutics*, 33, 167–174.

Ehrlich, B.E., Kaftan, E., Bezprozvannaya, S. and Bezprozvanny, I. (1994) The pharmacology of intracellular Ca^{2+}-release channels. *Trends in Pharmacological Sciences*, 15, 145–149.

El Yacoubi, M., Ledent, C., Menard, J.F., Parmentier, M., Costentin, J. and Vaugeois, J.M. (2000) The stimulant effects of caffeine on locomotor behaviour in mice are mediated through its blockade of adenosine A(2A) receptors. *British Journal of Pharmacology*, 129, 1465–1473.

Fastbom, J., Pazos, A., Probst, A. and Palacios, J.M. (1986) Adenosine A_1 receptors in human brain: characterization and autoradiographic visualization. *Neuroscience Letters*, 65, 127–132.

Foukas, L.C., Daniele, N., Ktori, C., Andersen, K.E., Jensen, J. and Shepherd, P.R. (2002) Direct effects of caffeine and theophylline on p110δ and other phosphoinositide kinases. Differential effects on lipid kinase and protein kinase activities. *Journal of Biological Chemistry*, 277, 37124–37130.

Fredholm, B.B. (1980) Are methylxanthine effects due to antagonism of endogenous adenosine? *Trends in Pharmacological Sciences*, 1, 129–132.

Fredholm, B.B., Arslan, G., Halldner, L., Kull, B., Schulte, G. and Wasserman, W. (2000) Structure and function of adenosine receptors and their genes. *Naunyn-Schmiedeberg's Archives of Pharmacology*, 362, 364–374.

Fredholm, B.B., Arslan, G., Johansson, B., Kull, B. and Svenningsson, P. (1997) Adenosine A_{2A} receptors and the actions of caffeine, in *Role of Adenosine in the Nervous System*, Okada, Y., Ed., Elsevier Science, Amsterdam, pp. 51–74.

Fredholm, B.B., Bättig, K., Holmén, J., Nehlig, A. and Zvartau, E. (1999) Actions of caffeine in the brain with special reference to factors that contribute to its widespread use. *Pharmacological Reviews*, 51, 83–153.

Fredholm, B.B., Ijzerman, A.P., Jacobson, K.A., Klotz, K.N. and Linden, J. (2001a) International Union of Pharmacology. XXV. Nomenclature and classification of adenosine receptors. *Pharmacological Reviews,* 53, 527–552.

Fredholm, B.B., Irenius, E., Kull, B. and Schulte, G. (2001b) Comparison of the potency of adenosine as an agonist at human adenosine receptors expressed in Chinese hamster ovary cells. *Biochemical Pharmacology,* 61, 443–448.

Galione, A. (1994). Cyclic ADP-ribose, the ADP-ribosyl cyclase pathway and calcium signaling. *Molecular and Cellular Endocrinology*, 98, 125–131.

Guerrero, A., Fay, F.S. and Singer, J.J. (1994) Caffeine activates a Ca^{2+}-permeable, nonselective cation channel in smooth muscle cells. *Journal of General Physiology*, 104, 375–394.

Hague, F., Matifat, F., Brûlé, G. and Collin, T. (2000) Caffeine exerts a dual effect on capacitative calcium entry in *Xenopus* oocytes. *Cellular Signalling*, 12, 31–35.

Halldner-Henriksson, L.M., Johansson, B., Dahlberg, V., Åden, U. and Fredholm, B.B. (2002) Maintained biphasic effect of caffeine in the adenosine A_1 receptor knock-out mouse. *Society of Neuroscience Abstracts,* No. 783.20.

Haller, C.A., Jacob, P., III and Benowitz, N.L. (2002) Pharmacology of ephedra alkaloids and caffeine after single-dose dietary supplement use. *Clinical Pharmacology and Therapeutics*, 71, 421–432.

Harvey, J. and Lacey, M.G. (1997) A postsynaptic interaction between dopamine D1 and NMDA receptors promotes presynaptic inhibition in the rat nucleus accumbens via adenosine release. *The Journal of Neuroscience,* 17, 5271–5280.

Henry, J.P. and Stephens, P.M. (1980) Caffeine as an intensifier of stress-induced hormonal and pathophysiological changes in mice. *Pharmacology, Biochemistry and Behavior*, 13, 719–727.

Hess, S. (2001) Recent advances in adenosine receptor antagonist research. *Expert Opinion in Therapeutics Patents*, 11, 1533–1561.

Ikeuchi, Y., Nishizaki, T., Mori, M. and Okada, Y. (1996) Adenosine activates the K^+ channel and enhances cytosolic Ca^{2+} release via a P_{2Y} purinoceptor in hippocampal neurons. *European Journal of Pharmacology*, 304, 191–199.

Islam, M.S. (2002). The ryanodine receptor calcium channel of β-cells. Molecular regulation and physiological significance. *Diabetes*, 51, 1299–1309.

Islam, M.S., Larsson, O., Nilsson, T. and Berggren, P.O. (1995) Effects of caffeine on cytoplasmic free Ca^{2+} concentration in pancreatic β-cells are mediated by interaction with ATP-sensitive K^+ channels and L-type voltage-gated Ca^{2+} channels but not the ryanodine receptor. *Biochemical Journal*, 306, 679–686.

Islam, M.S., Leibiger, I., Leibiger, B., Rossi, D., Sorrentino, V., Ekstrom, T.J. et al. (1998) *In situ* activation of the type 2 ryanodine receptor in pancreatic beta cells requires cAMP-dependent phosphorylation. *Proceedings of the National Academy of Sciences of the USA*, 95, 6145–6150.

Ito, K., Lim, S., Caramori, G., Cosio, B., Chung, K.F., Adcock, I.M. and Barnes, P.J. (2002) A molecular mechanism of action of theophylline: induction of histone deacetylase activity to decrease inflammatory gene expression. *Proceedings of the National Academy of Sciences of the USA*, 99, 8921–8926.

Jarvis, M.F., Jackson, R.H. and Williams, M. (1989) Autoradiographic characterization of high affinity adenosine A_2 receptors in the rat brain. *Brain Research,* 484, 111–118.

Jarvis, M.F., Jacobson, Y.V. and Williams, M. (1987) Autoradiographic localization of adenosine A1 receptors in rat brain using [3H]XCC, a functionalized congener of 1,3-dipropylxanthine. *Neuroscience Letters*, 81, 69–74.

Ji, X.D., von Lubitz, D., Olah, M.E., Stiles, G.L. and Jacobson, K.A. (1994) Species differences in ligand affinity at central A_3-adenosine receptors. *Drug Development Research*, 33, 51–59.

Jiang, X., Lim, L.Y., Daly, J.W., Li, A.H., Jacobson, K.A. and Roberge, M. (2000) Structure-activity relationship for G2 checkpoint inhibition by caffeine analogs. *International Journal of Oncology*, 16, 971–978.

Jin, X., Shepherd, R.K., Duling, B.R. and Linden, J. (1997) Inosine binds to A3 adenosine receptors and stimulates mast cell degranulation. *Journal of Clinical Investigation,* 100, 2849–2857.

Johansson, B., Halldner, L., Dunwiddie, T.V., Masino, S.A., Poelchen, W., Giménez-Llort, L. et al. (2001) Hyperalgesia, anxiety, and decreased hypoxic neuroprotection in mice lacking the adenosine A_1 receptor. *Proceedings of the National Academy of Sciences of the USA*, 98, 9407–9412.

Kardos, J. and Blandl, T. (1994) Inhibition of a gamma aminobutyric acid A receptor by caffeine. *Neuroreport*, 5, 1249–1252.

Keijzers, G.B., DeGalan, B.E., Tack, C.J. and Smits, P. (2002) Caffeine can decrease insulin sensitivity in humans. *Diabetes Care*, 25, 364–369.

Kenakin, T. (1995) Agonist-receptor efficacy I: mechanisms of efficacy and receptor promiscuity. *Trends in Pharmacological Sciences*, 16, 188–192.

Kogure, A., Sakane, N., Takakura, Y., Umekawa, T., Yoshioka, K., Nishino, H. et al. (2002) Effects of caffeine on the uncoupling protein family in obese yellow KK mice. *Clinical and Experimental Pharmacology and Physiology*, 29, 391–394.

Kotsias, B.A. and Venosa, R.A. (2001) Caffeine-induced depolarization in amphibian skeletal muscle fibres: role of Na^+/Ca^{2+} exchange and K^+ release. *Acta Physiologica Scandinavica*, 171, 459–466.

Kramer, R.H., Mokkapatti, R. and Levitan, R.S. (1994) Effects of caffeine on intracellular calcium, calcium current and calcium-dependent potassium current in anterior pituitary GH_3 cells. *Pflügers Archives*, 426, 12–20.

Larach, M.G. (1989) Standardization of the caffeine halothane muscle contracture test. *Anesthesia and Analgesia*, 69, 511–515.

Ledent, C., Vaugeois, J.M., Schiffmann, S.N., Pedrazzini, T., El Yacoubi, M., Vanderhaeghen, J.J. et al. (1997) Aggressiveness, hypoalgesia and high blood pressure in mice lacking the adenosine A2A receptor. *Nature*, 388, 674–678.

Lee, J.L., Ha, J.H., Kim, S., Oh, Y. and Kim, S.W. (2002) Caffeine decreases the expression of Na^+/K^+-ATPase and the type 3 Na^+/H^+ exchanger in rat kidney. *Clinical and Experimental Pharmacology and Physiology*, 29, 559–563.

Lindskog, M., Svenningsson, P., Pozzi, L., Kim, Y., Fienberg, A.A., Bibb, J.A. et al. (2002) Involvement of DARPP-32 phosphorylation in the stimulant action of caffeine. *Nature*, 418, 774–778.

Lopez, F., Miller, L.G., Greenblatt, D.J., Kaplan, G.B. and Shader, R.I. (1989) Interaction of caffeine with the $GABA_A$ receptor complex: alterations in receptor function but not ligand binding. *European Journal of Pharmacology*, 172, 453–459.

Lu, Y.-P., Lou, Y.-P., Xie, J.-G., Peng, Q.-Y., Liao, J., Yang, C.S. et al. (2002) Topical applications of caffeine or (–)-epigallocatechin gallate (EGCG) inhibit carcinogenesis and selectively increase apoptosis in UVB-induced skin tumors in mice. *Proceedings of the National Academy of Sciences of the USA*, 99, 12455–12460.

Marangos, P.J., Paul, S.M., Parma, A.M., Goodwin, F.K., Syapin, P. and Skolnick, P. (1979) Purinergic inhibition of diazepam binding to rat brain (in vitro). *Life Sciences*, 24, 851–858.

Masino, S.A., Diao, L., Illes, P., Zahniser, N.R., Larson, G.A., Johansson, B. et al. (2002) Modulation of hippocampal glutamatergic transmission by ATP is dependent on adenosine A_1 receptors. *Journal of Pharmacology and Experimental Therapeutics*, 303, 356–363.

McPherson, P.S., Kim, Y.-K., Valdivia, H., Knudson, C.M., Takekura, H., Franzini-Armstrong, C. et al. (1991) The brain ryanodine receptor: a caffeine-sensitive calcium release channel. *Neuron*, 7, 17–25.

Mendoza-Fernandez, V., Andrew, R.D. and Barajas-Lopez, C. (2000) ATP inhibits glutamate synaptic release by acting at P2Y receptors in pyramidal neurons of hippocampal slices. *Journal of Pharmacology and Experimental Therapeutics*, 293, 172–179.

Missiaen, L., Parys, J.B., DeSmedt, H., Himpens, B. and Casteels, R. (1994) Inhibition of inositol trisphosphate-induced calcium release is prevented by ATP. *Biochemical Journal*, 300, 81–84.

Missiaen, L., Taylor, C.W. and Berridge, M.J. (1992) Luminal Ca^{2+} promoting spontaneous Ca^{2+} release from inositol triphosphate-sensitive stores in rat hepatocytes. *Journal of Physiology*, 455, 623–640.

Mukhopadhyay, S. and Poddar, M.K. (1998) Is GABA involved in the development of caffeine tolerance? *Neurochemical Research*, 23, 63–68.

Mukhopadhyay, S. and Poddar, M.K. (2000) Long-term caffeine inhibits Ehrlich ascites carcinoma cell-induced induction of central GABAergic activity. *Neurochemical Research*, 25, 1457–1463.

Nehlig, A. and Debry, G. (1994) Effects of coffee on the central nervous system, in *Coffee and Health*, Debry, G. Ed., John Libbey, London, pp. 157–249.

Ohta, T., Wakade, A.R., Yonekubo, K. and Ito, S. (2002) Functional relation between caffeine- and muscarine-sensitive Ca^{2+} stores and no Ca^{2+} releasing action of cyclic adenosine diphosphate-ribose in guinea pig adrenal chromaffin cells. *Neuroscience Letters*, 326, 167–170.

Parker, I. and Ivorra, I. (1991) Caffeine inhibits inositol triphosphate-mediated liberation of intracellular calcium in *Xenopus* oocytes. *Journal of Physiology*, 433, 229–240.

Parkinson, F.E. and Fredholm, B.B. (1990) Autoradiographic evidence for G-protein coupled A_2-receptors in rat neostriatum using [^{3}H]-CGS 21680 as a ligand. *Naunyn-Schmiedeberg's Archives of Pharmacology*, 342, 85–89.

Qi, W., Qiao, D. and Martinez, J.D. (2002) Caffeine induces TP53-independent G_1-phase arrest and apoptosis in human lung tumor cells in a dose-dependent manner. *Radiation Research*, 157, 166–174.

Reisser, M.A., D'Souza, T. and Dryer, S.E. (1996) Effects of caffeine and 3-isobutyl-1-methylxanthine on voltage-activated potassium currents in vertebrate neurons and secretory cells. *British Journal of Pharmacology*, 118, 2145–2151.

Robertson, D., Frölich, J.C., Carr, R.K., Watson, J.T., Hollifield, J.W., Shand, D.G. and Oates, J.A. (1978) Effects of caffeine on plasma renin activity, catecholamines and blood pressure. *New England Journal of Medicine*, 298, 181–186.

Rosin, D.L., Robeva, A., Woodard, R.L., Guyenet, P.G. and Linden, J. (1998) Immunohistochemical localization of adenosine A2A receptors in the rat central nervous system. *Journal of Comparative Neurology*, 401, 163–186.

Sardão, V.A., Oliveira, P.J. and Moreno, A.J.M. (2002) Caffeine enhances the calcium-dependent cardiac mitochondrial permeability transition: relevance for caffeine toxicity. *Toxicology and Applied Pharmacology*, 179, 50–56.

Sattin, A. and Rall, T.W. (1970) The effect of adenosine and adenine nucleotides on the adenosine 3',5'-phosphate content of guinea pig cerebral cortex slices. *Molecular Pharmacology*, 69, 13–23.

Sawynok, J. (1995) Pharmacological rationale for the clinical use of caffeine. *Drugs*, 49, 37–50.

Sawynok, J. and Yaksh, T.L. (1993) Caffeine as an analgesic adjuvant: a review of pharmacology and mechanisms of action. *Pharmacological Reviews*, 45, 43–85.

Schroder, R.L., Jensen, B.S., Strobaek, D., Olesen, S.-P. and Christophersen, P. (2000) Activation of the human, intermediate-conductance, Ca^{2+}-activated K^+ channel by methylxanthines. *Pflügers Archives*, 440, 809–818.

Schulte, G. and Fredholm, B.B. (2000) Human adenosine A_1, A_{2A}, A_{2B}, and A_3 receptors expressed in Chinese hamster ovary cells all mediate the phosphorylation of extracellular-regulated kinase 1/2. *Molecular Pharmacology*, 58, 477–482.

Schwarzschild, M.A., Chen, J.-F. and Ascherio, A. (2002) Caffeinated clues and the promise of adenosine A_{2A} antagonists in PD. *Neurology*, 58, 1154–1160.

Sei, Y., Gallagher, K.L. and Daly, J.W. (2001) Multiple effects of caffeine on Ca^{2+} release and influx in human B-lymphocytes. *Cell Calcium*, 29, 149–160.

Seo, J.T., Sugiya, H., Lee, S.I., Steward, M.C. and Elliott, A.C. (1999) Caffeine does not inhibit substance P-evoked intracellular Ca^{2+} mobilization in rat salivary acinar cells. *American Journal of Physiology*, 276, C915–C922.

Serafin, W.E. (1996) Drugs used in the treatment of asthma, in *Goodman & Gilmans's The Pharmacological Basis of Therapeutics*, Hardman, J.G. and Limbird, L.E., Eds., McGraw Hill, New York, pp. 659–682.

Shi, D., Nikodijevic, O., Jacobson, K.A. and Daly, J.W. (1993) Chronic caffeine alters the density of adenosine, adrenergic, cholinergic, GABA, and serotonin receptors and calcium channels in mouse brain. *Cellular and Molecular Neurobiology*, 13, 247–261.

Shi, D., Padgett, W.L. and Daly, J.W. (2003) Caffeine analogs: effects on ryanodine-sensitive calcium-release channels and $GABA_A$ receptors. *Cellular and Molecular Neurobiology*, 23, 33–347.

Shinozuka, K., Bjur, R.A. and Westfall, D.P. (1988) Characterization of prejunctional purinoceptors on adrenergic nerves of rat caudal artery. *Naunyn-Schmiedeberg's Archives of Pharmacology*, 338, 221–227.

Silinsky, E.M. and Ginsborg, B.L. (1983) Inhibition of acetylcholine release from preganglionic frog nerves by ATP but not adenosine. *Nature*, 305, 327–328.

Squires, R.F. and Saederup, E. (1987) $GABA_A$ receptor blockers reverse the inhibitor effect of GABA on brain-specific [^{35}S]TBPS binding. *Brain Research*, 414, 357–364.

Svenningsson, P., Hall, H., Sedvall, G. and Fredholm, B.B. (1997a) Distribution of adenosine receptors in the postmortem human brain: an extended autoradiographic study. *Synapse*, 27, 322–335.

Svenningsson, P., Le Moine, C., Kull, B., Sunahara, R., Bloch, B. and Fredholm, B.B. (1997b) Cellular expression of adenosine A_{2A} receptor messenger RNA in the rat central nervous system with special reference to dopamine innervated areas. *Neuroscience*, 80, 1171–1185.

Svenningsson, P., Le Moine, C., Aubert, I., Burbaud, P., Fredholm, B.B. and Bloch, B. (1998) Cellular distribution of adenosine A_{2A} receptor mRNA in the primate striatum. *Journal of Comparative Neurology,* 399, 229–240.

Svenningsson, P., Le Moine, C., Fisone, G. and Fredholm, B.B. (1999a) Distribution, biochemistry and function of striatal adenosine A2A receptors. *Progress in Neurobiology*, 59, 355–396.

Svenningsson, P., Nomikos, G.G. and Fredholm, B.B. (1995) Biphasic changes in locomotor behavior and in expression of mRNA for NGFI-A and NGFI-B in rat striatum following acute caffeine administration. *The Journal of Neuroscience*, 15, 7612–7624.

Svenningsson, P., Nomikos, G.G. and Fredholm, B.B. (1999b) The stimulatory action and the development of tolerance to caffeine is associated with alterations in gene expression in specific brain regions. *The Journal of Neuroscience*, 19, 4011–4022.

Teramoto, N., Yunoki, T., Tanaka, K., Takano, M., Masaki, I., Yonemitsu, Y. et al. (2000) The effects of caffeine on ATP-sensitive K^+ channels in smooth muscle cells from pig urethra. *British Journal of Pharmacology*, 131, 505–513.

Teraoka, H., Akiba, H., Takai, R., Taneike, T., Hiraga, T. and Ohga, A. (1997) Inhibitory effects of caffeine on Ca^{2+} influx and histamine secretion independent of cAMP in rat peritoneal mast cells. *General Pharmacology*, 28, 237–243.

Thong, F.S.L. and Graham, T.E. (2002) Caffeine-induced impairment of glucose tolerance is abolished by β-adrenergic receptor blockade in humans. *Journal of Applied Physiology*, 92, 2347–2352.

Ticku, M.K. and Burch, T. (1980) Purine inhibition of [^{3}H]-γ-aminobutyric acid receptor binding to rat brain membranes. *Biochemical Pharmacology*, 29, 1217–1220.

Toescu, E.C., O'Neill, S.C., Petersen, O.H. and Eisner, D.A. (1992) Caffeine inhibits the agonist-evoked cytosolic Ca^{2+} signal in mouse pancreatic acinar cells by blocking inositol triphosphate production. *Journal of Biological Chemistry*, 267, 23467–23470.

Ufret-Vincenty, C.A., Short, A.D., Alfonso, A. and Gill, D.L. (1995) A novel Ca^{2+} entry mechanism is turned on during growth arrest induced by Ca^{2+} pool depletion. *Journal of Biological Chemistry*, 270, 26790–26793.

Uhde, T.W., Boulenger, J.-P., Jimerson, D.C. and Post, R.M. (1984) Caffeine: relationship to human anxiety, plasma MHPG and cortisol. *Psychopharmacology Bulletin*, 20, 426–430.

Uneyama, H., Harata, N. and Akaike, N. (1993) Caffeine and related compounds block inhibitory amino acid-gated Cl currents in freshly dissociated rat hippocampal neurons. *British Journal of Pharmacology*, 109, 459–465.

Weaver, D.R. (1996) A1-adenosine receptor gene expression in fetal rat brain. *Developmental Brain Research*, 94, 205–223.

Yoshino, M., Matsufuji, Y. and Yabu H. (1996) Voltage-dependent suppression of calcium current by caffeine in single smooth muscle cells of the guinea-pig urinary bladder. *Naunyn-Schmiedeberg's Archives of Pharmacology*, 353, 334–341.

Yoshioka, K., Saitoh, O. and Nakata, H. (2001) Heteromeric association creates a P2Y-like adenosine receptor. *Proceedings of the National Academy of Sciences of the USA*, 98, 7617–7622.

Zhang, W.Y. (2001) A benefit-risk assessment of caffeine as an analgesic adjuvant. *Drug Safety*, 24, 1127–1142.

2 Effects of Caffeine on Sleep and Wakefulness: An Update

Jan Snel, Zoë Tieges, and Monicque M. Lorist

CONTENTS

Introduction 13
Sleep 14
    Alertness Cycles 15
Motives for Consumption 16
Regular and Irregular Sleep–Wake Schedules...17
    Simulated Real-Life Situations 18
    Real-Life Work Situations 20
Aids to Caffeine 21
    Bright Light..21
    Naps 21
    Slow-Release Caffeine 22
Methodological Comments..24
    Awake or Less Sleepy...24
    Measuring Caffeine Intake Assessment: Underreporting...............................24
        Self-Report 24
        Sources of Caffeine 25
        Subjective and Objective Assessment 25
    Expectancy, Instruction, and Placebo 26
        Withdrawal Effects 27
        Blaming Coffee, the Placebo Effect 28
Discussion and Conclusion..28
References 29

INTRODUCTION

It is a daily observation that in public transport, at home in the evening, and at times when people are expected to be fully awake, they suffer from a continuous sleep deprivation and too low a level of wakefulness. Data from laboratory studies show that a shortage of nocturnal sleep by as little as 1.3 to 1.5 h for one night results in a one third reduction of daytime objective alertness (Bonnet and Arand, 1995). Other studies show that 17 to 57% of healthy young adults have sleep onset latencies (SOL) during daytime of <5.5 min (±50% of the normal SOL) and that about 28% of young adults as a rule sleep less than 6.5 h each night of the week. In general, there exists a significant sleep loss in at least one third of all adults. For this reason it is not amazing that fatigue is a factor in 57% of traffic accidents, resulting in many casualties and an estimated loss of $56 billion in the U.S. alone (Bonnet and Arand, 1995). It is no surprise that people look for ways to compensate for a shortage of sleep and to stay awake when necessary. Caffeine-containing beverages such as coffee might be of help. Unfortunately most studies, especially those conducted before the 1990s, have been focused on disturbing sleep and wakefulness by giving caffeine shortly before sleep. Hence, the conclusion from a review (Snel, 1993) was that caffeine induced a restless sleep, predominantly in the first half of the sleep. Effects of caffeine on sleepiness were assessed mainly by measuring sleep latency, mood, and task performance. With doses of caffeine up to 400 mg, sleep latency increased and task performance improved on easy tasks but tended to be impaired on complex tasks. More recent studies also adhere to the tradition of giving caffeine shortly before going to sleep (Landolt et al., 1994; Lin et al., 1997; Hindmarch et al., 2000) or even administer caffeine (5 mg/kg) intravenously during sleep (Lin et al., 1997). Such studies make it difficult to appraise the influence of coffee on sleep and wakefulness in everyday life.

The general conclusion was that caffeine, corrected for the influence of age, gender, personality, and consumption habits, modulates arousal level and that, depending on this interaction, divergent and even contradictory effects on sleep and waking have been found.

Herein an attempt is made to emphasize in particular the effects of caffeine on sleep and wakefulness assessed in more real-life situations. A MEDLINE search using the terms *coffee*, *caffeine*, *sleep*, and *wakefulness*, covering the period 1993 to 2002, was conducted to determine whether the more recent literature offers support for this attempt.

A short introduction discussing what sleep is will be followed by the proper subject of this chapter: the role of caffeine on sleep and wakefulness in real-life settings.

SLEEP

About one third of our lives is spent in sleeping, but the reason we sleep is still unknown. Mostly, sleep is described as a part of the 24-h endogenous arousal cycle with its peak in the afternoon (postlunch dip of arousal) and its trough around 3:00 A.M. and a low shortly after noon. The behavioral manifestation of the circadian arousal cycle, which has to do with the underlying endogenous variations of adenosine and its metabolites (Chagoya de Sánchez, 1995), is expressed as sleep and wakefulness. The best-known adenosine-receptor antagonist, caffeine, and adenosine form an important subject in sleep research.

Adenosine can be seen as a sleep-inducing factor (Porkka-Heiskanen, 1999). Its concentration is higher during wakefulness than during sleep, it accumulates in the brain during prolonged wakefulness, and local perfusions as well as systemic administration of adenosine and its agonists induce sleep and decrease wakefulness. Adenosine receptor antagonists, caffeine and theophylline, are widely used as stimulants of the central nervous system to induce vigilance and increase the time spent awake. Caffeine is an antidote of sleep or an antihypnotic. Van Dongen et al. (2001) concluded from their study that caffeine was efficacious in overcoming sleep inertia by its occupation of adenosine receptors in the brain.

Recording brain activity with an electroencephalogram (EEG) is useful to follow the periodic fluctuations in arousal that are characteristic of sleep. The recorded sleep structure is used to describe the quality and depth of sleep, ranging from stage 1 through stage 4 to the rapid eye movement (REM) stage. Stages 1 and 2 together form light sleep. Stage 2 is the transition from the period of falling asleep to deep sleep and is used as an objective criterion to measure sleepiness. Stages 3 and 4 together represent deep sleep or slow-wave sleep (SWS). Stages 1 to 4 are called non-REM-sleep (NREM-sleep). When stage 4 is reached, there is a quick return via stages 3, 2, and 1 to a state in which REM-sleep occurs. Physiological characteristics of REM-sleep, contrary to NREM-sleep, are an irregular heart and respiration rate, absent muscle tonus of the extremities, a higher threshold to awaken, and the relatively easy reporting of detailed dreams. In the first half of the night more NREM-sleep, especially more SWS, is found; in the second half increasingly more REM-sleep and light sleep are found. The period needed to change from NREM to REM is called a sleep cycle. Although sleep as a biological rhythm is determined largely by endogenous physiological factors with a free-running length of about 25 h, exogenous factors, so-called Zeitgebers, overrule this free-running 25-h period and force it to a 24-h sleep–wake rhythm. Important Zeitgebers are the succession of light and dark and social factors such as the scheduling of work and leisure activities. Disturbing these Zeitgebers by responding to the demands of the 24-h economy can only have serious consequences for sleep and wakefulness.

Alertness Cycles

In addition to the 24-h sleep–wake cycle, smaller Ultradian 90-min cycles exist during sleep, occurring about six times during a normal sleep period. Some authors speculate that this 90-min rhythm also exists during waking and manifests itself in fluctuations of arousal and alertness, the so-called basic rest activity cycles. If true, this explains why in particular the postlunch dip in attention is compensated so well by coffee (Brice and Smith, 2002) and why the enjoyment of coffee may be distributed over the day in a specific pattern to counteract sleepiness. The literature offers little information on specific diurnal trends in patterns of caffeine consumption. Dekker et al. (1993) found in 365 families that about 90% of all coffee is drunk early in the morning, at the morning break, at lunch, late in the afternoon, and early in the evening. A study done by Bättig (1991) shows a similar, but more detailed, picture for 338 20- to 40-year-old women. Twenty-seven percent drank coffee at wake-up, 73% at breakfast, 60% at the morning break, 23% late in the morning, 52% with lunch, 48% at the afternoon break, 32% in the late afternoon, 18% at dinner, and 43% after dinner. Remarkably, the consumption of decaffeinated coffee increased throughout the day from hardly 1% at breakfast to 12.6% after dinner. This may reflect the shift over the day in reasons why people enjoy their coffee and also the unconscious preparation for sleep. Corresponding data were found for caffeine intake in 691 undergraduate students (Shohet and Landrum, 2001). From morning (6:00 A.M. to 12:00 P.M.) through the afternoon (12:00 P.M. to 6:00 P.M.) and evening (6:00 P.M. to 12:00 A.M.), the average consumption decreased from 534.2 ± 1218.7 to 488.2 ± 552.4 to 473.1 ± 532.2 mg. During the night (12:00 A.M. to 6:00 A.M.) the average consumption was 86.8 ± 281.2 mg. It remains difficult to deduce from this data whether ultradian cycles are involved and, if so, whether they are masked by the influence of cultural, situational, social, work-related, and personal factors such as health attitudes, sensitivity, diurnal type, and age.

According to Akerstedt and Ficca (1997), the disturbance of sleep in everyday situations seems negligible even for high doses up to 6 to 7 mg/kg (about six to seven cups of coffee per day). In other words, in the majority of the population, up to 3 mg/kg of coffee hardly influences sleep. That the last coffee is drunk shortly after dinner supports this point (Bättig, 1991). Nevertheless, Alford et al. (1996) found in six healthy volunteers averaging 23.8 years old that, of two doses given, a 4-mg/kg dose given 20 min before bedtime resulted only in a doubling of the SOL. An 8-mg/kg dose, however, decreased sleep efficiency 17%, tripled the number of awakenings to 11.1%, decreased SWS 4.2%, and decreased NREM-sleep 8.2 to 58.6%. In spite of this ecologically invalid procedure of offering high doses of caffeine 20 min before going to bed, the point to stress is that even after a relatively high dose of 4 mg/kg, sleep structure was hardly influenced.

In general, these studies indicate that doses ranging from 2 to 4 mg/kg, comparable to normal use in everyday life, may cause a slight postponement of falling asleep (5 to 10 min of increased sleep latency) (Rosenthal et al., 1991; Penetar et al., 1993). Nevertheless, a critical look at such findings is advised. After an abstinence period of 3 d, of which 2 d were spent in the laboratory, nine healthy students who on average consumed 1.5 cups of coffee daily received 200 mg of caffeine at 7:10 A.M. (Landolt et al., 1995). In the night that followed, the EEG showed that compared to the two previous baseline nights, sleep quality was significantly lower: Total sleep time (TST) ($p < .05$) and sleep efficiency ($p < .05$) decreased and sleep latency to stage 2 increased ($p < .05$) as revealed by 42 tests. A closer look at the absolute values showed that compared with baseline nights 1 and 2, the TST was diminished by 2.5 and 2.2%, respectively, resulting in a sound sleep lasting 440 ± 5.3 min. Sleep efficiency was 2.4 and 2.1% lower than the normal 91.6%, and sleep latency to stage 2 only increased in comparison to the second baseline night 2. Correction for capitalization on chance (Bonferroni correction) would have resulted in nonsignificant results.

The disadvantages of this kind of study are that it includes a caffeine abstinence period, which makes it unclear whether caffeine ameliorates withdrawal effects, it uses subjects not accustomed to caffeine (Landolt et al., 1995), and it does not take the clinical significance of the findings into consideration.

In order to give a more valid indication of the effect of caffeine on sleep in everyday situations, we suggest studying the usual practice, that is, taking the last caffeine of the day 3 to 4 h before bedtime. Because caffeine has an average half-life of 5 h in adults, the effects of caffeine on sleep will then hardly be found.

Engleman et al. (1990) gave 11 medical students a total dose of 5×200 mg of caffeine every 2 h between 7:00 A.M. and 5:00 P.M. after a maximum night's sleep of 3 h. This regular caffeine intake during the day, the latest at 5:00 P.M., did not substantially affect nighttime sleep.

In a study aimed at assessing the influences of caffeine use on the experience of low back pain, information was gathered on sleep onset latency and the numbers of awakenings (Currie et al., 1995). The 64 male and 67 female patients with mean age of 42.1 years and an average pain history of 6.1 years gave detailed information on their daily use of coffee, tea, and cola drinks. There were no differences in sleep quality among the groups that consumed low (mean = 33.7 ± 36.0 mg daily), medium (mean = 226.1 ± 87.8 mg), and high (mean = 562.1 ± 179.6 mg) amounts of caffeine.

Whether coffee hampers sleep quality in everyday, more natural settings were investigated by Janson et al. (1995) in a random population of 2202 subjects aged 20 to 45 years. In this three-country study (Iceland, Sweden, and Belgium), information was gathered on problems falling asleep, nightmares, nocturnal and early awakenings, and the use of psychoactive substances including coffee. Caffeine was not found to be a risk factor for difficulties inducing sleep or other sleep disturbances when making adjustments for age, gender, smoking, country, or seasonal variation. For those who consumed at least six cups per day, however, there was a negative correlation with nocturnal awakenings (Janson et al., 1995). Habitual caffeine consumption during daytime in a regular sleep–wake cycle has no deteriorating effects on sleep quality.

MOTIVES FOR CONSUMPTION

Early in the morning coffee is taken mostly to awaken. During the day coffee is taken more for conviviality (17%) and relaxation (34%) rather than for stimulation (14%); only 7% take coffee to cope with stress (Harris Research Centre, 1996). Support for this comes from a study done by Höfer et al. (1993) in which 120 students were put on a strict abstinence regimen, after which they received caffeine during 12 complete days. Although caffeine abstinence caused moderate and transient withdrawal effects, there was no so-called titration of caffeine, that is, coffee consumers did not consume more when the coffee contained less caffeine. Apparently, caffeine itself is a minor reason for coffee consumption, although the studies by Hughes' team repeatedly show that abstained coffee drinkers prefer caffeinated coffee above decaffeinated coffee (Hughes et al., 1995).

These motives to drink coffee, essentially all of a positive nature, imply that the disturbing effects of coffee on sleep are confounded by other aspects. Illustrative of this view is research by De Groen et al. (1993), who studied snoring and anxiety dreams in 98 veterans from World War II. Fifty-five of them suffered from current posttraumatic stress disorder. The outcome showed that the association between snoring and anxiety dreams was independent of many factors that were expected to be related, one of which was coffee consumption. A comparable study was done in 14,800 male twins, born between 1939 and 1995, who served the army in Vietnam between 1964 and 1975 (Fabsitz et al., 1997). Responses were collected from 8870 men on the frequency of their sleep problems as reported on the Jenkins sleep questionnaire, which inventories the prevalence of at least one sleep problem per month. Sixty-seven percent of the respondents awoke often, 61.5% awoke tired or worn out, 48.1% experienced trouble falling asleep, and 48.6% awoke early. It appeared that of the 11 conditions inventoried, coffee consumption of at least eight cups per day vs. up to seven cups per day was related only to awaking tired (odds ration [OR] 1.32), while heavy alcohol use and type A behavior were associated with a higher risk for all sleep problems. The conclusion was that a number of the risk factors associated with these sleep problems came from lifestyle characteristics or stress.

The same conclusion can be drawn from a study of locomotive engineers and their spouses (Dekker et al., 1993). Twenty-seven engineers who were working irregular work schedules and their spouses completed daily logs for 30 d. These logs were divided into workdays and nonworkdays. Workday sleep length was significantly shorter than nonworkday sleep length for both subject groups. The number of cups of coffee consumed on workdays was higher (2.75 cups per day) than on nonworking days (2.17 cups per day), but only for the locomotive engineers. The authors concluded that increased coffee consumption was correlated with longer sleep latency, increased negative mood, and decreased positive mood on both work and nonwork days. Driving a locomotive is a taxing task that demands continuous vigilance; the stress of this combined with the frequent intake of coffee to compensate for this stress may have caused this decrease in sleep quality and feelings of well-being.

The same conclusion may apply to a study by Ohayon et al. (1997), who researched the prevalence of snoring and breathing pauses during sleep in 2894 women and 2078 men aged 15 to 100 years, a representative sample of the U.K. population. Forty-five percent of this sample reported snoring regularly, which was associated with the male sex, aged 25 years or more, and consuming at least 6 cups/d (OR 1.4, $p < .002$). Since snoring was also associated with obesity, daytime sleepiness or naps, nighttime awakenings, and smoking, it could be that, as found in the former studies, an inadequate lifestyle was the causal factor of the sleep-related problems, and not caffeine itself.

The same line of reasoning goes for the restless legs syndrome and periodic limb movement disorder (PLMD), two other sleep-impairing disorders. Cross-sectional studies in the U.K., Spain, Italy, Portugal, and Germany among 18,980 subjects, 15 to 100 years old, revealed that caffeine intake was not associated with restless legs syndrome, although it was with PLMD (Ohayon and Roth, 2002). The specific factors associated with PLMD included being a shift or night worker, snoring, daily caffeine intake, use of hypnotics, and stress.

Depression may lead to bad sleep, but stress is not always the causative factor. Chang et al. (1997) followed 1053 men in a prospective study to assess the relationship between self-reported sleep disturbance and subsequent clinical depression and psychiatric distress over a median follow-up period of 34 years. The relative risk for depression was greater for those who reported a bad sleep at the start of the follow-up period. Coffee, however, had no influence. In this case, sleep disturbances reflected a vulnerability for depression, since even after resolution of the depressive period, sleep EEG abnormalities remained. It is unlikely that coffee as a mood enhancer and cognitive stimulant has anything to do with a genetic predisposition to vulnerability for bad sleep and depression.

Although these results may shed light on studies reporting impaired sleep quality due to caffeine intake, they may only count for those who use sedative hypnotics, which may hinder a refreshing sleep.

In general, it can be said that coffee drinking is often associated with a cluster of factors that are representative of a stressful and risky lifestyle. It is these factors that might be responsible for certain sleep–wake problems, and not coffee.

REGULAR AND IRREGULAR SLEEP–WAKE SCHEDULES

Regular sleep is an important requisite of a good sleep and should result in low levels of daytime sleepiness. Manber et al. (1996) evaluated prospectively the effects of two manipulations of sleep–wake schedules on subjective ratings of daytime sleepiness in 39 17- to 22-year-old students. Subjects in the sleep–only and in the regularity groups were given a 7.5-h limit for total sleep time. Those in the regularity group were instructed to stick to a regular sleep schedule. After a 12-d baseline period, the experimental conditions were introduced and lasted four weeks. Five weeks after this experimental phase, a follow-up phase of one week started. The findings were that when nocturnal sleep was not deprived, regularization of sleep–wake schedules was associated with less sleepiness. Subjects in the regular schedule group reported greater and longer-lasting improvements in alertness and improved sleep efficiency compared with subjects in the sleep-only group. As for coffee consumption, there were no differences between the groups, suggesting that subjects are able to attune their coffee consumption to the way they live their lives.

Although regular work schedules mostly imply working in close harmony with the sleep–wake cycle, and hence may not cause trouble with keeping awake, the use of caffeine in the work situation is a universal phenomenon. To assess the effect of caffeine on neuroendocrine stress responses in the workplace during the daytime, Lane (1994) studied 14 habitual coffee drinkers (two to seven cups/d; mean 3.4 ± 1.7 cups) doing their normal work-related activities. Catecholamine and cortisol levels were measured in 2 d in a 4-h interval from morning until noon. After overnight caffeine abstinence, 300 mg of caffeine or placebo was administered in a blind study between 8:00 and 8:30 A.M. Scores of mood (POMS) from a symptom checklist were collected at the end of each morning. Caffeine elevated adrenaline levels during work by 37% but did not affect norepinephrine or cortisol levels. The subjective reports suggested that caffeine abstinence was associated with symptoms of caffeine withdrawal by the end of the morning. Effects included higher ratings of sleepiness, lethargy, and headache and a reduced desire to socialize. Apparently, after caffeine abstinence, caffeine may increase the activity of the sympathetic adrenal–medullar system during everyday activities in the work environment. It indicates an acceleration of the increase of arousal, a normal stage of the circadian rhythm early in the morning. Its acceleration by coffee may help to attain a habitual level of functioning sooner.

Irregular patterns of life, voluntarily chosen (leisure activities) or imposed by work (health care, traveling intercontinentally, security industry), form a health risk due to excessive sleepiness, disturbed sleep, and accidents. There are many ways people could use to compensate for the consequences of irregularity, such as exposure to bright light, taking naps or a break, improving work scheduling, or manipulating their sleep. The role of coffee in sleep deprivation due to irregular sleep–wake schedules has recently been assessed in a few field studies.

SIMULATED REAL-LIFE SITUATIONS

The majority of studies on caffeine have investigated its effects on performance of typical computerized laboratory tasks that represent single basic functions underlying real-life performance. Exceptions are those studies with tasks that are found in situations like that in the following study.

In a study on simulated driving (Brice and Smith, 2001), the effect of 3 mg/kg of caffeine, comparable to the everyday practice of consuming two cups of coffee per occasion, was assessed. Participants were 24 healthy students, all nonsmokers and habitual consumers of regular coffee and with an average 4.63 years of driving experience. The subjects were not required to abstain from caffeine-containing beverages before the experiment. The results showed that in the group that consumed coffee, in addition to greater alertness and more hits on a repeated digit memory task (54.1% compared to 48.8%), steering variability was significantly less (95.5 to 101%; mean percentage change from baseline) and continued throughout the 1-h drive. Previous studies have found similar beneficial effects of a 150-mg dose of caffeine on driving performance (Horne and Reyner, 1996), but in sleep-deprived, hence fatigued, subjects. In a more recent study (Reyner and Horne, 2000), sleep deprivation was added. Eight male and eight female students, 23 years old, were sleep deprived until midnight or for the whole night, then had to drive continuously for 2 h (6:00 to 8:00 A.M.) on a dull, monotonous roadway. A dose of 200 mg of caffeine improved driving performance in a dose-dependent fashion, resulting in fewer incidents and less subjective sleepiness. Caffeine taken as regular coffee effectively reduces early morning driver sleepiness for about half an hour following total sleep deprivation for one night and for around 2 h after a short sleep of

5 h. The best way to counteract sleepiness appeared to be caffeine, because of its more consistent alerting effects; taking a break alone proved ineffective. Brice and Smith's (2001) study confirmed that caffeine works beneficially in monotonous tasks, also in nonsleep-deprived subjects. It supports the many studies showing that caffeine is quite useful in compensating for the fatigueing effect of performing a task uninterruptedly, the so-called time-on-tasks effect, especially while performing tasks that are simple, monotonous, and not intrinsically interesting enough to keep fatigue away (see Chapter 4 of this book).

In a simulated shift-work situation including seven men and five women, 19 to 36 years old, all of whom did not drink coffee or were moderate users (~ 2 cups/d), the influence of a 200-mg dose of caffeine was studied. Work started at 5:30 in the evening and continued until 10:00 the next morning. During the 1-h rest period from 1:30 to 2:30 A.M., the participants performed four computer tests lasting 90 to 95 min. Caffeine alone had a general beneficial effect on performance during the night, which was ascribed to the suppression of the melatonin level or to a delay of the melatonin rhythm. Indeed, Shilo et al. (2002) found in six volunteers who drank decaffeinated coffee or regular coffee in a double-blind study that caffeinated coffee caused a decrease of 6-MT, the main metabolite of melatonin, in urine. Wright et al. (1997a) found that caffeine delayed the melatonin rhythm dose-dependently. This explains why sleepiness during the daytime may occur after caffeine is administered during sleep deprivation the night before. It also implies that individuals who suffer from sleep abnormalities or claim to be sensitive to caffeine should avoid drinking coffee late in the evening.

Modern life may require that individuals sacrifice their regular sleep schedules, with the consequence of partial or complete loss of sleep. To determine the effects during and after a 62-h period of prolonged wakefulness, Kamimori et al. (2000) studied 50 healthy, nonsmoking males (aged 18 to 32 years) who did not take more than 300 mg/d of caffeine. This study resembled their 1993 study with the same design, doses of 150, 300, or 600 mg/70 kg, and had similar results (Penetar et al., 1993). In their 2000 study, the caffeine doses (2.1, 4.3, and 8.6 mg/kg) were given as before, double-blind, in a sweetened lemon juice drink, and only once after 49 h of wakefulness. Plasma caffeine concentrations were measured every half hour until 12 h after intake. Caffeine had no significant effect on noradrenaline, but adrenaline was significantly increased 1 and 4 h postintake, but only in the high-dose group. Sleepiness scores as assessed with the MSLT showed dose-related responses that were nonsignificant for the 2.1-mg/kg dose. This result was ascribed to possible tolerance effects in these regular coffee consumers. The high dose was especially effective in stimulating adrenaline and almost tripled the SOL in the MSLT over the first 4 h postintake. Interesting was the significant and disproportional increase in the dose-normalized caffeine AUC and the significant decrease in its clearance rate with increasing dose (Kamimori et al., 1995). Also, the paraxanthine:caffeine ratio significantly decreased with increasing dose. Apparently, with increasing dose, the metabolism of caffeine slows down, suggesting a capacity-limited metabolism (Kamimori et al., 1995).

The effect of caffeine in simulation of jet-lag situations was studied by Moline et al. (1994) in five healthy men between 37 and 58 years of age, all moderate caffeine intakers (~ 3 caffeinated beverages). They were recruited for two 16-d sessions in the laboratory. The first 5 d were based on habitual sleep–wake schedules, followed by the sixth night from which they were awakened 6 h before their usual arising time. This was repeated during the subsequent sleep periods. Throughout the study they received, double-blind, 200-mg caffeine or placebo tablets at breakfast. Prior to one of the scheduled shifts, the caffeine was replaced by placebo. Subjective alertness remained approximately constant following the shift in the drug condition, whereas it declined in the no-caffeine condition ($p < .05$). The preshift alertness was the same in both conditions. The summarizing conclusion was that using a single dose of 200 mg of caffeine, taken at breakfast, prevented a decline in subjective alertness following a 6-h time advance in middle-aged male subjects.

A comparable study was done by Muehlbach and Walsh (1993) on simulated night work with 2 mg/kg of caffeine in eight males and seven females with a mean age of 24.7 years. Dependent variables were physiological sleepiness (MSLT), performance, mood (POMS), and subsequent daytime sleep. Caffeine appeared to be beneficial in improving alertness during three successive night shifts without impairing mood and daytime sleep. Nighttime performance, however, was not significantly improved and sleepiness at the circadian trough remained at weak levels (see also Lee, K.A., 1992; Walsh et al., 1995). In a replication study (Muehlbach and Walsh, 1995), one of the objectives was to determine the efficacy of caffeine in maintaining performance and alertness during the circadian trough. Ten healthy young adults were totally sleep-deprived for 54.5 h. After 41.5 h awake, the subjects received, double-blind, 600 mg of caffeine followed by hourly testing. Performance and alertness assessed every 2 h were significantly improved by caffeine. Caffeine maintained performance and alertness during the early morning hours, when the combined effects of sleep loss and the circadian morning trough of performance and alertness were manifest.

In sum, caffeine is an effective and cheap means of improving performance and alertness during sleep loss in normal, healthy adults in conditions comparable to those of real-life work situations.

REAL-LIFE WORK SITUATIONS

Greenwood et al. (1995) were interested in the role of caffeine, alcohol, and tobacco use, exercise, activities upon going to bed, and sleep-enhancing measures in minimizing sleep disturbance in shift-work situations. Subjects were 72 workers (nurses, computer operators working for a bank, and industrial workers at a chemical factory) who worked on rotating-shift systems for periods of at least three consecutive nights or days. The 40 men and 32 women, on average 30.5 years old, 61% married, had a mean shift-work experience of 8.8 ± 6.5 years and worked on average 39.5 h/week (24 to 80 h). Sixty percent of the workers consumed caffeine from different sources in the 6 h preceding sleep on all night shifts, only 5% more than on day shifts. To protect their sleep quality, they were recommended to abstain from caffeine during the 6 h before sleep. In general there was little evidence of workers changing their caffeine consuming habits according to the shift they worked. Correlating the consumption of caffeine during the 6-h period before going to bed with the length of sleep for only those who consumed caffeine resulted in no significant correlations, ranging from –0.31 to 0.10. Also, there was no correlation between the time they took the last caffeine-containing beverage and sleep duration on any of the three night- and day-shift days (Greenwood et al., 1995). Further, sleep at a time common for night workers was not markedly disrupted by the moderate caffeine consumption at night (Muehlbach and Walsh, 1995). In sum, shift workers do not adjust their coffee intake with the explicit intention of improving sleep quality, and coffee has hardly any effect on sleep quality.

The question of whether moderate doses of caffeine (100, 200, or 300 mg) or placebo given after 72 h of sleep deprivation would reduce adverse effects was posed by Lieberman et al. (2002). The 68 U.S. Navy Sea–Air-level trainees were tested 1 to 8 h after sleep deprivation on several cognitive tasks, mood, and performance (marksmanship). The sleep deprivation and the stress of the simulated combat situation adversely affected performance and mood. However, caffeine, dose-dependently (200 and 300 mg) improved visual vigilance, reaction time, and alertness. Marksmanship was not affected by caffeine. The greatest effect of caffeine was found 1 h after intake, but significant effects persisted for 8 h. The conclusion was that when cognitive performance is critical and must be maintained during severe stress under sleep deprivation conditions, caffeine, especially the 200-mg dose, provides a significant advantage.

The foregoing studies (simulated and real-life) indicate that caffeine taken at appropriate times reduces sleepiness and compensates performance decrements during nighttime work hours. These findings confirm the series of studies done by Smith (1994). A summary from his five studies showed that 1.5- and 3.0-mg/kg doses of caffeine (the usual amount people drink at one occasion) are beneficial for aspects of mood and performance when a person's arousal level is reduced, as in sleep deprivation. A secondary advantage is that this increase in alertness and improvement of performance reduces problems with safety and loss of performance efficiency. Whether the systematic use of coffee is feasible for all occupational settings is difficult to decide. Caffeine certainly is used to improve alertness during work, but the use is spontaneous and *ad hoc*, and there is still a lack of data on its systematic application. Particularly, the optimal amount and pattern of administration need elucidation, and also its effect combined with other means people use to stay awake, among them bright light, naps, and, more recently, the use of slow-release caffeine.

AIDS TO CAFFEINE

Bright Light

The most important Zeitgeber in regulating the 24 sleep–wake cycle is light. Therefore, the abundance of light in our 24-h economy, sometimes termed *light pollution*, will disturb the sleep–wake cycle. On the other hand, the increase of light intensity could be used in situations to keep people awake. To check this option, some studies have been done with bright light only, more recently also combined with caffeine (Wright et al., 1997b; Babkoff et al., 2002). Such a study during nighttime hours across 45.5 h of sleep deprivation was done under four conditions (dim light-placebo, dim light-caffeine, bright light-placebo, and bright light-caffeine) (Wright et al., 1997b). Subjects were 46 healthy male volunteers, aged 18 to 25 years, and the study was conducted during the nighttime hours. Measures were alertness and performance. A caffeine dose of 200 mg was given at 8:00 P.M. and at 2:00 A.M.; bright light-exposure (> 2000 lux) was from 8:00 P.M. to 8:00 A.M. The combined treatment of caffeine and all-night bright light enhanced performance to a larger degree than either the dim light-caffeine or the bright light-placebo condition. Beneficial effects of the treatments on performance were largest after the trough in the arousal level at 2:00 A.M. At that time, performance in the dim light-placebo condition was the worst. Notably, the bright light-caffeine condition was successful in overcoming the circadian drop in performance for most tasks measured. Both caffeine conditions improved wakefulness. Taken together, the results suggest that the combined treatment of bright light and caffeine is effective in enhancing alertness and performance during sleep loss. A similar experiment was done by Babkoff et al. (2002) with the same treatment as in the study by Wright et al. (1997b), except that the brightness was 3000 lux. Subjects were seven men and five women who were on average 24.6 years old. The simulated work condition started at 5:30 P.M. and ended 10:00 A.M. Exposure to bright light was for 1 h at 1:30 A.M., combined with a 200-mg dose of caffeine. Performance on spatial discrimination, working memory, and a letter cancellation task was maintained throughout the reminder of the night and morning; however, the exposure to light alone without the caffeine actually degraded the performance.

The most effective way to compensate for the increase in fatigue was the combination of 200 mg of caffeine and the exposure to bright light (3000 lux), followed by caffeine, and then by bright light.

Naps

Interesting from the point of view of everyday practice are those studies that combine caffeine with naps to counteract fatigue from a shortage of sleep, such as those done by Bonnet and Arand (1995) and by Reyner and Horne (1997). Intuitively, naps and caffeine do not match; naps are meant to induce sleep, while caffeine is used to awaken. Nevertheless, previous studies have shown that performance during sleep loss is improved by prophylactic naps, dose-dependently with nap length. The study by Bonnet and Arand (1995) compared the effects of repeated vs. single-dose administration of caffeine and varying amounts of sleep taken prior to sleep loss on performance, mood, and physiological measures during two nights and days of sleep loss. A total of 140 young adult males (recruits and college students) participated. No data were reported on their coffee consumption. Ninety-eight subjects were randomly assigned to one of four nap conditions (0, 2, 4, or 8 h) and 42 subjects were assigned to one of the following caffeine conditions: a single 400-mg dose of caffeine at 1:30 A.M. each night or repeated doses of 150 or 300 mg every 6 h starting at 1:30

A.M. on the first night of sleep loss. The placebo control group did not nap and received placebo every 6 h on the repeated caffeine schedule. After a normal baseline night of sleep and morning baseline tests of performance, mood, and nap latency, subjects in the nap groups returned to bed at 12:00 P.M., 4:00 P.M., 6:00 P.M., or not at all. Bedtimes were varied so that all naps ended at 8:00 P.M. (Bonnet and Arand, 1995). The MSLT was administered every 3 h starting at 10:00 P.M. and the visual vigilance every 6 h starting at 11:30 P.M. on the initial sleep loss night. The results confirmed earlier findings that alertness and performance during sleep loss are directly proportional to prophylactic nap length. Naps in general were superior to caffeine and resulted in longer and less-graduated changes in performance and alertness. Caffeine displayed peak effectiveness and lost its effect within 6 h. The combination of short naps and small, repetitive doses of caffeine (150 mg), however, did maintain alertness and performance during sleep loss better than no naps, caffeine, or large single doses (300 and 400 mg). However, neither nap nor caffeine could preserve function at baseline levels beyond 24 h, after which alertness and functioning approached placebo levels.

Reyner and Horne (1997) combined naps with 150 mg of caffeine to assess the combined effect on driver sleepiness and driving safety on a simulated dull and tedious motorway in 12 healthy, 20- to 30-year-old, medication-free, good sleepers; all were infrequent nappers. The drive started at 2:00 P.M., followed by a 30-min break and then a 2-h drive. Decaffeinated coffee with 150 mg of caffeine was taken 5 min before the break, followed by a nap of 15 min that ended 5 min before the 2-h drive. The event frequencies for incidents during caffeine + nap and caffeine were, respectively, 0.09 and 0.34, compared to 1.0 for placebo. Caffeine + nap was additive in its decrease of subjective sleepiness. The EEG trends for all conditions resembled those for sleepiness. Taken altogether, it appeared that the combined treatment was additive with respect to incidents and sleepiness. During the 2-h postcaffeine treatment, compared with caffeine only, the combination of caffeine + naps reduced the number of incidents a further three- to fourfold.

Slow-Release Caffeine

Slow-release caffeine (SRC) is a relatively recent means of caffeine delivery. SRC appears especially suitable in long work or shift-work schedules that necessitate an elevated and prolonged level of vigilance and performance and in which fatigue and sleepiness, which may impair efficiency, should be avoided. Use of SRC could have implications for civilian life, for physicians and nurses on duty, pilots, truckers, rescue workers, and perhaps even for the chronically sleep-deprived general public.

A work-simulated study, quite interesting because of the use of SRC, was done in a laboratory situation with a group of 12 young adults (Bonnet and Arand, 1993). The aim was to compare the relatively best effect of either napping for four periods of 1 h each or one nap period of 4 h in combination with or without a 200-mg SRC dose. Addition and logical reasoning were improved during the night with the combination of the 4-h nap before the shift and caffeine. Performance after the 1-h naps in the beginning of the night was very poor, probably due to the fact that 60% of the naps ended in SWS compared with 10% of the prophylactic naps, the main advantage of this method to stay awake. Apparently, the best strategy to stay awake during shift work is to take a nap before starting to work and using SRC (200 mg) or two doses of 200 mg at 1:30 and 7:30 A.M. (Bonnet and Arand, 1994a,b).

The effect of partial sleep derivation induced by a short night of 4.5 h was measured with a driving simulator for 45 min with or without a 300-mg SRC dose. The subjects were seven male and five female students, with a mean age of 22.5 years. They were all moderate to normal sleepers and had at least 2 years of driving experience. After the normal sleep condition (7.5 h), caffeine decreased lane drifting, while after the 4.5-h night, the SRC resulted in less lane drifting, smaller speed deviation, and less accident liability. Subjectively, SCR resulted in less fatigue and more vigor, but only in the short 4.5-h sleep condition. Especially when there is no opportunity to take a nap, as is the case in most industrial settings (continuous process and monitoring activities, transport, health care), SRC is indicated to decrease fatigue. Another reason for considering using

SRC instead of caffeine-containing beverages is to prevent the risk of unwanted side effects from repeated administration of caffeine over long periods.

Beaumont et al. (2001) investigated the efficacy of 600 mg of SRC on sustained attention and cognitive performance during a 64-h continuous wakefulness period. Sixteen healthy, nonsmoking male volunteers participated in this double-blind, two-way crossover study. The subjects were nonconsumers or low consumers of xanthine-based coffee, tea, or cola. Twice a day, at 9:00 P.M. and 9:00 A.M., during a 64-h sleep deprivation (SD) period, a 300-mg dose of SRC or placebo was given. From continuous EEG recording, vigilance and sleepiness was objectively assessed. Cognitive functions (information processing and working memory) and selective and divided attention were determined with computerized tests. Attention was measured also with the symbol cancellation task and Stroop's test and with visual analog scales. Measurements were done during the low vigilance periods from 2:00 to 4:00 A.M. and from 2:00 to 4:00 P.M. and during the high vigilance periods from 10:00 A.M. to 12:00 P.M. and from 10:00 P.M. to 12:00 A.M. In comparison with those given the placebo, the SRC subjects were less sleepy from the onset ($p = .001$) to the end of SD ($p < .0001$). Some cognitive functions were improved until the 33rd hour of SD, others were ameliorated through all the SD period, and alertness was better from the 13th h of SD as shown by Stroop's test ($p < .05$). In conclusion, a dose of 300 mg of SRC given twice daily is able to counteract the impairment of vigilance and cognitive functions produced by a 64-h SD.

Patat et al. (2000) used 600 mg of SRC during a 36-h SD period. The subjects are 12 healthy men, on average 28 years old. All subjects abstained from taking medication for 15 d before the study began. A variety of tests to assess psychomotor and cognitive functions were administered, in addition to EEG recordings, five times during the SD period. Similar to the Beaumont study (Beaumont et al., 2001), one single dose of 600 mg of SRC increased alertness, compensated for performance decrements throughout SD, and was able to reverse the deleterious effect of 36 h of SD for at least 24 h. The peak effect occurred 4 h after its intake.

Whether this means that in SD situations one high-dose, caffeinated beverage, SRC, or repeated, moderate caffeine doses should be taken is difficult to say. At any rate, in a well-rested state, a dose of 600 mg induces an awakening effect 5 h after intake. No significant effects on alertness or contentedness were reported; only calmness was influenced (Sicard et al., 1996). A single dose of 600 mg of SRC is well tolerated and its pharmacokinetics are not influenced by habitual caffeine use. In a well-rested state, however, it may disturb sleep onset and elicit a reduction in calmness. Since well-rested subjects may already be close to their optimal level of alertness, SRC in such a state may increase alertness beyond the optimum and thereby impair mood and performance. According to Sicard et al. (1996), in well-rested subjects it is sufficient to use lower SRC doses or to stick to the frequent enjoyment of regular coffee. Since SRC is designed to be used in chronic sleep-deprivation situations where fatigued subjects are more responsive to caffeine's effects, more studies should be done with different doses.

To find out at which dose of SRC is optimal (maximum effect without side effects), Lagarde et al. (2000) compared three doses (150, 300, and 600 mg) to a placebo in 24 young people in a 32-h sleep deprivation experiment. The subjects were moderate caffeine consumers (≤ 3 cups/d). Wakefulness was assessed with the MSLT, questionnaires, and analog visual scales. Several aspects were measured: attention, grammatical reasoning, spatial recognition, mathematical processing, visual tracking, memory search, and a dual task. Motor activity was evaluated by actimeters. There was a significant effect of the three doses of SRC vs. placebo on vigilance and performance when subjects became tired. This was found particularly in the number of errors on the four-letter memory search task and the visual tracking task. Women appeared to be more sensitive to the lowest dose, while in men the most obvious effects were found with 300 and especially with 600 mg of caffeine. Remarkably, the effects of SRC lasted 13 h after treatment. Considering all results together, the authors concluded that the SRC doses of 300 and 600 mg were efficacious but the optimal dose for both men and women was 300 mg. SRC (300 mg) seems to be an efficient and safe substance to maintain a good level of vigilance and performance during sleep deprivation. SRC may permit a long, good-quality wakefulness not only in laboratory situations but also in real-life, such as in jet-lag and simulated driving conditions.

METHODOLOGICAL COMMENTS

The foregoing studies showed that caffeine in simulated and real-life work situations is effective in counteracting sleepiness and improving performance and has almost no effect on sleep quality. The results could possibly be more salient and revealing if proper attention were paid to a number of factors: sleepiness or staying awake, total amount of caffeine, age and gender, expectancy, instruction, and placebo effects.

AWAKE OR LESS SLEEPY

In SD studies the subject of interest is mainly sleepiness. This is mostly done with the MSLT as the instrument and by using EEG parameters to assess the tendency to fall asleep in an objective way. In real-life situations, to remain awake and perform is of more practical concern than one's ability to fall asleep. To assess the ability to remain awake without assistance, the maintenance of wakefulness test (MWT) is used. The MWT is a procedure that uses EEG measures to determine the ability of the subject to remain awake while sitting in a quiet, darkened room. The test consists of 20-min trials conducted four times at 2-h intervals, commencing 2 h after awakening from a night of sleep (Mitler et al., 1998). In Kelly's 64-h SD study, 300 mg of caffeine or a placebo was given to 11 and 14 nonsmoking males who reported an average daily caffeine intake of no more than 250 mg (Kelly et al., 1997). The doses were given double-blind at 11:20 P.M., 4:50 A.M., 11:20 A.M., 5:20 P.M., 11:20 P.M., 4:50 A.M., and 5:20 P.M. The instruments were performance tests, the MSLT, and one single MWT trial each day at 1:30 P.M. Ability to go to sleep and ability to stay awake during SD appear to be affected differently by caffeine. The MSLT showed only an effect of caffeine the 1st day, but neither a group effect nor a group × day effect. The MWT showed a group effect ($p < .005$) and a day × group interaction ($p < .03$), but no day effect, since on both days the MWT showed a positive effect of caffeine on the ability to stay awake. In sum, the MSLT was sensitive to caffeine's effects only during the first 24 h without sleep; the MWT demonstrated that caffeine improved the ability to remain awake even after two nights of SD. Performance testing may fail to detect this stimulant effect because it often happens that experimenters prevent subjects from nodding off during testing, an external support not available to subjects during the MWT and also not available in many real-world work environments. In that sense, the MWT is more ecologically valid and more practical than the MSLT. In addition, the MWT was more sensitive to stimulant amelioration of SD effects than the MSLT.

MEASURING CAFFEINE INTAKE ASSESSMENT: UNDERREPORTING

Self-Report

Most studies on caffeine and sleep have been done on young students, but there is little evidence of age differences in overall levels of consumption. Brice and Smith (2002) refer to a study done in 1989 in the U.K. in which no age-related differences were found. A study of an American sample revealed that those between 60 and 69 years of age consumed the most caffeine per day (472.3 mg), whereas those between 20 and 29 years had the lowest consumption levels (284.6 mg). It was not clear from this paper whether these figures also included caffeine from sources other than coffee. Figures from 1996 (Barone and Roberts, 1996), across a range of groups, including children, teenagers, adults, and pregnant women, showed that for the adult population in Denmark, the mean total daily intake from all products containing caffeine was on average 7.0 mg/kg; in the U.K. and U.S. intake was found to be 4 mg/kg. Per capita for the whole U.S. population was found to be

2.4 mg/kg (Mandel, 2002). For those caffeine consumers in the U.S. who are under 18 years of age, daily intake is about 1 mg/kg.

More recent figures (Brice and Smith, 2002) showed that in a student sample caffeine intake from coffee alone was 778.5 mg/week; tea, hot chocolate, soft drinks, and other products contributed an additional 386.5 mg/week. Hence, inquiring about coffee consumption as a representative measure for total caffeine intake implies underreporting of total caffeine intake by about one third (31.6%). Since sleep is affected by caffeine, especially when taken shortly before going to sleep (Snel, 1993), it should be normal practice to gather information on caffeine intake from other sources as well. Gathering reliable and valid data on caffeine consumption is essential for a correct interpretation of results. Estimates are that there is an underreporting of caffeine consumption ranging from 25 to 40% (Brown et al., 2001).

Sources of Caffeine

Daily caffeine consumption is, almost without exception, reported retrospectively, but in general screening in the laboratory is refrained from. This is rather surprising for two reasons. Most people view their coffee intake as their only source of caffeine and forget or do not know that other products, such as tea, soft and energy drinks, hot chocolate, certain food products such as cakes and candies, and over-the-counter (OTC) medications, including analgesics and cold remedies, contain caffeine as well (Brice and Smith, 2002). Data on daily caffeine intake from foods, beverages, and medications were also collected through mailed questionnaires from 481 30- to 75-year-old men and women in Ontario, Canada (Brown et al., 2001). The mean total daily caffeine intake from coffee alone, with large variances, for men 30 to 44 years old, 324 mg (4.3 mg/kg), 45 to 59 years old, 426 mg (5.7 mg/kg), and 60 to 75 years old, 359 mg (4.8 mg/kg). For women these figures were 288 mg (4.2 mg/kg), 322 mg (4.6 mg/kg), and 314 mg (4.5 mg/kg), respectively. The percentage of caffeine intake from tea in men within the three age groups was 9.0, 12.4, and 18.7% and for the women was 23.6, 23.0, and 33.8%, respectively. The caffeine intake from soft drinks, medications, and chocolate decreased over the three age groups from 16.9 to 10.9% in men; a similar decrease from 15.9 to 8.9% was seen in the women. Caffeine from medications was a stable 6 mg in men and in women averaged 9 mg. The percentage of caffeine intake from coffee, soft drinks, and tea over the three age groups in men was 72, 13, and 12%, respectively; in women these percentages were 60, 9.5, and 27%.

In the U.S., based on figures from 1996, the estimate is that coffee accounts for about three fourths of the caffeine that is consumed. Tea makes up 15%, soft drinks 10%, and chocolate about 2% (NAH, 1996). Whether these data are reliable is questionable. Among 13- to 17-year-old U.S. teenagers, information from a drug-use questionnaire on the daily intake of caffeine from only caffeinated beverages collected over a 3-d period (Bernstein et al., 2002) showed a total average daily intake of 244.4 ± 173 mg, or 3.2 ± 2.0 mg/kg/d. The figures showed that 61.8% of the total caffeine intake came from soft drinks, 34.9% from coffee, and 3.3% from tea.

Subjective and Objective Assessment

There is a large interindividual variability in the half-life of caffeine due to age-related changes in the smell and taste of caffeine (Gilmore and Murphy, 1989; Murphy and Gilmore, 1989), body weight (BW) (Goldstein et al., 1965; Abernathy et al., 1985), stage in the menstrual cycle, habitual use of caffeine, etc. Since saliva caffeine levels peak at similar times for coffee (42 ± 5 min) and cola (39 ± 5 min) but later for caffeine in capsules (67 ± 7 min; $p = .004$), the vehicle in which caffeine is present may have consequences for self-reporting scales and sleep parameters. Also, perceived differences in the effects of coffee vs. other caffeinated products may be due to differences in dose, time of day, added sweetener or other substances, environmental context, or chance occurrences (Liguori et al., 1997).

However, the associations between self-reporting of caffeine use and laboratory screening are confusing for a correct classification of subjects. James et al. (1989) obtained saliva samples of 142 first- and second-year medical students and tested them for caffeine and paraxanthine levels. These levels correlated 0.31 and 0.42, respectively, with self-reported consumption data. Also, in a study of 181 healthy community dwellers, estimates of caffeine intake from coffee, tea, and other sources did not correlate with plasma caffeine concentrations (Curless et al., 1993).

The summarizing conclusion from these caffeine consumption figures is that there are huge differences between men and women and different age groups. It also means that in general underestimation ranges from 33.3 to 48.4%. The implication is that among noncoffee drinkers, who are sometimes used as control subjects in coffee studies, 32.3% would have underestimated their caffeine intake by at least 100 mg, and more than 6% of the nondrinkers are misclassified by at least 300 mg. About 2% of the noncoffee drinkers ingest at least 500 mg of caffeine (Schreiber et al., 1988).

Studies on effects of caffeine may suffer from this inaccuracy in information on caffeine intake. There are many factors that play a role. Subjects may simply forget. The majority of studies gathers information exclusively on coffee use only and does that retrospectively. Also, in the literature, it is not always reported which sources of caffeine have been included. Subjects may vary in their rate of metabolism due to age, gender, habitual use, or use of other psychoactive substances as well.

EXPECTANCY, INSTRUCTION, AND PLACEBO

Nocturnal sleep quality, objectively measured, is not necessarily correlated with subjective daytime sleepiness, and caffeine studies are no exception. Johnson et al. (1990) found that nocturnal sleep was associated with objectively measured sleepiness (MSLT), but not with subjective sleepiness during daytime. Apparently, the subject's expectancy of the efficacy of caffeine determines his or her perception of wakefulness and alertness and may explain the discrepancies found with more objective wakefulness/sleepiness measures. This discrepancy between objective sleep parameters and reported sleep or wakefulness may be a source of invalid conclusions. Caffeine treatment is often given double-blind in order not to influence the experimenter or the subject. Whether such a procedure is justifiable is questionable and may lead to spurious conclusions. Kirsch and Weixel (1988) made subjects believe that they were assigned to the caffeine condition, while in the double-blind condition, subjects knew they might receive a placebo. In both conditions only decaffeinated coffee was present. Both procedures produced different, and in some instances even opposite, effects on pulse rate, systolic blood pressure, and mood.

Christensen et al. (1990) studied the influence of expectancy on the reporting of caffeine-related symptoms in 62 undergraduates. In the expectancy condition with specific instructions on the effects of caffeine, the subjects received a cellulose-filled gelatin capsule that ostensibly was filled with caffeine; in the nonexpectancy condition, this was a placebo. The subjects in the expectancy group reported higher alertness and more caffeine-related symptoms (all $p < .05$) and 90% of them remembered the instructions, compared with 50% among the nonexpectancy group. To discern the pharmacological and expectancy effects of caffeine, a balanced placebo design was used with 100 male psychology students between 18 and 35 years of age who were daily coffee drinkers (1 to 4 cups/d) (Lotshaw et al., 1996). The aim was to separate the active effects of caffeine from the expectancy to have consumed caffeine on mood, performance, and physiological measures. The four conditions concerned caffeinated/decaffeinated coffee. The subjects were told that they would get drug/receive drug; get drug/receive placebo; get placebo/receive drug; or get placebo/receive placebo. After a caffeine abstinence night, the manipulation of expectancies was highly effective on subjects' judgments of caffeine dosage, regardless of actual caffeine content (always 150 mg). Expectancy set and caffeine content were equally powerful and worked additively to affect the subjects' ratings of how much the coffee influenced their mood and performance (digit symbol and trail making). Main effects on blood pressure, pulse rate, and fatigue (POMS), however, were found only in those actually given caffeine, but not in those given no caffeine. An almost identical study was done by Mikalsen et al. (2001). They investigated whether administration of stimuli associated with caffeine elicited conditioned arousal and whether information that a drink contained or did not contain 2 mg/kg of caffeine modulated arousal. The conclusions roughly confirmed those of Lotshaw et al. (1996) that stimuli associated with caffeine increased arousal while information about the content of the drink modulated arousal in the direction indicated by the information. Placebo effects were strongest when both conditioned responses and expectancy-based responses acted in the same direction (Mikalsen et al., 2001). In conclusion, the way in which subjects are informed on aspects concerning the influence of caffeine or whether they think caffeine is at play may determine the experimental outcome.

Withdrawal Effects

Withdrawal effects from regular caffeine consumption is a continuously controversial subject in research on caffeine. The "syndrome" starts, on average, after 12 to 24 h of abstinence and has a peak between 20 and 48 h; it may result in several symptoms, including headaches, irritation, lethargy, anxiety, etc. (Dews et al., 2002); and it may start after a relatively short-term exposure from 6 to 15 d with doses ≥ 600 mg/d. It is puzzling that symptoms have been reported following low amounts of caffeine intake, as well as excess caffeine, and that the prevalence of symptoms reported in different studies covers a range from 11 to 100%. In a 1988 review (Griffiths and Woodson, 1988) only one of eight studies published before 1943 mentioned headache. Since the attention of researchers fell on severe headache upon withdrawal, from this time on reports of headache upon withdrawal became more frequent (Dews et al., 2002). Such complaints are given attention because of the ambivalent attitude to coffee in our culture. Since withdrawal effects may interact with the effects of caffeine treatment, in experiments that have the objective of assessing the effects of caffeine treatment, no instructions should be given concerning abstinence from caffeine before the experiment.

One might expect that in high-level caffeine consumers a change to abstinence would precipitate severe withdrawal symptoms. This need not to be the case, as the following study shows. In Guatemala, the sixth largest producer of coffee, coffee is one of the first drinks given to infants after breast milk. The average caffeine intake of Guatemalan infants is 9 mg/kg/d, which is three times the intake of American adults and over six times the intake of Americans 2 to 17 years of age (Engle et al., 1999). After a sudden change to a caffeine-free diet, followed by a period of 5 months without caffeine, no significant effect of a coffee substitute was found on behavior. As for sleep, the no-coffee group slept 0.4 h more during the night (9.9 vs. 10.3) and 0.5 h longer overall (night plus naps; 10.8 vs. 11.3 h) than children in the coffee group. No differences were found in sleep difficulty or times waking at night. In sum, the minor changes due to an abrupt change of heavy coffee use in 12- to 24-month-old infants may point to an attitude toward coffee that is quite different from that in countries where coffee is appreciated heavily for its flavor and taste but is suspected of having adverse effects on health.

Depending on the state of the consumer, caffeine may have effects contrary to the caffeine withdrawal syndrome in habitual drinkers. The hypnic "alarm clock" headache syndrome, a moderately severe, enduring headache at a consistent time of the night, has been reported by Dodick et al. (1998). This syndrome is a relatively rare, sleep-related, benign headache disorder; those with the syndrome are awakened regularly during sleep by a short-lasting headache. The physiological mechanism of hypnic headache is unknown, although the hypothesis is that it is a perturbation of the chronobiologic rhythms. Characteristically, patients awaken at a consistent time, usually between 1:00 and 3:00 A.M., the time of night just before the trough in the circadian arousal rhythm. Successful prophylaxis can be found by taking coffee by bedtime; if a hypnic headache occurs, a cup of coffee or a caffeine-containing medication may abort a single episode successfully and promote a good sleep (Dodick et al., 1998).

Blaming Coffee, the Placebo Effect

Health complaints, bad mood, and hampered functioning are often attributed to coffee, and also to the lack of coffee. In 8 out of 14 placebo nights, eight subjects estimated their average caffeine intake before bedtime as 275 mg (Mullin et al., 1933). On two of these eight placebo nights, subjects experienced extreme restlessness, which was blamed on a caffeine intake of up to 390 mg. Similar placebo effects were found (Levenson and Bick, 1977) on the auditory threshold. A similar phenomenon is the attribution of health complaints to coffee use. Such a "reverse placebo effect" was found in a study done by Goldstein (1964). When the subjects were told that caffeine was given 30 min before going to bed, wakefulness was minimal. Yet, identical amounts of caffeine, blindly offered, caused more wakefulness. The absence of that part of the effects that is caused by the placebo may be found when subjects are not aware that caffeine is involved. Being aware of having received a substance could be sufficient to induce a placebo effect, but this effect might be absent in those more accustomed to receiving or taking medication. Twelve patients with a history of sleeping problems who routinely received daily medication were studied in a nursing home (Ginsburg and Weintraub, 1976). Treatment with a placebo or 48, 138, or 228 mg of caffeine did not alter sleep, as rated by the nursing staff. Even more surprising was the fact that three patients improved their sleep length after the 138-mg dose. Although this result may give the impression that caffeine may be beneficial to sleep in subjects who are not aware of receiving caffeine and are accustomed to taking medication as a daily routine, in such a situation effects of interactions with other drugs on sleep and alertness cannot be ruled out. Nevertheless, the suggestion is strong that if the objective is to determine the pharmacological effects of caffeine alone it should be given covertly.

DISCUSSION AND CONCLUSION

The objective of this chapter was to focus on the effects of caffeine on sleep and wakefulness as presented in the literature from the past 10 years.

Remarkably, the major part of the research concerned the role of caffeine in naturalistic or simulated work situations. This development in scientific interest follows the change in working situations in our 24-h economy, which now include a large variety of flex-, irregular, and temporary work.

Also, the nature of the research has changed. Caffeine is seen more as an aid to regulate sleep quality than as a substance that disturbs sleep and counters sleepiness. Consistent with this change is the fact that the subject of research has moved from sleepiness or the ability to fall asleep to wakefulness. The ability to remain awake over a long time is of vital importance in work situations that induce partial or total sleep deprivation. Hence, the use of slow-release caffeine, especially in long sleep deprivation periods, is a promising development because of its long efficacy and the almost complete absence of side effects.

The striking feature of the studies done on caffeine in real-life situations is that caffeine has only a few negative effects, which may be explained by the ambivalent attitude to coffee that exists in our culture. In spite of that, in everyday working situations, those who enjoy coffee or other caffeinated beverages have learned to use caffeine in such a way that their functioning and sleep quality benefit from it.

The combination of caffeine and bright light and slow-release caffeine are promising developments and could possibly be implemented in certain working situations that by their nature induce sleep deprivation. Since in such conditions to remain awake is more relevant than combating falling asleep, these opportunities should be preferred over the combination of caffeine and napping.

In addition, more attention should be paid to differences in effects of caffeine on sleep and wakefulness due to age and gender. It is well known that general metabolism and the sleep–wake pattern in the very young and the aged differ from those in the adult and middle-aged. Nevertheless, hardly any attempt has been made to assess the role of gender and lifestyle of the different age-groups in the effects of caffeine on sleep. Most studies are still done in young adults, mostly students.

One factor is a bias in gathering information and reporting on caffeine use; most studies focus exclusively on coffee use. There are only a few studies that include consumption of tea, and hardly any that gather information on chocolate, energized soft drinks, and OTC medication. Medications, although their absolute caffeine content is low, might be important for sleep quality and wakefulness, most notably in middle-aged and elderly people who use medications shortly before going to bed. The consequence might be that sleep complaints are erroneously attributed to coffee and may possibly be exaggerated due to the existing ambivalent attitude toward coffee (Knibbe and De Haan, 1998).

Differences in sensitivity to caffeine may contribute to differences in outcome and are hardly mentioned in the recent research. Fredholm et al. (1999) reported that for the same amount of caffeine ingested, the plasma concentration of methylxanthine can vary among individuals by a factor of 15.9. Mandel (2002) reported on studies that found that a consumption of up to about 6 cups/d resulted in plasma caffeine levels of 2 to 6 µg/l. It is hardly conceivable that such large differences would have no effect on sleep quality and functioning. Also, there is no explanation yet for the evidence that the effects of caffeine are different for those who complain of a bad sleep and for those who enjoy a good sleep or should be different for those who consume hardly any caffeine or are excessive users. Also, the effects of the diurnal pattern of coffee drinking on sleep and waking are not clear, not to mention the influence of interindividual differences such as genetic factors, personality, cultural influences, socioeconomic status, and certain aspects of behavior, such as physical exercise and diet.

Since caffeine is taken often for divergent reasons, positive as well as negative, coffee use could be associated with a complex of factors, positive (good quality of life, success) or negative (stress, tension, ill health), and these factors that may affect sleep more strongly than caffeine itself (Fredholm et al., 1999). The effects of caffeine on sleep and wakefulness should be evaluated for the influence of these factors.

Other factors, alone or in combination, are situational or contextual factors that go together with coffee consumption (Lee, 1992b). Coffee drinking is more than taking caffeine. Coffee use is embedded in a context of rituals, conviviality, social activity, and enjoyment of aroma, taste, and warmth, which form a "gestalt" that could be part of a specific lifestyle. Consequently, depriving people of this gestalt, as is done in most caffeine deprivation studies, will result in worsened sleep, bad mood, and impaired performance (Lane and Phillips-Bute, 1998). To view coffee drinking as a gestalt may offer an alternative explanation for the so-called withdrawal effects of caffeine deprivation, which in fact could be abstinence of all aspects associated with the joy of coffee drinking. Thus, in real-life work situations slow-release caffeine may not be accepted in spite of its obvious advantages over the frequent consumption of coffee.

Caffeine might be considered a stimulant of choice, since it is universally available in different beverages that suit one's taste. It is legal, socially accepted, and has hardly any side effects or abuse potential. It depends on the context whether caffeine is used predominantly to stay awake and alert or whether it is used for social, relaxation, and pleasure reasons, or for all these reasons together. In the former case, one could consider using slow-release caffeine since it has a proven long-term efficacy; in the second case, a good cup of coffee would be preferable; while in the third case, two cups of coffee might suffice.

REFERENCES

Abernathy, D.R., Todd, E.L. and Schwartz, J.B. (1985) Caffeine disposition in obesity. *British Journal of Clinical Pharmacology,* 20, 61–66.

Akerstedt, T. and Ficca, G. (1997) Alertness-enhancing drugs as a countermeasure to fatigue in irregular work hours. *Chronobiology International,* 14, 145–158.

Alford, C., Bhatti, J., Leigh, T., Jamnieson, A. and Hindmarch, I. (1996) Caffeine-induced sleep disruption: effects on waking the following day and its reversal with an hypnotic. *Human Psychopharmacology,* 11, 185–198.

Babkoff, H., French, J., Whitmore, J. and Sutherlin, R. (2002) Single-dose bright light and/or caffeine effects on nocturnal performance. *Aviation, Space and Environmental Medicine,* 73, 341–350.

Barone, J.J. and Roberts, H. (1996) Consumption of caffeine. *Food and Chemical Toxicology,* 34, 119–129.

Bättig, K.A. (1991) Cross-sectional study: coffee consumption, life-style, personality and cardiovascular reactivity. *Internal Report, ETH-Zentrum, Zürich,* pp. 1–27.

Beaumont, M., Batejat, D., Pierard, C., Coste, O., Doireau, P., Van Beers, P. et al. (2001) Slow release caffeine and prolonged (64-h) continuous wakefulness: effects on vigilance and cognitive performance. *Journal of Sleep Research,* 10, 265–276.

Bernstein, G.A., Carrol, M.E., Thuras, P.D., Cosgrove, K.P. and Roth, M.E. (2002) Caffeine dependence in teenagers. *Drug and Alcohol Dependence,* 66, 1–6.

Bonnet, M.H. and Arand, D.L. (1993) Prophylactic naps and caffeine or nocturnal naps for the night shift? *Sleep Research,* 22 (Abstr.), 324.

Bonnet, M.H. and Arand, D.L. (1994a) Impact of naps and caffeine on extended nocturnal performance. *Physiology and Behavior,* 56, 103–109.

Bonnet, M.H. and Arand, D.L. (1994b) The use of prophylactic naps and caffeine to maintain performance during a continuous operation. *Ergonomics,* 37, 1009–1020.

Bonnet, M.H. and Arand, D.L. (1995) We are chronically sleep deprived. *Sleep,* 18, 908–911.

Bonnet, M.H., Gomez, S., Wirth, O. and Arand, D.L. (1995) The use of caffeine versus prophylactic naps in sustained performance. *Sleep,* 18, 97–104.

Brice, C. and Smith, A. (2001) The effects of caffeine on simulated driving, subjective alertness and sustained attention. *Human Psychopharmacology,* 16, 523–531.

Brice, C.F. and Smith, A. (2002) Factors associated with caffeine consumption. *International Journal of Food Sciences and Nutrition,* 53, 55–64.

Brown, J., Kreiger, N., Darlington, G.A. and Sloan, M. (2001) Misclassification of exposure: coffee as a surrogate for caffeine intake. *American Journal of Epidemiology,* 153, 815–820.

Chagoya de Sánchez, V. (1995) Circadian variations of adenosine and of its metabolism. Could adenosine be a molecular oscillator for circadian rhythms? *Canadian Journal of Physiology and Pharmacology,* 73, 339–355.

Chang, P.P., Ford, D.E., Mead, L.A., Cooper-Patrick, L. and Klag, M.J. (1997) Insomnia in young men and subsequent depression: the Johns Hopkins Precursors Study. *American Journal of Epidemiology,* 146, 105–114.

Christensen, L., White, B., Krietsch, K. and Steele, G. (1990) Expectancy effects in caffeine research. *The International Journal of Addictions,* 25, 27–31.

Curless, R., French, J.M., James, O.F. and Wynne, H.A. (1993) Is caffeine a factor in subjective insomnia of elderly people? *Age and Ageing,* 22, 41–45.

Currie, S.R., Wilson, K.G. and Gauthier, S. T. (1995) Caffeine and chronic low back pain. *The Clinical Journal of Pain,* 11, 214–219.

De Groen, J.H., Op den Velde, W., Hovens, J.E., Falger, P.R., Schouten, E.G. and van Duijn, H. (1993) Snoring and anxiety dreams. *Sleep,* 16, 35–36.

Dekker, D.K., Paley, M.J., Popkin, S.M. and Tepas, D.I. (1993) Locomotive engineers and their spouses: coffee consumption, mood, and sleep reports. *Ergonomics,* 36, 233–238.

Dekker, P., Van het Reve, C. and Den Hartog, A.P. (1993) Koffieverbruik en koffiegewoonten in Nederland [Coffee consumption and coffee habits in the Netherlands]. *Voeding,* 54, 6–9.

Dews, P.B., O'Brien, C.P. and Bergman, J. (2002) Caffeine: behavioral effects of withdrawal and related issues. *Food and Chemical Toxicology,* 40, 1257–1261.

Dodick, D.W., Mosek, A.C. and Campbell, J.K. (1998) The hypnic ("alarm clock") headache syndrome. *Cephalalgia,* 18, 152–156.

Engle, P.L., VasDias, T., Howard, I., Romero-Abal, M.E., Quan de Serrano, J., Bulux, J. et al. (1999) Effects of discontinuing coffee intake on iron deficient Guatemalan toddlers' cognitive development and sleep. *Early Human Development,* 53, 251–269.

Engleman, H., Ronald, P. and Shapiro, C.M. (1990) Effect of caffeine and sleep deprivation on daytime performance, paper presented at Sleep '90, Proceedings of the Tenth European Congress on Sleep Research, Strasbourg, France.

Fabsitz, R.R., Sholinsky, P. and Goldberg, J. (1997) Correlates of sleep problems among men: the Vietnam Era Twin Registry. *Journal of Sleep Research,* 6, 50–56.

Fredholm, B.B., Bättig, K., Holmén, J., Nehlig, A. and Zvartau, E.E. (1999) Actions of caffeine in the brain with special reference to factors that contribute to its widespread use. *Pharmacoogical Reviews,* 51, 83–133.

Gilmore, M.M. and Murphy, C. (1989) Aging is associated with increased Weber ratios for caffeine, but not for sucrose. *Perception and Psychophysics,* 46, 555–559.

Ginsburg, R. and Weintraub, M. (1976) Caffeine in the "sundown syndrome": report of negative results. *Journal of Gerontology,* 31, 419–420.

Goldstein, A. (1964) Wakefulness caused by caffeine. *Naunyn–Schmiedebergs' Archives of Experimental Pathologie und Pharmakologie,* 248, 269–278.

Goldstein, A., Warren, R. and Kaizer, S. (1965) Psychotropic effects of caffeine in man. I. Individual differences in sensitivity to caffeine-induced wakefulness. *Journal of Pharmacology and Experimental Therapeutics,* 149, 156–159.

Greenwood, K.M., Rich, W.J. and James, J.E. (1995) Sleep hygiene practices and sleep duration in rotating shift-workers. *Work and Stress,* 9, 262–271.

Griffiths, R.R. and Woodson, P.P. (1988) Caffeine physical dependence: a review of human and laboratory animal studies. *Psychopharmacology,* 94, 437–451.

Harris Research Centre (1996) The value of pleasure and the question of guilt: international data tabulations, in *Research Report JN66029,* Simpson, T. and Bellchambers, L., Eds., p. 220.

Hindmarch, I., Rigney, U., Stanley, N., Quinlan, P., Rycroft, J. and Lane, J. (2000) A naturalistic investigation of the effects of day-long consumption of tea, coffee and water on alertness, sleep onset and sleep quality. *Psychopharmacology,* 149, 203–216.

Höfer, I., Kos, J. and Bättig, K. (1993) Effects of caffeinated vs. decaffeinated coffee: a blind field study, Proceedings 15th International Scientific Colloquium on Coffee, Vol. II, pp. 470–475.

Horne, F.A. and Reyner, L.A. (1996) Counteracting driver sleepiness: effects of napping, caffeine and placebo. *Psychophysiology,* 33, 306–309.

Hughes, W.K., Olivetto, A.H., Bickel, W.K., Higgins, S.T. and Badger, G.J. (1995) The ability of low doses of caffeine to serve as reinforcers in humans: a replication. *Experimental and Clinical Psychopharmacology,* 3, 358–363.

James, J.E., Bruce, M.S., Lader, M.H. and Scott, N.R. (1989) Self-report reliability and symptomatology of habitual caffeine consumption. *British Journal of Clinical Pharmacology,* 27, 507–514.

Janson, C., Gislason, T., De Backer, W., Plaschke, P., Bjornsson, E., Hetta, J. et al. (1995) Prevalence of sleep disturbances among young adults in three European countries. *Sleep,* 18, 589–597.

Johnson, L.C., Spinweber, C.L. and Gomez, S.A. (1990) Benzodiazepines and caffeine: effect on daytime sleepiness, performance and mood. *Psychopharmacology,* 101, 160–167.

Kamimori, G., Lugo, S.I., Penetar, D.M., Chamberlain, A.C., Brunhart, G.E., Brunhart, A.E. and Eddington, N.D. (1995) Dose-dependent caffeine pharmacokinetics during severe sleep deprivation. *International Journal of Clinical Pharmacology and Therapeutics,* 33, 182–186.

Kamimori, G.H., Penetar, D.M., Headley, D.B., Thorne, D.R., Otterstetter, R. and Belenky, G. (2000) Effect of three caffeine doses on plasma catecholamines and alertness during prolonged wakefulness. *European Journal of Clinical Pharmacology,* 56, 537–544.

Kelly, T.L., Mitler, M.M. and Bonnet, M.H. (1997) Sleep latency measures of caffeine effects during sleep deprivation. *Electroencephalography and Clinical Neurophysiology,* 102, 397–400.

Kirsch, I. and Weixel, L.J. (1988) Double-blind versus deceptive administration of a placebo. *Behavioral Neuroscience,* 102, 319–323.

Knibbe, R.A. and De Haan, Y.T. (1998) Coffee consumption and subjective health: interrelations with tobacco and alcohol, in *Nicotine, Caffeine and Social Drinking: Behaviour and Brain Function,* Snel, J. and Lorist, M.M., Eds., Harwood Academic Publishers, London, pp. 229–243.

Lagarde, D., Batejat, D., Sicard, B., Trocherie, S., Chassard, D., Enslen, M. and Chauffard, F. (2000) Slow-release caffeine: a new response to the effects of a limited sleep deprivation. *Sleep,* 23, 651–661.

Landolt, H.P., Werth, E., Borbely, A.A. and Dijk, D.J. (1994) Effect of caffeine (200 mg) administered in the morning on sleep and EEG power spectra at night. *Sleep Research,* 23, 68 (Abstract).

Landolt, H.P., Werth, E., Borbely, A.A. and Dijk, D.J. (1995) Caffeine intake (200 mg) in the morning affects human sleep and EEG power spectra at night. *Brain Research,* 675, 67–74.

Lane, J.D. (1994) Neuroendocrine responses to caffeine in the work environment. *Psychosomatic Medicine,* 56, 267–270.

Lane, J.D. and Phillips-Bute, B.G. (1998) Caffeine deprivation affects vigilance performance and mood. *Physiology and Behavior,* 65, 171–175.

Lee, K.A. (1992) Self-reported sleep disturbances in employed women. *Sleep,* 15, 493–498.

Lee, S. (1992) Does coffee really cause insomnia? *Tea and Coffee Trade JNY,* 164, 3–4.

Levenson, H.S. and Bick, E.C. (1977) Psychopharmacology of caffeine, in *Psychopharmacology in the Practice of Medicine,* Jarvik, M.E., Ed., Plenum Press, New York, pp. 451–467.

Lieberman, H.R., Wurtman, R.J., Emde, G.G., Roberts, C., Coviella, I.L., Tharion, W.J. et al. (2002) Effects of caffeine, sleep loss, and stress on cognitive performance and mood during U.S. Naval SEAL training. *Psychopharmacology,* 164, 250–261.

Liguori, A., Hughes, J.R. and Grass, J.A. (1997) Absorption and subjective effects of caffeine from coffee, cola and capsules. *Pharmacology, Biochemistry and Behavior,* 58, 721–726.

Lin, A.S., Uhde, T.W., Slate, S.O. and McCann, U.D. (1997) Effects of intravenous caffeine administered to healthy males during sleep. *Depression and Anxiety,* 5, 21–28.

Lotshaw, S.C., Bradley, J.R. and Brooks, L.R. (1996) Illustrating caffeine pharmacological and expectancy effects utilizing a balanced placebo design. *Journal of Drug Education,* 26, 13–24.

Manber, R., Bootzin, R.R., Acebo, C. and Carskadon, M.A. (1996) The effects of regularizing sleep-wake schedules on daytime sleepiness. *Sleep,* 19, 432–441.

Mandel, H.G. (2002) Update on caffeine consumption, disposition and action. *Food and Chemical Toxicology,* 40, 1231–1234.

Mikalsen, A., Bertelsen, B. and Flaten, M.A. (2001) Effects of caffeine, caffeine-associated stimuli, and caffeine related information on physiological and psychological arousal. *Psychopharmacology,* 157, 373–380.

Mitler, M.M., Walsleben, J., Sangal, R.B. and Hirshkowitz, M. (1998) Sleep latency on the maintenance of wakefulness test (MWT) for 530 patients with narcolepsy while free of psychoactive drugs. *Electroencephalography and Clinical Neurophysiology,* 107, 33–38.

Moline, M.L., Zendell, S.M., Pollak, C.P. and Wagner, D.R. (1994) Acute effects of caffeine on subjective alertness in simulated jet-lag. *Sleep Research,* 23 (Abstr.), 505.

Muehlbach, M.J. and Walsh, J.K. (1993) The effects of caffeine on simulated night shift work. *Sleep Research,* 22, 412.

Muehlbach, M.J. and Walsh, J.K. (1995) The effects of caffeine on simulated night-shift work and subsequent daytime sleep. *Sleep,* 18, 22–29.

Mullin, F.J., Kleitman, N. and Cooperman, N. (1933) Studies on the physiology of sleep X. The effects of alcohol and caffeine on motility and body temperature during sleep. *American Journal of Physiology,* 106, 478–487.

Murphy, C. and Gilmore, M.M. (1989) Quality-specific effects of aging on the human taste system. *Perception and Psychophysics,* 46, 121–128.

NAH (1996) Caffeine: the inside scoop — the good, the bad, and the myth. Nutrition Action Health Letter, www.cspinet.org/nah/good bad.htm, Retrieved Nov. 13, 2002.

Ohayon, M.M. and Roth, T. (2002) Prevalence of restless legs syndrome and periodic limb movement disorder in the general population. *Journal of Psychosomatic Research,* 53, 547–554.

Ohayon, M.M., Guilleminault, C., Priest, R.G. and Caulet, M. (1997) Snoring and breathing pauses during sleep: telephone interview survey of a United Kingdom population sample. PG-860-3. *British Medical Journal,* 314, 860–863.

Patat, A., Rosenzweig, P., Enslen, M., Trocherie, S., Miget, N., Bozon, M.C. et al. (2000) Effects of a new slow release formulation of caffeine on EEG, psychomotor and cognitive functions in sleep-deprived subjects. *Human Psychopharmacology,* 15, 153–170.

Penetar, D.M., McCann, U.D., Thorne, D.R., Kamimori, G., Galinski, C.L., Singh, H. et al. (1993) Caffeine reversal of sleep deprivation effects on altertness and mood. *Psychophysiology,* 112, 359–365.

Porkka-Heiskanen, T. (1999) Adenosine in sleep and wakefulness. *Annals of Medicine,* 31, 125–129.

Reyner, L.A. and Horne, J.A. (1997) Caffeine combined with a short nap effectively counteracts driver sleepiness. *Sleep Research,* 269 (Abstr.), 625.

Reyner, L.A. and Horne, J.A. (2000) Early morning driver sleepiness: effectiveness of 200 mg caffeine. *Psychophysiology,* 37, 252–256.

Rosenthal, L., Roehrs, T., Zwyghuizen-Doorenbos, A., Plath, D. and Roth, T. (1991) Alerting effects of caffeine after normal and restricted sleep. *Neuropsychopharmacology,* 4, 103–108.

Schreiber, G.B., Maffeo, C.E., Robins, M., Masters, M.N. and Bond, A.P. (1988) Measurement of coffee and caffeine intake: implications for epidemiologic research. *Preventive Medicine,* 17, 280–294.

Shilo, L., Sabbah, H., Hadari, R., Kovatz, S., Weinberg, U., Dolev, S. et al. (2002) The effects of coffee consumption on sleep and melatonin secretion. *Sleep Medicine,* 3, 271–273.

Shohet, K.L. and Landrum, R.E. (2001) Caffeine consumption questionnaire: a standardized measure for caffeine consumption in undergraduate students. *Psychonomic Reports,* 89, 521–526.

Sicard, B.A., Perault, M.C., Enslen, M., Chauffard, F., Vandel, B. and Tachon, P. (1996) The effects of 600 mg of slow release caffeine on mood and alertness. *Aviation, Space and Environmantal Medicine,* 67, 859–862.

Smith, A.P. (1994) Caffeine, performance, mood and status of reduced alertness. *Pharmacopsychoecologia,* 7, 75–86.

Snel, J. (1993) Coffee and caffeine: sleep and wakefulness, in *Caffeine, Coffee and Health,* Garattini, S., Raven Press, New York, pp. 255–290.

Van Dongen, H.P., Price, N.J., Mullington, J.M., Szuba, M.P., Kapoor, S.C. and Dinges, D.F. (2001) Caffeine eliminates psychomotor vigilance deficits from sleep inertia. *Sleep,* 24, 813–819.

Walsh, J.K., Muehlbach, M.J. and Schweitzer, P. (1995) Hypnotics and caffeine as countermeasures for shiftwork-related slepiness and sleep distrurbances. *Journal of Sleep Research,* 4, 80–83.

Wright, K.P., Badia, P., Myers, B.L., Plenzler, S.C. and Drake, C.L. (1997a) Caffeine and bright light effects on nighttime melatonin and body temperature in women during sleep deprivation. *Sleep Research,* 26 (Abstr.), 636.

Wright, K.P.J., Badia, P., Myers, B.L. and Plenzler, S.C. (1997b) Combination of bright light and caffeine as a countermeasure for impaired alertness and performance during extended sleep deprivation. *Journal of Sleep Research,* 6, 26–35.

3 Arousal and Behavior: Biopsychological Effects of Caffeine

Barry D. Smith, Amanda Osborne, Mark Mann, Heather Jones, and Thom White

CONTENTS

Introduction 35
Genetic and Neurophysiological Substrates 36
A Theoretical Model of Arousal and Caffeine 36
Cognition and Performance 37
    Psychomotor Functions..37
    Vigilance 37
    Selective Attention 38
    Memory 38
Effects of Caffeine on Mood 39
    Alertness..39
    Positive Mood States 39
    Complications and Caveats...40
Caffeine Abuse and Dependence 40
    Caffeine Dependence 41
    Epidemiology 41
    Caffeine Intoxication 42
    Caffeine and Anxiety Disorders 42
    The Role of Caffeine in Other Disorders..43
    Psychopharmacological Treatment: Caffeine Interactions 44
Caffeine and Pain..44
References 45

INTRODUCTION

The biobehavioral effects of caffeine on psychological functioning and behavior are mediated by underlying genetic mechanisms and their neurophysiological expressions. Primarily by blocking adenosine A_{2A} receptors, caffeine affects cognitive and psychomotor functions and mood states. A multidimensional, biobehavioral arousal model explains a number of these effects. The drug tends to decrease reaction times and improve vigilance performance, though its effects on memory are poorly understood. Caffeine increases alertness, the psychological representation of physiological arousal, and, in low-to-moderate doses, increases positive mood states. High doses can produce negative affect, but there is considerable individual variability in caffeine sensitivity, based on genetic predispositions and prior experience. Available statistics suggest that caffeine dependence and abuse may be quite widespread, in some cases producing caffeine intoxication (caffeinism) and exacerbating some anxiety disorders. Caffeine effects are also seen in depression, schizophrenia, eating disorders, attention deficit hyperactivity disorder (ADHD), and restless legs syndrome (RLS).

Caffeine has been shown to affect several aspects of psychological and related physiological functioning. We will begin by considering the genetic and neural mechanisms that underlie the biobehavioral effects of caffeine, then address specific areas of psychological functioning that are affected, including cognition, mood state, anxiety disorders, caffeine intoxication (caffeinism), other psychiatric disorders, psychopharmacological treatment, and pain perception. Both beneficial and detrimental effects of both acute and chronic caffeine ingestion are considered.

GENETIC AND NEUROPHYSIOLOGICAL SUBSTRATES

The behavioral effects of caffeine are strongly influenced by genetic mechanisms and their neurophysiological expressions. Both the extent of habitual caffeine consumption and the magnitude of responses to single doses appear to be subject to a hereditary predisposition (Carrillo et al., 1998). The effect likely occurs through cytochrome P450 CYP1A2, which is involved in caffeine metabolism. It is governed principally by genetic mechanisms, with a heritability estimate of 0.725 (Rasmussen et al., 2002). The genetic predisposition is expressed, at least in part, as the bimodally distributed ability to acetylate molecules possessing an amino functional group. The principal direct genetic effect is likely on the adenosine A_{2A} receptors, which are inhibited by caffeine (Lindskog et al., 2002). These receptors are found in the kidneys, digestive system, bronchial tree, heart and peripheral vasculature, and in the brain.

In the brain, adenosine acts as a neurotransmitter. It is synthesized in glial cells and neurons, and its release into extracellular space is enhanced during states of fatigue and sleep (Adrien, 2001). The adenosine A_{2A} receptors act, in part, by inhibiting the N-methyl-D-aspartate (NMDA) component of excitatory synaptic currents (Gerevich et al., 2002), and the neural distribution of these receptors suggests some probable sites of action of caffeine in the brain. Included are the striatum and medulla, as well as portions of the basal forebrain, the mesopontine area, and the sleep-regulating preoptic nucleus of the hypthothalamus (Boros et al., 2002). The nucleus acubens (Solinas et al., 2002) and the lateral amygdala (Svenningsson et al., 1999) may also be involved. Caffeine probably produces its stimulatory effect, in part, by blocking the A_{2A} receptors that activate the GABAergic neurons populating the inhibitory tracts to the striatal dopaminergic reward system (Daly and Fredholm, 1998). The neural distribution of adenosine receptors is also consistent with the apparent neuroprotective effects of caffeine in Parkinson's disease (Chen et al., 2001), where it may act to prevent dopamine deficits, and in Alzheimer's disease (Maia and de Mendonca, 2002).

A THEORETICAL MODEL OF AROUSAL AND CAFFEINE

The neural inhibitory action of caffeine on adenosine receptor substrates has numerous downstream effects on psychological functioning and behavior. We have cast these varied effects within the context of a set of theoretical models. The most encompassing of these models is dual-interaction theory, which postulates that both physiological functioning and behavior are ultimately products of the characteristics of the person, the situational context, and the interaction of these two determinants (Smith et al., 1999, 2002; Mann et al., 2002). Our multidimensional model of the arousal component of the dual-interaction theory postulates that biological and environmental background factors contribute separately and interactively to both chronic and acute arousal. Chronic exposure to an arousal agent, such as caffeine, contributes to arousal traits and thereby affects arousal states, while acute exposure contributes directly to the current arousal state. Both traits and states are multidimensional, and there are three broad dimensions of trait arousal: intensity, type (emotional or general), and individual differences. Arousal traits and states then contribute separately and interactively to current physiological functioning and to behavior. We have elsewhere outlined four general theoretical principles and five specific hypotheses that deal with the effects of caffeine (Smith et al., 1999), and these have received support from our own work and that of others (Quinlan et al., 2000; Smith et al., 2000, 2001; Wilken et al., 2000).

COGNITION AND PERFORMANCE

The alerting properties of caffeine and its possible beneficial effects on motor and cognitive performance have long been touted. Legend has it, in fact, that an observant goatherd named Kaldi discovered coffee in Ethiopia somewhere between about 300 and 800 A.D. (Smith et al., 2002). He noticed that his goats did not sleep at night after eating coffee berries. He took the berries to a local abbot, who brewed the first batch of coffee, noting its effects on arousal and cognition. While we do not know whether the Kaldi story is fact or fable, the effects of caffeine on cognitive and motor performance have certainly long been noted. In what follows, we briefly review the effects of both acute and habitual caffeine consumption on a number of psychomotor and cognitive functions.

Psychomotor Functions

Research has demonstrated that caffeine often decreases both simple and choice reaction times (RTs). These effects have been seen in the evening (Babkoff et al., 2002), in elderly people (Bryant et al., 1998), at normal room temperatures but not at unusually low temperatures (Kruk et al., 2001), and across a variety of tasks, conditions, and groups (Durlach, 1998; Smit and Rogers, 2000). There are, however, contrary findings, showing no effect at higher dosage levels (Roache and Griffiths, 1987), in elderly participants (Swift and Tiplady, 1988), or in self-paced RT tasks, particularly when there has been no prior sleep deprivation (Ruijter et al., 2000). In fact, one recent study demonstrated a reversal, in which caffeine actually increased RT (Hespel et al., 2002). The variability in results appears to reflect differences in groups, tasks, dosage levels, and amount of sleep deprivation, and more research will be required if these are to be fully addressed.

Vigilance

Studies of vigilance and sustained attention have been more consistent, typically showing that caffeine improves these forms of cognitive performance. On prolonged vigilance tasks, subjects display improved performance levels (Loke and Meliska, 1984; Linde, 1994; Stein et al., 1996; Lin et al., 1997; Rogers and Dernoncourt, 1998; Rees et al., 1999; Gilbert et al., 2000; Beaumont et al., 2001; Van Dongen et al., 2001). Some work suggests that 300 mg of caffeine improves vigilance performance (Gilbert et al., 2000), while 600 mg produces no further increment (Lagarde et al., 2000), and improvements in this dosage range may be greater in older than in younger groups (Rees et al., 1999). The latter study employed a newer, slow-release caffeine preparation that may prove useful in certain work settings (Sicard et al., 1996). Positive effects have also been seen at lower doses. Lieberman et al. (1987), for example, have shown that even 32 mg can increase the number of correct hits in a vigilance task without affecting error rate. Further studies show caffeine-induced increments even on brief tasks (Temple et al., 2000) and demonstrate that caffeine deprivation impairs performance in regular users (Lane and Phillips-Bute, 1998).

Although some earlier work suggested that caffeine had negative effects on cognitive functioning in children, more recent work fails to support this view (Stein et al., 1996). Kupietz and Winsberg (1977) found that some children exhibit improved functioning in both auditory vigilance and sustained attention tasks. The beneficial effects of caffeine have also been seen in adolescents (Tynjala et al., 1997), and Loke (1988) found that moderate doses of the drug (200 mg) decrease self-reported feelings of boredom, possibly providing a partial explanation for the positive effects on vigilance and sustained-attention tasks. These same results were seen in a population of middle-aged subjects (Landolt et al., 1996). In fact, a reviewer a few years ago noted an improvement in vigilance performance in 14 of 17 studies of adolescents (Koelega, 1993). A few studies have found no caffeine effect (Loke and Meliska, 1984; Linde 1995) or mixed effects (Lieberman et al., 1987; Smith, 1994), but a vast majority are positive.

Despite quite consistent findings, the mechanism of the caffeine effect on vigilance performance remains somewhat elusive. The dominant attribution, however, has been to the increased arousal level produced by the drug (Lieberman et al., 1987; Akerstedt and Ficca, 1997; Giam, 1997; Kelly et al., 1997; Smith et al., 1999). It appears that the effect of caffeine is to increase processing speed, rather than reducing distraction or affecting output (Lorist and Snel, 1997; Streufert et al., 1997; Smith et al., 2001; Smith, 2002). Consistent with this conclusion is the observation that the size of the arousal increment greatly influences performance (Brown et al., 1995; Landolt et al., 1996; Smith et al., 1999), which tracks dosage up to a point. However, it also appears that further increases in dose beyond that may actually cause performance to asymptote or deteriorate. Indeed, some studies have shown that either large doses of caffeine or the combination of caffeine and environmental stressors can reverse the positive effects of caffeine on performance (Frewer and Lader, 1991; Smith, et al., 1991). Smith et al. (1999) have hypothesized that these findings further support the theoretical inverted U-shaped arousal function.

SELECTIVE ATTENTION

Caffeine also enhances performance on selective attention tasks (Lorist et al., 1995; Warburton, 1995). For example, Lorist and Snel (1997) report improved response times with 3 mg/kg caffeine, with no decrease in accuracy. Other studies have shown that the drug improves performance on divided attention tasks (Pons et al., 1988) and increases the ability to self-focus (Debrah et al., 1996). Studies involving evoked potential responses have isolated the P300 component of the electroencephalogram (EEG) as a measure of visual or auditory attention, demonstrating that large P300 waves are associated with more intensive attention (Lorist et al., 1995; Wijers et al., 1997). Using the oddball paradigm, in which subjects attend to a stimulus in an effort to notice any change in that stimulus, researchers have shown that the drug increases the size of the P300 component, evidence that caffeine can enhance attention (Kenemans and Lorist, 1995; Kawamura et al., 1996).

MEMORY

Far less consistent are studies of the effects of caffeine on memory. Some investigators have shown enhanced performance on delayed recall (Warburton, 1995), recognition memory (Bowyer et al., 1983), semantic memory (Oborne and Rogers, 1983), working memory (Sawyer et al., 1982), and verbal memory tasks (Jarvis, 1993). Other work has demonstrated improved performance on short- and long-term memory retrieval and on encoding efficiency (Riedel et al., 1995; Smith et al., 1999). Additional investigators have reported increases in memory performance on both easy (Anderson and Revelle, 1983) and difficult (Loke, 1988) memory tasks and with both lower and higher doses following exercise (Hogervorst et al., 1999). Finally, caffeine counteracts the detrimental effects of aging on general memory performance (Riedel and Jolles, 1996) and reduces performance decrements over the course of the day in older subjects (Ryan et al., 2002).

Despite these findings, many studies have failed to demonstrate the positive effects of caffeine on memory. In some studies, caffeine has had no effect on immediate (Arnold et al., 1987; Landrum et al., 1988; Smith, 1994; Linde, 1995; Warburton, 1995) or delayed (Loke et al., 1985) recall or on short-term memory (Clubley et al., 1979; Hindmarch et al., 1998; Lieberman, 1988), long-term memory (Lieberman, 1988), verbal learning (Clubley et al., 1979; File et al., 1982; Herz, 1999), working memory (Smith et al., 1999), implicit or incidental memory (Turner, 1993), verbal or nonverbal memory (Warburton et al., 2001), or spatial learning (Battig et al., 1984). Indeed, caffeine has even been shown to decrease immediate or delayed word list recall under some circumstances (Loke et al., 1985; Terry and Phifer, 1986; Roache and Griffiths, 1987; Lieberman, 1988; Loke, 1988).

The discrepancies among memory studies are difficult to explain because the studies vary from one to another in memory assessment method, time frame (immediate vs. delayed), sex of subjects, and age of subjects. One possible explanation lies in the inverted U-shaped arousal-performance function (Smith et al., 1999). Supporting this conceptualization, Anderson and Revelle (1983), for example, reported that caffeine enhanced performance on low-load memory tasks and impaired it on high-load tasks. Similarly, a study in our laboratories showed that arousal increments produced by novel stimuli and white noise improved performance on a backward recall task, while the further addition of caffeine-induced arousal decreased performance (Davidson and Smith, 1991). Kaplan et al. (1997) found that low doses enhanced working memory performance, while higher doses decreased it.

EFFECTS OF CAFFEINE ON MOOD

Research on the mood effects of caffeine is complicated by differing definitions of mood, variations in drug dose, prior caffeine deprivation conditions, habitual caffeine usage levels of participants, and individual differences in caffeine sensitivity. The terminology and methodology used in defining mood states varies considerably from one laboratory to another. Some have used such formal definitions as those provided by the Profile of Mood States (POMS; Lorr and Wunderlich, 1998), the Activation-Deactivation Adjective Check List (Thayer, 1978), or the Positive Affect Negative Affect Schedule (PANAS; Watson et al., 1988). Others have employed Likert scales assessing a variety of mood terms (e.g., happy, sad, depressed, anxious, pleasant, calm, contented), and still others have used a set of factor analytic dimensions (Smit and Rogers, 2002) that accord with those proposed by Thayer (1978). They include Energetic Arousal, defined by such adjectives as "active," "elated," "enthusiastic," and "excited"; Hedonic Tone (pleasantness, high positive affect); and Tense Arousal, defined by such descriptors as "tense," "distressed," and "jittery." Despite the variability in mood definitions, research has provided a fairly consistent picture of the effects of caffeine on mood state.

ALERTNESS

It can be argued that the "purest" or most basic psychological effect of caffeine is an increase in feelings of alertness produced by the arousing effects of the drug as it blocks adenosine receptor action. In effect, the physiological state of arousal produces the psychological state of alertness. While some might argue that alertness is not, strictly speaking, a mood state, it certainly is a psychological state, and increases in alertness are among the most consistent effects of caffeine (Leathwood and Pollet, 1982–1983; Warburton, 1995; Hindmarch et al., 1998; Kamimori et al., 2000). Although it does occur under normal waking conditions, the effect of caffeine on alertness is more pronounced under low arousal conditions, as may occur early in the morning (Smith et al., 1992) or at night (Smith et al., 1993) and following sleep deprivation (Brauer et al., 1994). More generally, the drug may have greater alerting effects in fatigued subjects (Rusted and Yeomans, 2002) and under demanding performance conditions (Rusted, 1999).

POSITIVE MOOD STATES

In low to moderate doses, caffeine has been quite consistently shown to increase positive affect. Reflecting the varied definitions of positive affect in the literature, the drug has been found to increase feelings of well-being (Smith et al., 1991; Schuh and Griffiths, 1997), calmness (Yeomans et al., 2002), contentedness (Mikalsen et al., 2001), energetic arousal (Liguori et al., 1997; Quinlan et al., 1997, 2000; Robelin and Rogers, 1998; Herz, 1999; Smit and Rogers, 2000; Lieberman,

2001), and hedonic tone (Liguori et al., 1997; Quinlan et al., 1997, 2000) and to decrease uncertainty (Yeomans et al., 2002). Further evidence comes from studies showing decreases in feelings of friendliness (Jones et al., 2002), energetic arousal (Robelin and Rogers, 1998; Evans and Griffiths, 1999), and hedonic tone (Phillips-Bute and Lane, 1997; Evans and Griffiths, 1999) when regular users are deprived of caffeine.

COMPLICATIONS AND CAVEATS

Although numerous studies support the positive effects of caffeine on mood, high doses of the drug can produce negative mood state effects. Studies suggest that caffeine in high doses increases anxiety (Green and Suls, 1996; Sicard et al., 1996), tense arousal (Penetar et al., 1993), and negative mood state more generally (Silverman and Griffiths, 1992; Brauer et al., 1994; Liguori et al., 1997). Reports of negative affect at high doses are further supported by multiple-dose studies showing an inverted U-shaped relationship, in which positive affect shows a small dose-response relationship up to a point, with further dosage increases producing more negative mood states (Quinlan et al., 2000). This observation is consistent with our multidimensional arousal model (Smith et al., 1999; Mann et al., 2002).

If some doses of caffeine produce positive and others negative moods, one obvious issue is the dosage at which the change takes place. Lieberman (1992) suggested that positive effects appear to occur up to about 300 mg, then negative effects above that. This estimate of the peak of the inverted-U function would appear to be quite reasonable for the average study participant. However, other research points to substantial individual differences in caffeine sensitivity (Silverman and Griffiths, 1992; Mumford et al., 1994) and therefore in the response to a given dose of the drug. One study, for example, showed that subjects who chose to receive a high dose of caffeine reported an increase in positive affect, whereas those who chose not to receive the caffeine reported negative mood changes when the drug was administered (Stern et al., 1989). Anecdotally, participants in our laboratory occasionally refuse consent because they believe they are highly sensitive to caffeine. These individual differences in responses to the drug may well reflect the genetic underpinnings of caffeine sensitivity and preference noted above. In addition, personal history of caffeine consumption may influence current responses to the drug.

Overall, it appears that low doses of caffeine tend to improve mood states and that high doses are associated with negative affective change. Mood alteration in either direction may be influenced by the individual's arousal state at the time of drug administration, with lower preexisting arousal levels yielding larger drug effects. If the dose is "moderate" (with doses somewhat variable from study to study), individual differences in caffeine sensitivity very likely play a more substantial role in determining the positive or negative mood effects of the drug.

CAFFEINE ABUSE AND DEPENDENCE

As the most widely used psychoactive drug in the world, caffeine has considerable potential for abuse. Its adenosine receptor-mediated effects on the central nervous system serve to establish it as potentially addictive (Gilliland and Bullock, 1984; Nehlig, 1999; Griffiths and Chausmer, 2000), and the well-documented behavioral reinforcing properties of the drug further increase that probability (Griffiths and Chausmer, 2000). Caffeine displays both positive reinforcing characteristics, exemplified by the temporary enhancement of cognitive performance (Ryan et al., 2002), and negative reinforcing attributes, seen in its ability to relieve withdrawal symptoms (Bernstein et al., 2002). Both the powerful stimulant properties of caffeine and its abuse potential have peaked the interest of many scientists and led to the conduct of extensive research focused on the role of the drug in physiology and behavior, including that involved in psychopathology. As one indication of this escalating empirical effort, a search of Medline reveals a total of 1094 publications between 1971 and 1980, 4743 between 1981 and 1990, and 6476 between 1991 and 2000. By way of comparison, this latter number is higher than those for marijuana (2704), amphetamine (5683), or heroin (2991).

CAFFEINE DEPENDENCE

In the *Diagnostic and Statistical Manual of Mental Disorders — Fourth Edition* (DSM-IV; American Psychiatric Association, 1994), a diagnosis of substance abuse includes substance-related occupational, interpersonal, social, and psychological consequences. Dependence criteria include tolerance, withdrawal, a strong desire or unsuccessful attempt to stop usage, spending a great deal of time with the drug, using more than intended, use despite knowledge of harm, and foregoing other activities to use. The *International Classification of Diseases — Tenth Revision* (ICD-10; World Health Organization, 1992) uses the term *harmful use* instead of *abuse* and classifies it as a pattern of use that is health-damaging. Its dependence criteria overlap with those in the DSM-IV but also include a "compulsive use" criterion. Despite these formal definitions of the terms *abuse* and *dependence*, the terms are often used interchangeably in the literature (Griffiths et al., 1996; Holtzman, 1990).

EPIDEMIOLOGY

Though firm prevalence rates have yet to be established, caffeine dependence has been examined by measuring the endorsement of the DSM-IV dependence criteria (Hughes et al., 1992, 1998; Strain et al., 1994). Hughes et al. (1998) tested 162 random caffeine users and found that the most commonly endorsed criterion was a strong desire or unsuccessful attempt to stop usage. Another investigation employed a structured clinical interview based on the dependence criteria in the DSM and tailored specifically for caffeine (Strain et al., 1994). Results showed that the most frequently endorsed dependence criteria were the presence of withdrawal symptoms and continued use despite psychological or physiological harm. A third study recently investigated the presence of caffeine dependence in a small sample of teenagers. Using the Diagnostic Interview Scale for Children, with substance dependence criteria modified for caffeine use, it was found that withdrawal criteria were met by 77.8% of those consuming caffeine and tolerance criteria by 41.7% of respondents (Bernstein et al., 2002). Based on these and other studies, current estimates suggest prevalence rates ranging from 9 to 30% in caffeine consumers (Griffiths and Chausmer, 2000).

Given these prevalence rates, consider that about 80% of U.S. adults consume an average of 4.2 mg/kg/day of caffeine, equivalent to about four cups of coffee (Denaro et al., 1990), and many consume more than 15 mg/kg/day (Mandel, 2002). Even if the entire population, including non-coffee drinkers, is taken into account, average daily consumption is estimated to be 2.4 mg/kg/day (Chou, 1992; Mandel, 2002). Moreover, even when consumed in moderate amounts, the drug has a half-life of 4 to 5 h (Kaplan et al., 1997). These statistics suggest that there may be a very large number of caffeine-dependent users (Mandel, 2002).

Further evidence comes from the documentation of caffeine abuse in selected populations, including athletes and inpatients. Both professional and amateur athletes frequently consume caffeine, sometimes to the point of abuse, in order to take competitive advantage of its stimulating properties. Among a large sample of Canadian teenagers, for instance, 27% admitted to consuming caffeine to improve their performance in various sports (Melia et al., 1996). Although the drug does tend to enhance performance, particularly during endurance events, deleterious effects, such as precompetition anxiety and psychological dependence, may also be experienced by these athletes as a result of caffeine abuse (Sinclair and Geiger, 2000). In addition to athletes, caffeine abuse in clinical populations has been studied. In a sample of 60 hospital inpatients, MacKay and Rollins (1989) found that 47% consumed more than 750 mg of caffeine per day. It is likely that these patients, if queried, would endorse the diagnostic criteria for a caffeine-related disorder. Though those in an inpatient population may present themselves to a mental health provider complaining primarily of anxiety and depressive symptomatology, these patients often also consume excessive amounts of caffeine (Greden et al., 1978; Rihs et al., 1996).

Caffeine Intoxication

Caffeine intoxication is a syndrome involving psychological and physical distress caused by chronic or acute overconsumption of caffeine, usually in excess of 500 to 600 mg daily (James and Stirling, 1983). It is included in DSM-IV and ICD-10 and appeared in earlier editions of these nomenclatures as *caffeinism*, a term still used in the literature. The syndrome is often manifested by such somatic complaints as diuresis, tachycardia, and tremulousness, as well as by anxiety (Kendler and Prescott, 1999). Thus, the central nervous system, gastrointestinal system, and cardiovascular system are all affected by excessive caffeine consumption. Caffeinism may be related to an addiction to other licit substances, such as nicotine and alcohol, and to lower academic performance (Bradley and Petree, 1990; Kozlowski et al., 1993).

There is limited research on the treatments available for those with caffeinism. Fox and Rubinoff (1979) used a behavioral method involving self-monitoring and rewards to decrease daily caffeine intake. Baseline data were collected on three coffee drinkers suffering from caffeinism. Participants then switched from brewed to instant coffee, thereby gradually decreasing the amount of caffeine consumed. Treatment goals (decrease baseline level to 600 mg) were developed, and participants then invested $20 in the treatment, with portions of this deposit returned if progress occurred. This behavioral approach was successful in decreasing coffee-drinking behavior by 69%, and an average 67% reduction was maintained at a 10-month follow-up. In a similar treatment, focused on the reduction of caffeine intake to a safe level, James and Stirling (1983) also found significant reductions in caffeine use in a sample of 27 excessive users. Thus, there is some evidence that caffeinism can be effectively treated.

Caffeine and Anxiety Disorders

A substantial body of research has linked caffeine use to anxiety disorders. In fact, evidence suggests that caffeine is not only a contributing factor in anxiety and the anxiety disorders but can also precipitate the onset and exacerbate the symptoms of some of these disorders. Research in this area has centered on panic disorder (PD) and generalized anxiety disorder (GAD), but obsessive-compulsive disorder (OCD), social phobia, and post-traumatic stress disorder (PTSD) have also been addressed. Patients diagnosed with GAD or PD are negatively affected by caffeine. They show hypersensitivity to the drug (Boulenger et al., 1984; Bruce et al., 1992) and exhibit improvement in anxiety symptoms when they abstain from consuming this powerful stimulant (Bruce and Lader, 1989; Bruce et al., 1992). Self-ratings of anxiety also increase in PD patients when caffeine is consumed, and the drug can even trigger panic attacks (Charney et al., 1985; Lee et al., 1988; Bruce, 1990).

Caffeine consumption also impacts other anxiety disorders. Those diagnosed with social phobia also have a hypersensitivity to caffeine, though to a lesser extent than is the case for GAD and PD patients (Nutt et al., 1998; den Boer, 2000). Other studies show that caffeine abuse is common in OCD patients with comorbid major depressive disorder (Perugi et al., 1997). Some evidence suggests that the drug may be involved in the pathogenesis of PTSD (Iancu et al., 1996). In fact, some have proposed that the use of decaffeinated coffee in military settings could reduce the prevalence of anxiety reactions and perhaps of PTSD itself (Iancu et al., 1996). A further indication is the finding of Solursh and Solursh (1994) that reduction in caffeine use results in improvement of sexual functioning in some Vietnam combat veterans diagnosed with chronic PTSD.

A final note on the role of caffeine in anxiety disorders concerns the currency of the available studies. It is notable that most of the studies done to date appear in literature prior to 1990. This earlier literature provides the basic information that caffeine affects patients with anxiety disorders, but much more work is needed to determine the exact nature and extent of that impact. Additional studies in this area would certainly be welcome.

THE ROLE OF CAFFEINE IN OTHER DISORDERS

Caffeine has also been implicated in a number of other disorders, including depression, schizophrenia, bipolar disorder, eating disorders, ADHD in children, and restless legs syndrome. In depressive patients, caffeine is often used as a self-medication against the depressive mood state, including that seen in seasonal affective disorder (Krauchi et al., 1997; Abbott and Fraser, 1998). Its use is also higher in both clinically depressed patients and adolescents who report a high number of depressive symptoms (Leibenluft et al., 1993; Worthington et al., 1996; Bernstein et al., 2002). One unfortunate side effect of such self-medication with caffeine is that it exacerbates insomnia and parasomnias in many depressed patients, thereby further decreasing quality of life (Neylan, 1995). Thus, any short-term stimulant relief that depressed patients might receive from the use of caffeine may be outweighed by its deleterious effects.

Schizophrenia is also associated with the high levels of caffeine use (Donnelly et al., 1996; Van Ammers et al., 1997). Evidence concerning actual amounts of caffeine intake in this population is largely lacking because use is often neither regulated nor monitored. However, early researchers estimated that between 17 and 71% of schizophrenic inpatients and outpatients use more than 500 mg of caffeine per day (Furlong, 1975; Winstead, 1976), and more recent evidence suggests that 38% of schizophrenics use more than 500 mg per day (Mayo et al., 1993). Not surprisingly, then, members of the general population who score higher on a schizotypy scale also report greater caffeine intake than those who score low (Larrison et al., 1999). One explanation for the substantial use of caffeine in the schizophrenic population is that the adenosine A_{2A} antagonistic effects of the drug may produce an effect similar to increased dopaminergic neurotransmission in the ventral striatum (Ferre, 1997). An alternative suggestion is that caffeine may counteract some negative effects of neuroleptic medications (Kruger, 1996).

Despite the fact that many schizophrenic patients choose to consume caffeine, it is not clear whether the drug has primarily advantageous or deleterious effects on symptoms. In fact, some studies have found caffeine use to increase subjective distress and psychotic symptoms (Hamera et al., 1995; Hyde, 1990; Lucas et al., 1990; Ferre, 1997). Moreover, schizophrenics who are heavy caffeine users appear to need larger doses of antipsychotic medications than do nonusers (Hyde, 1990). Contrary findings show, however, that switching patients to decaffeinated coffee does not produce any amelioration of symptoms (Koczapski et al., 1989). One attempt to explain these discrepancies in results is the suggestion that certain subgroups of schizophrenics may be more highly sensitive to the possible psychotogenic effects of caffeine (Hyde, 1990).

Three final disorders to be considered are eating disorders, ADHD, and restless legs syndrome. Early case reports of excessive caffeine consumption among eating disorder patients (Sours, 1983) have been supported by a few studies showing caffeine abuse by bulimics (Kruger and Braunig, 1995), some suggesting higher levels of caffeine consumption in purgers than in restrictors (Haug et al., 2001). However, other recent research shows that caffeine consumption in these groups is about the same as or less than that of age-matched control samples (Stock et al., 2002). Research with ADHD children shows, not surprisingly, that caffeine may be better than no treatment in decreasing impulsivity, aggression, and both parents' and teachers' perceptions of symptom severity (Leon, 2000). In a second study, caffeine was better than placebo in decreasing hyperactivity and teacher-rated symptom severity and in improving executive functions. However, it was not as effective as methylphenidate and proved to have little or no effect on performance during tests of attention (Riccio et al., 2001). A final disorder is restless legs syndrome. Although there now appears to be a major genetic component in this disorder, perhaps transmitted on an autosomal dominant gene with multifactorial expression (Winkelmann, 2002), emotional and behavioral factors have also been identified. Symptoms of anxiety and depressive disorders, social alienation, and diminished cognitive focus have been reported as typical of RLS patients (Kuny, 1991; Aikens et al., 1999; Ulfberg et al., 2001), and caffeine has been implicated (Lutz, 1978; Paulson, 2000). In fact, one recent investigation showed that regular use of nonopioid analgesics, frequently containing caffeine, is a risk factor for RLS and is associated with increases in both psychiatric and medical comorbidity (Leutgeb, 2002).

Psychopharmacological Treatment: Caffeine Interactions

In addition to its observed effects on psychiatric symptomatology, caffeine interacts with some of the medications used to treat psychological disorders. One early investigation showed that caffeine can interfere with the action of benzodiazepines used in treating the anxiety disorders (Greden et al., 1981). More recent work demonstrates that it can increase the risk of clozapine toxicity, such as sedation, seizures, and hypotension (Carrillo et al., 1998; Patton and Beer, 2001), and increase the excretion of lithium, thereby leading to treatment failure in lithium patients (Jefferson, 1988). In a recent review, Patton and Beer (2001) summarized the effects of caffeine on antipsychotics, benzodiazopines, and tricyclic antidepressants. Most commonly, treatment failure and increased risk of toxicity occur through reduced sedative, anxiolytic, or anticonvulsant effects or reduced metabolism of the drug.

CAFFEINE AND PAIN

Pain clearly has psychological, as well as physical, consequences for the patient, and many studies show that caffeine can be effective in the management of pain. It is for that reason that the drug appears, in combination with such common analgesics as acetaminophen and aspirin, in many prescription and nonprescription headache and arthritis medications. Excedrin, Midol, Cafergot, and Butalbital are examples of analgesic medications that contain caffeine. Research demonstrates, for example, that combinations of acetaminophen and caffeine provide greater relief of tension-type headaches than acetaminophen alone (Diamond, 1999). In fact, caffeine may itself have analgesic properties. In one double-blind, placebo-controlled study, caffeine had an independent effect equivalent to that of acetaminophen in relieving nonmigraine headaches (Ward et al., 1991). Relief of the symptoms of migraines has also been found with caffeine-containing combination products, such as Excedrin. These products produce a decrease in pain intensity ratings and relieve photophobia, phonophobia, functional disability, and nausea symptoms in menstruating and non-menstruating women (Silberstein et al., 1999). The analgesic effects of caffeine have also been studied in relation to pain management in cancer, though the role of the drug here is different. One study showed that caffeine was superior to placebo in reducing the frequently observed cognitive and psychomotor effects of morphine in cancer patients (Mercadante et al., 2001).

Although caffeine clearly plays a role in the relief of acute pain, its effects on chronic pain remain uncertain. A preliminary study in this area recorded dietary caffeine intake in patients with chronic back pain and found no differences in low, medium, and high caffeine users with regard to pain severity (Currie et al., 1995). However, another study showed that patients suffering from chronic back pain consumed twice as much caffeine as controls (McPartland and Mitchell, 1997). Other research on chronic pain has demonstrated that analgesics, including caffeine, can actually cause chronic headaches and that reducing the use of analgesics can reduce headache severity and incidence (Berciano, 1993; Le Jeunne, 2001). The discrepancies among studies may reflect differences in dosage, frequency of use, disorders, or patient populations. Clearly, considerable additional research is needed before we will fully understand the role of caffeine in chronic pain syndromes.

REFERENCES

Abbott, F.V. and Fraser, M.I. (1998) Use and abuse of over-the-counter analgesic agents. *Journal of Psychiatry Neuroscience,* 23, 13–34.

Adrien, J. (2001) Adenosine in sleep regulation. *Revue Neurologique,* 157, 7–11.

Aikens, J.E., Vanable, P.A., Tadimeti, L., Caruana-Montaldo, B. and Mendelson, W.B. (1999) Differential rates of psychopathology symptoms in periodic limb movement disorder, obstructive sleep apnea, psychophysiological insomnia, and insomnia with psychiatric disorder. *Sleep,* 22, 775–780.

Akerstedt, T. and Ficca, G. (1997) Alertness-enhancing drugs as a countermeasure to fatigue in irregular work hours. *Chronobiology International,* 14, 145–158.

American Psychiatric Association. (1994) *Diagnostic and Statistical Manual of Mental Disorders,* 4th ed., Washington, DC.

Anderson, K. and Revelle, W. (1983) The interactive effects of caffeine, impulsivity and task demands on visual search task. *Personality and Individual Differences,* 4, 127–132.

Arnold, M., Petros, T., Beckwith, B., Coons, G. and Gorman, N. (1987) The effects of caffeine, impulsivity and sex on memory for word lists. *Physiology and Behavior,* 41, 25–30.

Babkoff, H., French, J., Whitmore, J. and Sutherlin, R. (2002) Single-dose bright light and/or caffeine effect on nocturnal performance. *Aviation, Space, and Environmental Medicine,* 73, 341–350.

Battig, K., Buzzi, R., Martin, J.R. and Feierabend, J.M. (1984) The effects of caffeine on physiological functions and mental performance. *Experientia,* 40, 1218–1223.

Beaumont, M., Batejat, D., Pierard, C., Coste, O., Doireau, P., Van Beers, P. et al. (2001) Slow release caffeine and prolonged (64-h) continuous wakefulness: effects on vigilance and cognitive performance. *Journal of Sleep Research,* 10, 265–276.

Berciano, J. (1993) Gangliosides and the Guillain-Barré Syndrome. *Medicina Clinica,* 101, 758–759.

Bernstein, G.A., Carroll, M.E., Thuras, P.D., Cosgrove, K.P. and Roth, M.E. (2002) Caffeine dependence in teenagers. *Drug and Alcohol Dependence,* 66, 1–6.

Boros, I., Lengyel, Z., Balkay, L., Horvath, G., Szentmiklosi, A.J., Fekete, I. et al. (2002) In vivo investigation of the A2A adenosine receptor distribution using the 11C-CSC radioligand. *Orvosi Hetilap,* 143, 1317–1319.

Boulenger, J.P., Uhde, T.W., Wolff, E.A. and Post, R.M. (1984) Increased sensitivity to caffeine in patients with panic disorders. *Archives of General Psychiatry,* 41, 1067–1071.

Bowyer, P., Humphreys, M. and Revelle, W. (1983) Arousal and recognition memory: the effects of impulsivity, caffeine and time on a task. *Personality and Individual Differences,* 4, 41–45.

Bradley, J.R. and Petree, A. (1990) Caffeine consumption, expectations of caffeine-enhanced performance and caffeinism symptoms among university students. *Journal of Drug Education,* 20, 319–328.

Brauer, L.H., Buican, B. and de Wit, H. (1994) Effects of caffeine deprivation on taste and mood. *Behavioral Pharmacology,* 5, 111–118.

Brown, S.L., Harris, T.B., Wallace, R.B., Langlois, J.A., Corti, M.C., Foley, D.J. et al. (1995) Occult caffeine as a source of sleep problems in an older population. *Journal of the American Geriatrics Society,* 43, 860–984.

Bruce, M., Scott, N., Shine, P. and Lader, M. (1992) Anxiogenic effects of caffeine in patients with anxiety disorders. *Archives of General Psychiatry,* 49, 867–869.

Bruce, M.S. (1990) The anxiogenic effects of caffeine. *Postgraduate Medical Journal,* 66, 518–524.

Bruce, M.S. and Lader, M. (1989) Caffeine abstention in the management of anxiety disorders. *Psychological Medicine,* 19, 211–214.

Bryant, C.A., Farmer, A., Tiplady, B., Keating, J., Sherwood, R., Swift, C.G. and Jackson, S.H. (1998) Psychomotor performance: investigating the dose-response relationship for caffeine and theophylline in elderly volunteers. *European Journal of Clinical Pharmacology,* 54, 309–313.

Carrillo, J.A. and Benitez, J. (1994) Caffeine metabolism in a healthy Spanish population: N-acetylator phenotype and oxidation pathways. *British Journal of Clinical Pharmacology,* 38, 471–473.

Carrillo, J.A., Herriaz, A.G., Ramos, S.L. and Benitez, J. (1998) Effects of caffeine withdrawal from the diet on the metabolism of clozapine in schizophrenic patients. *Journal of Clinical Psychopharmacology,* 18, 311–316.

Charney, D.S., Heninger, G.R. and Jatlow, P.I. (1985) Increased anxiogenic effects of caffeine in panic disorders. *Archives of General Psychiatry,* 42, 233–243.

Chen, J.F., Xu, K., Petzer, J.P., Staal, R., Xu, Y.H., Beilstein, M. et al. (2001) Neuroprotection by caffeine and A(2A) adenosine receptor inactivation in a model of Parkinson's disease. *Journal of Neuroscience*, 21, RC143.

Chou, T. (1992) Wake up and smell the coffee: caffeine, coffee, and the medical consequences. *Western Journal of Medicine*, 157, 544–553.

Clubley, M., Bye, C.E., Henson, T.A., Peck, A.W. and Riddington, C.J. (1979) Effects of caffeine and cyclizine alone and in combination on human performance, subjective effects and EEG activity. *British Journal of Clinical Pharmacology*, 7, U57–U63.

Currie, S.R., Wilson, K.G. and Gauthier, S.T. (1995) Caffeine and chronic low back pain. *The Clinical Journal of Pain*, 11, 214–219.

Daly, J.W. and Fredholm, B.B. (1998) Caffeine: an atypical drug of dependence. *Drug and Alcohol Dependence*, 51, 199–206.

Davidson, R. and Smith, B. (1991) Caffeine and novelty: effects on electrodermal activity and performance. *Physiological Behavior*, 49, 1169–1175.

Debrah, K., Kerr, D., Murphy, J. and Sherwin, R.S. (1996) Effect of caffeine on recognition of and physiological responses to hypoglycemia in insulin-dependent diabetes. *Lancet*, 347, 19–24.

Denaro, C.P., Brown, C.R., Wilson, M., Jacob, P., III and Benowitz, N.L. (1990) Dose-dependency of caffeine metabolism with repeated dosing. *Clinical Pharmacology and Therapeutics*, 48, 277–285.

den Boer, J.A. (2000) Social anxiety disorder/social phobia: epidemiology, diagnosis, neurobiology, and treatment. *Comprehensive Psychiatry*, 41, 405–415.

Diamond, S. (1999) Caffeine as an analgesic adjuvant in the treatment of headache. *Headache Quarterly*, 10, 119–125.

Donnelly, C.L., Narasimhachari, N., Wilson, W.H. and McEnvoy, J.P. (1996) A study of the potential confounding effects of diet, caffeine, nicotine, and lorazepam, on the stability of plasma and urinary homovanillic acid levels in patients with schizophrenia. *Biological Psychiatry*, 40, 1218–1221.

Durlach, P.J. (1998) The effects of a low dose of caffeine on cognitive performance. *Psychopharmacology*, 140, 116–119.

Evans, S.M. and Griffiths, R.R. (1999) Caffeine withdrawal: a parametric analysis of caffeine dosing conditions. *Journal of Pharmacology and Experimental Therapeutics*, 289, 285–294.

Ferre, S. (1997) Adenosine-dopamine interactions in the ventral striatum: implications for the treatment of schizophrenia. *Psychopharmacology*, 133, 107–120.

File, S., Bond, A. and Lister, R. (1982) Interaction between effects of caffeine and lorazepam in performance and self-ratings. *Journal of Clinical Psychopharmacology*, 2, 102–106.

Fox, R.M. and Rubinoff, A. (1979) Behavioral treatment of caffeinism: reducing excessive coffee drinking. *Journal of Applied Behavior Analysis*, 12, 335–344.

Frewer, L. and Lader, M. (1991) The effects of caffeine on two computerized tests of attention and vigilance. *Human Psychopharmacology and Clinical Experimentation*, 6, 119–126.

Furlong, F.W. (1975) Possible psychiatric significance of excessive coffee consumption. *Canadian Psychiatric Association Journal*, 20, 577–583.

Gerevich, Z., Wirkner, K. and Illes, P. (2002) Adenosine a(2a) receptors inhibit the n-methyl-d-aspartate component of excitatory synaptic currents in rats' striatal neurons. *European Journal of Pharmacology*, 451, 161–164.

Giam, G.C. (1997) Effects of sleep deprivation with reference to military operations. *Annals of the Academy of Medicine, Singapore*, 26, 88–93.

Gilbert, D.G., Dibb, W.D., Plath, L.C. and Hiyane, S.G. (2000) Effects of nicotine and caffeine, separately and in combination, on EEG topography, mood, heart rate, cortisol, and vigilance. *Psychophysiology*, 37, 583–595.

Gilliland, K. and Bullock, W. (1984) Caffeine: a potential drug of abuse. *Addictive Behaviors*, 3, 53–73.

Greden, J.F., Fontaine, P., Lubetsky, M. and Chamberlin, K. (1978) Anxiety and depression associated with caffeinism among psychiatric inpatients. *American Journal of Psychiatry*, 135, 963–966.

Greden, J.F., Procter, A. and Victor, B. (1981) Caffeinism associated with greater use of other psychotropic agents. *Comprehensive Psychiatry*, 22, 565–571.

Green, P.J. and Suls, J. (1996) The effects of caffeine on ambulatory blood pressure, heart rate, and mood. *Journal of Behavioral Medicine*, 19, 111–128.

Griffiths, R.R. and Chausmer, A.L. (2000) Caffeine as a model drug of dependence: recent developments in understanding caffeine withdrawal, the caffeine dependence syndrome, and caffeine negative reinforcement. *Nihon Shinkei Seishin Yakurigaku Zasshi, 20,* 223–231.

Griffiths, R.R., Holtzman, S.G., Daly, J.W., Hughes, J.R., Evans, S.M. and Strain, E.C. (1996) Caffeine: a model drug of abuse. *NIDA Research Monogram, 162,* 73–75.

Hamera, E., Deviney, S. and Schneider, J.K. (1995) Alcohol, cannabis, nicotine, and caffeine use and symptom distress in schizophrenia. *The Journal of Nervous and Mental Disease, 183,* 559–565.

Haug, N.A., Heinberg, L.J. and Guarda, A.S. (2001) Cigarette smoking and its relationship to other substance use among eating disorder inpatients. *Eating and Weight Disorders, 6,* 130–139.

Herz, R.S. (1999) Caffeine effects on mood and memory. *Behaviour Research and Therapy, 37,* 869–879.

Hespel, P., Op't Eijnde, B. and Van Leemputte, M. (2002) Opposite actions of caffeine and creatine on muscle relaxation time in humans. *Journal of Applied Physiology, 92,* 513–518.

Hettema, J.M., Corey, L.A. and Kendler, K.S. (1999) A multivariate genetic analysis of the use of tobacco, alcohol, and caffeine in a population based sample of male and female twins. *Drug and Alcohol Dependence, 57,* 69–78.

Hindmarch, I., Quinlan, P.T., Moore, K.L. and Parkin, C. (1998) The effects of black tea and other beverages on aspects of cognition and psychomotor performance. *Psychopharmacology, 139,* 230–238.

Hogervorst, E., Riedel, W.J., Kovacs, E., Brouns, F. and Jolles, J. (1999) Caffeine improves cognitive performance after strenuous physical exercise. *International Journal of Sports Medicine, 20,* 354–361.

Holtzman, S.G. (1990) Caffeine as a model drug of abuse. *Trends in Pharmacological Sciences, 11,* 355–356.

Hughes, J.R., Oliveto, A.H., Helzer, J.E., Higgins, S.T. and Bickel, W.K. (1992) Should caffeine abuse, dependence, or withdrawal be added to DSM-IV and ICD-10? *American Journal of Psychiatry, 149,* 33–40.

Hughes, J.R., Oliveto, A.H., Liguouri, A., Carpenter, J. and Howard, T. (1998) Endorsement of DSM-IV dependence criteria among caffeine users. *Drug and Alcohol Dependence, 52,* 99–107.

Hyde, A. (1990) Response to 'Effects of caffeine on behaviour of Schizophrenic inpatients.' *Schizophrenia Bulletin, 15,* 371–372.

Iancu, I., Dolberg, O.T. and Zohar, J. (1996) Is caffeine involved in the pathogenesis of combat-stress reaction? *Military Medicine, 161,* 230–232.

James, J.E. and Stirling, K.P. (1983) Caffeine: a survey of some of the known and suspected deleterious effects of habitual use. *British Journal of Addiction, 78,* 215–258.

Jarvis, M.J. (1993) Does caffeine intake enhance absolute levels of cognitive performance? *Psychopharmacology, 110,* 45–52.

Jefferson, J.W. (1988) Lithium tremor and caffeine intake: two cases of drinking less and shaking more. *Journal of Clinical Psychiatry, 49,* 72–73.

Jones, N., Duxon, M.S. and King, S.M. (2002) Ethopharmacological analysis of the unstable elevated exposed plus maze, a novel model of extreme anxiety: predictive validity and sensitivity to anxiogenic agents. *Psychopharmacology, 161,* 314–323.

Kamimori, G.H., Penetar, D.M, Headley, D.B., Thorne, D.R., Otterstetter, R. and Belenky, G. (2000) Effect of three caffeine doses on plasma catecholamines and alertness during prolonged wakefulness. *European Journal of Clinical Pharmacology, 56,* 537–544.

Kaplan, G.B., Greenblatt, D.J., Ehrenberg, B.L., Goddard, J.E., Cotreau, N.M., Harmatz, J.S. et al. (1997) Dose-dependent pharmacokinetics and psychomotor effects of caffeine in humans. *Journal of Clinical Pharmacology, 37,* 693–703.

Kawamura, N., Maeda, H., Nakamura, J., Morita, A. and Nakazawa, Y. (1996) Effects of caffeine on event-related potentials: comparison of oddball with single-tone paradigm. *Psychiatry of Clinical Neuroscience, 50,* 217.

Kelly, T.L., Bonnet, M.H. and Mitler, M.M. (1997) Sleep latency measures of caffeine effects during sleep deprivation. *Electroencephalogy and Clinical Neurophysiology, 102,* 397–400.

Kendler, K.S. and Prescott, C.A. (1999) Caffeine intake, tolerance, and withdrawal in women: a population-based twin study. *American Journal of Psychiatry, 156,* 223–228.

Kenemans, J.L. and Lorist, M.M. (1995) Caffeine and selective visual processing. *Pharmacology, Biochemistry and Behavior, 52,* 461–471.

Koczapski, A., Paredes, J., Kogan, C., Ledwidge, B. and Higenbottam, J. (1989) Effects of caffeine on behavior of schizophrenic inpatients. *Schizophrenia Bulletin, 15,* 339–344.

Koelega, H.S. (1993) Stimulant drugs and vigilance performance: a review. *Psychopharmacology,* 111, 1–16.

Kozlowski, L.T., Henningfield, J.E., Keenan, R.M., Lei, H., Leigh, G., Jelinek, L.C. et al. (1993) Patterns of alcohol, cigarette, and caffeine and other drug use in two drug abusing populations. *Journal of Substance Abuse Treatment,* 10, 171–179.

Krauchi, K., Reich, S. and Wirz-Justice, A. (1997) Eating style in seasonal affective disorder: who will gain weight in winter? *Comprehensive Psychiatry,* 38, 80–87.

Kruger, A. (1996) Chronic psychiatric patients' use of caffeine pharmacological effects and mechanisms. *Psychological Reports,* 78, 915–923.

Kruger, S. and Braunig, P. (1995) Abuse of body weight reducing agents in Bulimia Nervosa. *Der Nervenarzt,* 66, 66–69.

Kruk, B., Chmura, J., Krzeminski, K., Ziemba, A.W., Nazar, K., Pekkarinen, H. et al. (2001) Influence of caffeine, cold and exercise on multiple choice reaction time. *Psychopharmacology,* 157, 197–201.

Kuny, S. (1991) Psychiatric catamnesis in patients with restless legs. *Schweizerische Medizinische Wochenschrift,* 121, 72–76.

Kupietz, S. and Winsberg, B. (1977) Caffeine and inattentiveness in reading-disabled children. *Perception and Motor Skills,* 44, 1238.

Lagarde, D., Batejat, D., Sicard, B., Trocherie, S., Chassard, D., Enslen, M. et al. (2000) Slow-release caffeine: a new response to the effects of a limited sleep deprivation. *Sleep,* 23, 651–661.

Landolt, H.P., Borbely, A.A., Dijk, D.J. and Roth, C. (1996) Late-afternoon ethanol intake affects nocturnal sleep and the sleep EEG in middle-aged men. *Journal of Clinical Psychopharmacology,* 16, 428–436.

Landrum, R., Meliska, C. and Loke, W. (1988) Effects of caffeine and task experience on task performance. *Psychologia: International Journal of Psychology, Orient,* 37, 801–805.

Lane, J.D. and Phillips-Bute, B.G. (1998) Caffeine-deprivation affects vigilance performance and mood. *Physiology and Behavior,* 65, 171–175.

Larrison, A.L., Briand, K.A. and Sereno, A.B. (1999) Nicotine, caffeine, alcohol and schizotypy. *Personality and Individual Differences,* 27, 101–108.

Leathwood, P.D. and Pollet, P. (1982–1983) Diet-induced mood changes in normal populations. *Journal of Psychiatric Research,* 17, 147–154.

Lee, M.A., Flegel, P., Greden, J.F. and Camerson, O.G. (1988) Anxiogenic effects of caffeine on panic and depressed patients. *American Journal of Psychiatry,* 145, 632–635.

Leibenluft, E., Fiero, P.L., Bartko, J.J., Moul, D.E. and Rosenthal, N.E. (1993) Depressive symptoms and the self-reported use of alcohol, caffeine, and carbohydrates in normal volunteers and four groups of psychiatric outpatients. *American Journal of Psychiatry,* 150, 294–301.

Le Jeunne, C. (2001) Analgesic-induced chronic headache. *Annals of International Medicine,* 152, 50–53.

Leon, R. (2000) Effects of caffeine on cognitive, psychomotor, and affective performance on children with attention-deficit/hyperactivity disorder. *Journal of Attention Disorders,* 4, 27–47.

Leutgeb, U. (2002) Regular intake of non-opioid analgesics is associated with an increased risk of restless leg syndrome in patients maintained on antidepressants. *European Journal of Medical Research,* 7, 368–378.

Lieberman, H.R. (1988) Beneficial effects of caffeine, in *Twelfth International Scientific Colloquium on Coffee*, ASIC, Paris.

Lieberman, H.R. (1992) Caffeine, in *Handbook of Human Performance,* vol. 2, Smith, A.P. and Jones, D.M., Eds., Academic Press, London, pp. 49–72.

Lieberman, H.R. (2001) The effects of ginseng, ephedrine, and caffeine on cognitive performance, mood and energy. *Nutrition Reviews,* 59, 91–102.

Lieberman, H.R., Wurtman, R.J., Emde, G.G., Roberts, C. and Coviella, I.L.G. (1987) The effects of low doses of caffeine on human performance and mood. *Psychopharmacology,* 92, 308–312.

Liguori, A., Hughes, J.R. and Grass, J.A. (1997) Absorption and subjective effects of caffeine from coffee, cola and capsules. *Pharmacology Biochemistry and Behavior,* 58, 721–726.

Lin, A.S., McCann, U.D., Slate, S.O. and Uhde, T.W. (1997) Effects of intravenous caffeine administered to healthy males during sleep. *Depression and Anxiety,* 5, 21–28.

Linde, L. (1994) An auditory attention task: a note on the processing of verbal information. *Perception and Motor Skills,* 78, 563–570.

Linde, L. (1995) Mental effects of caffeine in fatigued and non-fatigued female and male subjects. *Ergonomics,* 38, 864–885.

Lindskog, M., Svenningsson, P., Pozzi, L., Kim, Y., Fienberg, A.A., Bibb, J.A. et al. (2002) Involvement of DARPP-32 phosphorylation in the stimulant action of caffeine. *Nature*, 418, 774–778.

Loke, W.H. (1988) Effects of caffeine on mood and memory. *Physiology and Behavior*, 44, 367–372.

Loke, W.H., Hinrichs, J.V. and Ghoneim, M.M. (1985) Caffeine and diazepam: separate and combined effects on mood, memory, and psychomotor performance. *Psychopharmacology*, 87, 344–350.

Loke, W. and Meliska, C. (1984) Effects of caffeine use and ingestion of a protracted visual vigilance task. *Psychopharmacology*, 84, 54–57.

Lorist, M.M., Kok, A., Mulder, G. and Snel, J. (1995) Aging, caffeine, and information processing: an event-related potential analysis. *Electroencephalography and Clinical Neurophysiology*, 96, 453–467.

Lorist, M.M. and Snel, J. (1997) Caffeine effects on perceptual and motor processes. *Electroencepharlogr Clinical Neurophysiology*, 102, 401–413.

Lorr, M. and Wunderlich, R. (1988) Self-esteem and negative affect. *Journal of Clinical Psychology*, 44, 36–39.

Lucas, P.B., Pickar, D., Kelsoe, J., Rapaport, M., Pato, C. and Hommer, D. (1990) Effects of the acute administration of caffeine in patients with schizophrenia. *Biological Psychiatry*, 28, 35–40.

Lutz, E.G. (1978) Restless legs, anxiety and caffeinism. *Journal of Clinical Psychiatry*, 39, 693–698.

Mackay, D.C. and Rollins, J.W. (1989) Caffeine and caffeinism. *Journal of the Royal Navy Medical Services*, 75, 65–67.

Maia, L. and de Mendonca, A. (2002) Does caffeine intake protect from Alzheimer's disease? *European Journal of Neurology*, 9, 377–382.

Mandel, H.G. (2002) Update on caffeine consumption, disposition, and action. *Food and Chemical Toxicology*, 40, 1231–1234.

Mann, M., Smith, B.D., Tola, K. and Farley, L. (2002) Alcoholic tendency and EEG arousal in women: effects of family history and personality under emotional stimulation. *International Journal of Neuroscience*, 112, 639–661.

Mayo, K.M., Falkowski, W. and Jones, C.A. (1993) Caffeine: use and effects in long-stay psychiatric patients. *The British Journal of Psychiatry*, 162, 543–545.

McPartland, J.M. and Mitchell, J.A. (1997) Caffeine and chronic back pain. *Archives of Physical Medical Rehabilitation*, 78, 61–63.

Melia, P., Pipe, A. and Greenberg, L. (1996) The use of anabolic-androgenic steroids by Canadian students. *Clinical Journal of Sport Medicine*, 6, 9–14.

Mercadante, S., Serretta, R. and Casuccio, A. (2001) Effects of caffeine as an adjuvant to morphine in advanced cancer patients: a randomized, double-blind, placebo-controlled, crossover study. *Journal of Pain and Symptom Management*, 21, 369–372.

Mikalsen, A., Bertelsen, B. and Flaten, M.A. (2001) Effects of caffeine, caffeine-associated stimuli, and caffeine-related information on physiological and psychological arousal. *Psychopharmacology*, 157, 373–380.

Mumford, G.K., Evans, S.M., Kaminski, B.J., Preston, K.L., Sannerud, C.A., Silverman, K. et al. (1994) Discriminative stimulus and subjective effects of theobromine and caffeine in humans. *Psychopharmacology*, 115, 1–8.

Nehlig, A. (1999) Are we dependent upon coffee and caffeine?: a review of human and animal data. *Neuroscience and Biobehavioral Reviews*, 23, 563–576.

Neylan, T.C. (1995) Treatment of sleep disturbances in depressed patients. *Journal of Clinical Psychiatry*, 56, 56–61.

Nutt, D.J., Bell, C.J. and Malizia, A.L. (1998) Brain mechanisms of social anxiety disorder. *Journal of Clinical Psychiatry*, 59, 4–11.

Oborne, D.J. and Rogers, Y. (1983) Interactions of alcohol and caffeine on human reaction time. *Aviation, Space, and Environmental Medicine*, 54, 528–534.

Patton, C. and Beer, D. (2001) Caffeine: the forgotten variable. *International Journal of Psychiatry in Clinical Practice*, 5, 231–236.

Paulson, W.D. (2000) Prediction of hemodialysis synthetic graft thrombosis: can we identify factors that impair validity of the dysfunction hypothesis? *American Journal of Kidney Disorders*, 35, 973–975.

Penetar, D., McCann, U., Thorne, D., Kamimori, G., Galinski, C., Sing, H. et al. (1993) Caffeine reversal of sleep deprivation effects on alertness and mood. *Psychopharmacology*, 112, 359–365.

Perugi, G., Akiskal, H.S., Pfanner, C., Presta, S., Gemignani, A., Milanfranchi, A. et al. (1997) The clinical impact of bipolar and unipolar affective comorbidity on obsessive-compulsive disorder. *Journal of Affective Disorders*, 46, 15–23.

Phillips-Bute, B.G. and Lane, J.D. (1997) Caffeine withdrawal symptoms following brief caffeine deprivation. *Physiology and Behavior,* 63, 35–39.

Pons, L., Trenque, T., Bielcki, M. and Moulin, M. (1988) Attentional deficits of caffeine in man: comparison with drugs acting on performance. *Psychiatric Research,* 23, 329–333.

Quinlan, P., Lane, J. and Aspinall, L. (1997) Effects of hot tea, coffee and water ingestion on physiological responses and mood: the role of caffeine, water and beverage type. *Psychopharmacology,* 134, 164–173.

Quinlan, P., Lane, J., Moore, K.L., Aspen, J., Rycroft, J.A. and O'Brien, D.C. (2000) The acute physiological and mood effects of tea and coffee: the role of caffeine level. *Pharmacology, Biochemistry and Behavior,* 66, 19–28.

Rasmussen, B.B., Brix, T.H., Kyvik, K.O. and Brosen, K. (2002) The interindividual differences in the 3-Demthylation of caffeine alias CYP1A2 is determined by both genetic and environmental factors. *Pharmacogenetics,* 12, 473–478.

Rees, K., Allen, D. and Lader, M. (1999) The influences of age and caffeine on psychomotor and cognitive function. *Psychopharmacology,* 145, 181–188.

Riccio, C.A., Waldrop, J.J.M., Reynolds, C.R. and Lowe, P. (2001) Effects of stimulants on the Continuous Performance Test (CPT). *Journal of Neuropsychiatry and Clinical Neurosciences,* 13, 326–335.

Riedel, W.J. and Jolles, J. (1996) Cognition enhancers in age-related cognitive decline. *Drugs and Aging,* 8, 245–274.

Riedel, W.J., Jolles, J., Van Praag, H., Verhey, F., Leboux, R. and Hogervorst, E. (1995) Caffeine attenuates scopolamine-induced memory impairment in humans. *Psychopharmacology,* 122, 158–165.

Rihs, M., Muller, C. and Baumann, P. (1996) Caffeine consumption in hospitalized psychiatric patients. *European Archives of Psychiatry and Clinical Neuroscience,* 246, 83–92.

Roache, J. and Griffiths, R. (1987) Interaction of diazepam and caffeine: behavioral and subjective dose effects in humans. *Pharmacology, Biochemistry and Behavior,* 26, 801–812.

Robelin, M. and Rogers, P.J. (1998) Mood and psychomotor performance effects of the first but not of subsequent, cup-of-coffee equivalent doses of caffeine consumed after overnight caffeine abstinence. *Behavioral Pharmacology,* 9, 611–618.

Rogers, P.J. and Dernoncourt, C. (1998) Regular caffeine consumption: A balance of adverse and beneficial effects for mood and psychomotor performance. *Pharmacology, Biochemistry and Behavior,* 59, 1039–1045.

Ruijter, J., Lorist, M., Snel, J. and De Ruiter, M.B. (2000) The influence of caffeine on sustained attention: an ERP study. *Pharmacology, Biochemistry and Behavior,* 66, 29–37.

Rusted, J. (1999) Caffeine and cognitive performance: effects on mood or mental processing? in *Caffeine and Behavior: Current Views and Research Trends,* Gupta, B.S. and Gupata, U., Eds., CRC Press, Boca Raton, FL, pp. 221–230.

Rusted, J. and Yeomans, M.R. (2002) Caffeine, mood and performance: a selective review. Sussex University, Brighton, 15, 179–200.

Ryan, L., Hatfield, C. and Hofstetter, M. (2002) Caffeine reduces time-of-day effects on memory performance in older adults. *Psychological Science,* 13, 68–71.

Sawyer, D.A., Julia, H.L. and Turin, A.C. (1982) Caffeine and human behavior: arousal, anxiety, and performance effects. *Journal of Behavioral Medicine,* 5, 415–439.

Schuh, K.J. and Griffiths, R.R. (1997) Caffeine reinforcement: the role of withdrawal. *Psychopharmacology,* 130, 320–326.

Sicard, B.A., Tachon, P., Vandel, B., Chauffard, F., Enslen, M. and Perault, M.C. (1996) The effects of 600 mg of slow release caffeine on mood and alertness. *Aviation, Space, and Environmental Medicine,* 67, 859–862.

Silberstein, S.D., Armellino, J.J., Hoffman, H.D., Battikha, J.P., Hamelsky, S.W. et al. (1999) Treatment of menstruation-associated migraine with the nonprescription combination of acetaminophen, aspirin, and caffeine: results from three randomized, placebo-controlled studies. *Clinical Therapeutics,* 21, 475–490.

Silverman, K. and Griffiths, R.R. (1992) Low-dose caffeine discriminations and self reported mood effects in normal volunteers. *Journal of Experimental Animal Behavior,* 57, 91–107.

Sinclair, C.J. and Geiger, J.D. (2000) Caffeine use in sports: a pharmacological review. *Journal of Sports Medicine and Physical Fitness,* 40, 71–79.

Smit, H.J. and Rogers, P.J. (2000) Effects of low doses of caffeine on cognitive performance, mood and thirst in low and higher caffeine consumers. *Psychopharmacology,* 152, 167–173.

Smit, H.J. and Rogers, P.J. (2002) Effects of caffeine on mood. *Pharmacopsychoecologia,* 15, 231–257.
Smith, A. (1994) Caffeine, performance, mood and states of reduced alertness. Special issues: caffeine research. *Pharmacopsychoecologia,* 7, 75–86.
Smith, A. (2002) Effects of caffeine on human behavior. *Food and Chemical Toxicology,* 40, 1243–1255.
Smith, A.P., Brockman, P., Flynn, R., Maben, A. and Thomas, M. (1993) Investigation of the effects of coffee on alertness and performance during the day and night. *Neuropsychobiology,* 27, 217–223.
Smith, A.P., Clark, R. and Gallagher, J. (1999) Breakfast cereal and caffeinated coffee: effects on working memory, attention, mood, and cardiovascular function. *Physiology and Behavior,* 67, 9–17.
Smith, A.P., Kendrick, A.M. and Maben, A.L. (1992) Effects of breakfast and caffeine on performance and mood in the late morning and after lunch. *Neuropsychobiology,* 26, 198–204.
Smith, B.D., Cranford, D. and Green, L. (2001) Hostility and caffeine: cardiovascular effects during stress and recovery. *Personality and Individual Differences,* 30, 1125–1137.
Smith, B.D., Cranford, D., and Mann, M. (2000) Gender, cynical hostility, and cardiovascular function: implications for differential cardiovascular disease risk? *Personality and Individual Differences,* 29, 659–670.
Smith, B.D., Kinder, N., Osborne, A.L. and Trotman, A.J. (2002) The arousal drug of choice: major sources of caffeine, *Pharmcopsychoecologia,* 15, 1–34.
Smith, B.D., Rafferty, J., Lindgren, K., Smith, D. and Nespor, A. (1991) Chronic and acute effects of caffeine: testing a biobehavioral model. *Physiology and Behavior,* 51, 131–142.
Solinas, M., Ferre, S., You, Z.B., Karcz-Kubicha, M., Popoli, P. and Goldberg, S.R. (2002) Caffeine induces dopamine and glutamate release in the shell of the nucleus accumbens. *Journal of Neuroscience,* 22, 6321–6324.
Solursh, L.P. and Solursh, D.S. (1994) Male erectile disorders in Vietnam combat veterans with chronic post-traumatic stress disorder. *International Journal of Adolescent Medicine and Health,* 7, 119–124.
Sours, J.A. (1983) Case reports of anorexia nervosa and caffeinism. *American Journal of Psychiatry,* 140, 235–236.
Stein, M.A., Bender, G.B., Phillips, W., Leventhal, B.L. and Krasowski, M. (1996) Behavioral and cognitive effects of methylxanthines: a meta-analysis of theophylline and caffeine. *Archives of Pediatric and Adolescent Medicine,* 150, 284–288.
Stern, K.N., Chait, L.D. and Johnson, C. (1989) Reinforcing and subjective effects of caffeine on normal human volunteers. *Psychopharmacology,* 98, 81–88.
Stock, S.L., Goldberg, E., Corbett, S. and Katzman, D.K. (2002) Substance use in female adolescents with eating disorders. *Journal of Adolescent Health,* 21, 176–182.
Strain, E.C., Mumford, G.K., Silverman, K. and Griffiths, R.R. (1994) Caffeine dependence syndrome: Evidence from case histories and experimental evaluations. *Journal of the American Medical Association,* 272, 1043–1048.
Streufert, S., Satish, U., Pogash, R., Gingrich, D., Landis, R., Roache, J. et al. (1997) Excess coffee consumption in simulated complex work settings: detriment or facilitation of performance? *Journal of Applied Psychology,* 82, 774–782.
Svenningsson, P., Nomikos, G.G. and Fredholm, B.B. (1999) The stimulatory action and the development of tolerance to caffeine is associated with alterations in gene expression in specific brain regions. *Journal of Neuroscience,* 19, 4011–4022.
Swift, C.G. and Tiplady, B. (1988) The effects of age on the response to caffeine. *Psychopharmacology,* 94, 29–31.
Temple, J.G., Warm, J.S., Dember, W.N., Jones, K.S., LaGrange, C.M. and Matthews, G. (2000) The effects of signal salience and caffeine on performance, workload, and stress in an abbreviated vigilance task. *Human Factors,* 42, 183–194.
Terry, W.S. and Phifer, B. (1986) Caffeine and memory performance on the AVLT. *Journal of Clinical Psychology,* 42, 860–863.
Thayer, R.E. (1978) Factor analytic and reliability studies on the Activation-Deactivation Adjective Check List. *Psychological Reports,* 42, 747–756.
Turner, J. (1993) Incidental information processing: effects of mood, sex and caffeine. *International Journal of Neuroscience,* 72, 1–14.
Tynjala, J., Levalahti, E. and Kannas, L. (1997) Perceived tiredness among adolescents and its association with sleep habits and use of psychoactive substances. *Journal of Sleep Research,* 6, 189–198.

Ulfberg, J., Nystrom, B., Carter, N. and Edling, C. (2001) Prevalence of restless legs syndrome among men aged 18 to 64 years: an association with somatic disease and neuropsychiatric symptoms. *Movement Disorders*, 16, 1159–1163.

Van Ammers, E.C., Mulder, R.T. and Sellman, J.D. (1997) Temperament and substance abuse in schizophrenia: is there a relationship? *The Journal of Nervous and Mental Disease*, 185, 283–288.

Van Dongen, H.P., Prince, N.J., Mullington, J.M., Szuba, M.P., Kapoor, S.C. and Kinges, D.F. (2001). Caffeine eliminates psychomotor vigilance deficits from sleeping inertia. *Sleep*, 24, 813–819.

Warburton, D.M. (1995) Effects of caffeine on cognition and mood without caffeine abstinence. *Psychopharmacology*, 119, 66–70.

Warburton, D.M., Bersellini, E. and Sweeney, E. (2001) An evaluation of a caffeinated taurine drink on mood, memory and information processing in healthy volunteers without caffeine abstinence. *Psychopharmacology*, 158, 322–328.

Ward, N., Whitney, C., Avery, D. and Dunner, D. (1991) The analgesic effects of caffeine in headache. *Pain*, 44, 151–155.

Watson, D., Clark, L.A. and Tellegen, A. (1988) Development and validation of brief measures of positive and negative affect: the PANAS scales. *Journal of Personality and Social Psychology*, 54, 1063–1070.

Wijers, A.A., Lange, J.J., Mulder, G. and Mulder, L.J. (1997) An ERP study of visual spatial attention and letter target detection for isoluminant and nonisoluminant stimuli. *Psychophysiology*, 34, 553–565.

Wilken, J.A., Smith, B.D., Tola, K. and Mann, M. (2000) Trait anxiety and prior exposure to non-stressful stimuli: effects on psychophysiological arousal and anxiety. *International Journal of Psychophysiology*, 37, 233–242.

Winkelmann, J. (2002) "Restless Legs" in "The Wedding Proposal" by Anton Chekhov. *Acta Neurologica Scandinavica*, 105, 349–350.

Winstead, D.K. (1976) Coffee consumption among psychiatric inpatients. *American Journal of Psychiatry*, 133, 1447–1450.

World Health Organization. (1992) *Draft International Classification of Diseases and Related Health Problems*, 10th ed., Geneva, Switzerland.

Worthington, J., Fava, M., Agustin, C., Alpert, J., Nierenberg, A.A., Pava, J.A. et al. (1996) Consumption of alcohol, nicotine, and caffeine among depressed outpatients: relationship with response to treatment. *Psychosomatics*, 37, 518–522.

Yeomans, M.R., Pryke, R. and Durlach, P.J. (2002) Effects of caffeine-deprivation on liking for a non-caffeinated drink. *Appetite*, 39, 35–42.

4 Coffee, Caffeine, and Cognitive Performance

Jan Snel, Monicque M. Lorist, and Zoë Tieges

CONTENTS

Introduction 54
Studies Used 54
Sensation and Perception 55
    Visual Modality 55
        Color Discrimination 55
        Critical Flicker Fusion 55
    Auditory Modality 55
    Auditory and Visual Senses Combined 56
Cognition 56
    Simple and Choice Reaction Times 56
    Speeded Decision Making 57
    Cancellation 57
    Substitution 57
    Other Cognitive Tasks 57
Learning and Memory 58
    Learning 58
        Paired-Associate Learning 58
        Serial Learning (Intentional Learning) 58
        Incidental Learning 59
    Memory 59
        Free Recall 59
        Delayed Free Recall and Recognition 60
        Task Load 61
        Display Load 61
        Memory Search 61
        Target Detection 61
        Memory Span 62
Attention 62
    Focused Attention 62
        Stroop Test 62
        "Filter" or Selective Attention Tasks 62
    Divided Attention 63
    Sustained Attention 63
Mental Fatigue 64
Conclusions 65
References 66

INTRODUCTION

Caffeine in doses equivalent to the contents of one to four cups of coffee has been found to facilitate several aspects of cognitive activity. In this chapter attention will be focused systematically on different aspects of human information processing: sensory activity, perceptual activities, evaluation of incoming information, decision making, central processing of newly received and already present knowledge in the brain, and response-related executive functions. Different types of tasks have been used to address these different cognitive subprocesses, with several doses of coffee or caffeine, in different types of subjects, at different times of day, and with different research protocols and procedures.

This chapter aims at providing relevant information with respect to coffee and caffeine's effects on different aspects of cognitive performance under different conditions.

STUDIES USED

Studies were used that: (a) examined acute effects of caffeine on aspects of thinking, that is, mental performance; (b) included a placebo condition; and (c) utilized healthy young human subjects (mostly students). These studies had the following characteristics:

- The self-reported habitual level of daily caffeine consumption was about 200 to 300 mg, in general, which is similar to the contents of two to four cups of coffee (85 mg/cup). These amounts may be somewhat stronger than most people habitually drink at one occasion and also are substantially larger than those amounts typically consumed in beverages (soft drinks), foods (chocolate or cookies), or over-the-counter (OTC) drugs (Lieberman, 1992).
- In nearly all studies the participants were instructed to abstain from caffeine-containing substances for a certain period of time (usually 10 to 12 h or more) prior to testing, which is similar to the condition when people get out of bed in the morning.
- In some studies the subjects were also asked to abstain from alcohol and/or to fast for some hours before the experiment. Since most people drink coffee especially during work or other activities later during the day in a nondeprived state, generalizations from laboratory findings to everyday situations should be made with caution.
- Strikingly, many studies did not assess caffeine levels in saliva or plasma. As a consequence, uncontrolled variations in baseline and/or achieved plasma or saliva caffeine concentrations, due to noncompliance with the abstinence instructions or to differences in caffeine metabolism, may have confounded some of the reported results.
- Experimental control was often, but certainly not always, exerted over a number of well-known factors associated with interindividual differences in caffeine metabolism, such as smoking, liver disease, and, for females, the use of oral contraceptives and pregnancy.
- The caffeine given to the subjects was mostly taken orally, as a fixed dose or as a dose of milligrams per kilogram body weight (mg/kg), in powder form (e.g., in gelatin capsules) or dissolved in a drink (e.g., fruit juice or decaffeinated coffee).

About half of the studies used a within-subjects (cross-over) design. The other half employed a between-subjects (independent groups) design. The potential advantage of the former over the latter design is its higher statistical power to detect true caffeine effects by preventing interindividual differences from contributing to the error variance. A disadvantage is the presence of a potential differential carryover effect of caffeine, which confounds estimates of its effects.

Several studies have examined caffeine's effects with a test battery purporting to sample a diversity of mental functions. A problem with this approach is that most tests in these batteries have no history of reliability and validity, which seriously hampers the interpretation of the results.

By comparison, other studies have evaluated caffeine's effects within an information-processing framework, for example Lorist and colleagues who used the Sanders' Additional Factor Method (AFM). This method typically employs single tasks, each representing a solid, theoretically based aspect of mental performance. Systematically manipulating a specific task variable, such as stimulus quality (degraded and intact stimuli) (Lorist, 1998), usually allows more robust and specific interpretations of caffeine's effects.

Although there exists no generally accepted taxonomy of human task performance, to facilitate generalizations of research findings the tasks are ordered, if possible, on the basis of their nature or structure and as similarly as possible to the sequential stages of information processing. Five broad but related aspects of mental functioning are distinguished: (1) sensation and perception, (2) cognition, (3) learning and memory, (4) attention, and (5) mental fatigue.

SENSATION AND PERCEPTION

Visual Modality

Diamond and Cole (1979) and Diamond and Smith (1974) showed that taking a 90- and 180-mg caffeine dose lowered (improved) the visual luminance threshold (measured as the amount of light emitted from a surface in a given direction) by 20 and 38%, respectively. This greater sensitivity to emitted light was maintained for the whole 70-min session, and moreover it also counteracted the decrement in sensitivity found later in the placebo session. In general, from the few studies done it can be said that caffeine-containing beverages sharpen and sensitize sensory modalities.

Color Discrimination

Böhme and Böhme (1985) and Diamond and Smith (1974) found that 100 and 200 mg of caffeine facilitated the ability to discriminate colors in men and women. For females, however, caffeine was found to impair color discrimination when oral contraceptives were also administered. Fine and McCord (1991) studied 43 female students who were not caffeine-deprived and were low caffeine users (zero to seven caffeinated beverages a week) or high users (greater than seven caffeinated beverages a week). The high-caffeine group had 57% fewer color discrimination errors than the low-caffeine group. A similar result was found by Kohen and Schneider (1986), who found in six men (18 to 39 years of age) receiving a relatively high dose of 7.5 mg/kg of body weight (BW) followed by a maintenance dose of 3.75 mg/kg (BW) 2 to 3 h after the initial dose, that the mean error score decreased by 23% from 35 to 27.

Critical Flicker Fusion

Several studies have evaluated caffeine's effects on the subject's perception threshold to the fusion of a flickering light source, referred to as the critical flicker fusion threshold (CFFT). The CFFT often is seen as an index of the state of the central nervous system's (CNS) arousal. The measurement of this threshold seems to be generally a reliable, valid, and pharmacosensitive technique. Six studies (Nicholson et al., 1984; Mattila et al., 1988; Swift and Tiplady, 1988; Yu et al., 1991; King and Henry, 1992; Fagan et al., 1994) out of eight failed to find effects of caffeine on the CFFT. Two studies had a trend to significant results (Kerr et al., 1991; Rees et al., 1999) showing a lower fusion threshold, that is, an impairment by caffeine.

Auditory Modality

Only a few data are available with regard to caffeine and sensory and perceptual processing of auditory information. Bullock and Gilliland (1993) recorded the human Brainstem Auditory Evoked Potential (BAEP) and observed that 1.5 and 3.0 mg/kg of caffeine in 18- to 24-year-old male students speeded up Wave V, quite an early sensory component, of the BAEP. Nicholson et al. (1984) did not find an effect of 300 mg of caffeine on the Auditory Evoked Potential (AEP) in six women who were measured during the night. Schicatano and Blumenthal (1994), Smith et al. (1993c), and Zahn and Rapoport (1987) reported that 4 mg of caffeine/kg delayed the habituation of the acoustic startle reflex in nine 18-year-old psychology students. This habituation depends on habitual coffee use. A 300-mg dose in "high" coffee users (> 200 mg/d) slowed habituation more than in low coffee users (Smith et al., 1991b). These findings suggest that caffeine keeps the auditory sensory pathways alert, presumably at the brainstem level.

AUDITORY AND VISUAL SENSES COMBINED

Tharion et al. (1993) used a setting in which both auditory and visual stimuli had to be processed under a placebo or 200 mg of caffeine. The 2-h visual vigilance task, which was the *main* task, required 18 military men (19 to 28 years of age) to respond to a small, dim rectangular target that appeared on a computer screen at random time intervals. The importance of this task was explicitly stressed, and they were asked to perform to the best of their ability. During performance of the vigilance task, there was a constant and invariant auditory stimulus; this was the secondary, or *distracting,* task. Due to the instruction, it was expected that the focus of attention would be on the vigilance task only. Auditory evoked potentials (ERPs) obtained during the performance of the visual vigilance task showed significant increases in the N1 and P2 latencies and decreases in amplitude voltage by caffeine for the nontargets over the whole 2 h of the experimental session. The interpretation was that caffeine helps one to ignore distracting or irrelevant stimulation.

COGNITION

SIMPLE AND CHOICE REACTION TIMES

Several caffeine studies have used a simple reaction time (SRT) and/or a choice reaction time (CRT) type of task. In SRT tasks, the subject is required as rapidly as possible to make a fixed response to a single stimulus, for example, pressing a knob. CRT tasks are similar to SRT tasks, except that the subject is exposed to *different* stimuli, each of which requires a different response.

In about half of the experiments, utilizing either visual or auditory stimuli, caffeine was found to reduce reaction time (RT) in both SRT and CRT. An effect of caffeine on RT was absent in other studies (Münte et al., 1984; Bruce et al., 1986; Kuznicki and Turner, 1986; Lieberman et al., 1987a,b; Zahn and Rapoport, 1987; Fagan et al., 1988; Landrum et al., 1988; Swift and Tiplady, 1988; Smith et al., 1991a,b; Bullock and Gilliland, 1993; Rogers and Dernoncourt, 1998).

Response accuracy (more hits/fewer errors) was found to be either improved by caffeine (Lieberman et al., 1987a,b; Roache and Griffiths, 1987; Swift and Tiplady, 1988; Kerr et al., 1991; King and Henry, 1992; Smith et al., 1993a, 1994a,b; Riedel, 1995; Smit and Rogers, 2000) or not affected (Smith et al. 1991a,b, 1993b; Rogers and Dernoncourt, 1998). These caffeine effects were observed utilizing a wide range of doses (32 to 600 mg). Remarkably, Jacobson and Edgley (1987) reported that even 600 mg of caffeine exhibited no effect on response accuracy.

Possible reasons for these positive and null findings found with caffeine may lie in differences between experiments in dose, protocol, task, and/or subject variables (Lieberman, 1992). In addition, most studies fail to distinguish between the "decision" and "motor" component of the task. The former component is believed to index central cognitive processes in the brain that have to do with discriminatory and decisional processes, while the latter is assumed to index only peripheral, motor output, or execution processes. Accordingly, it may be the case that caffeine's effects on RT are selective in that only the relatively minor, more peripheral aspects of the reaction process, the motor execution of the response, are affected by caffeine (Swift and Tiplady, 1988; Kerr et al., 1991;

King and Henry, 1992; Lorist, 1998), although Smith et al. (1977) and Zahn and Rapoport (1987) found that both decision time and movement time were impaired by 225 mg of caffeine.

Studies by Lorist et al. (1994a), applying the Additive Factor Method (AFM; Sanders, 1983) to assess main and interaction effects of task variables and caffeine on visual CRT, confirmed and extended the finding that peripheral motor processes benefit from caffeine.

SPEEDED DECISION MAKING

In studies on verbal reasoning subjects typically are shown statements about the order of the letters *A* and *B*, each sentence being followed by a pair of letters, *AB* or *BA* (e.g., *A* follows *B*. *BA* right? yes/no?). In such "logical reasoning" tasks, the subject's task is to read the statement, look at the pair of letters, and then decide, as fast as possible, whether the statement is true or false. In 9 out of 12 studies (Borland et al., 1986; Rogers et al., 1989; Smith et al., 1991a, 1992b, 1993a, 1994a,b; Mitchell and Redman, 1992; Bonnet and Arand, 1994b; Linde, 1994; Smith, 1994; Warburton, 1995), all done with young participants, caffeine was found to improve the speed or accuracy of logical reasoning.

Caffeine also improved speeded semantic processing in tasks in which the subjects were shown a series of sentences referring to general knowledge (e.g., canaries have wings) and had to decide whether the sentence was true or false (Smith et al., 1993a, 1994a; Smith, 1994).

CANCELLATION

In performance on a cancellation task, subjects usually are presented sheets with digits, letters, or symbols. The task is to cancel as many specified target items as fast as possible. In six studies caffeine improved task performance (Bättig et al., 1984; Loke, 1988; Frewer and Lader, 1991; Bullock and Gilliland, 1993; Anderson, 1994; Marsden and Leach, 2000), but five studies found no effects (Loke et al., 1985; Bruce et al., 1986; Rogers et al., 1989; Loke, 1990; Rees et al., 1999). The benefits seen with caffeine, however, were modified by dose (Frewer and Lader, 1991), time on task (Bättig et al., 1984), and the memory load of the task (Borland et al., 1986; Frewer and Lader, 1991), signifying that caffeine improves cancellation performance only when relatively few target items have to be retained in memory.

SUBSTITUTION

Digit or symbol substitution tasks, which require subjects as rapidly as possible to replace symbols with digits or letters (or vice versa), do not seem to be sensitive to caffeine (Clubley et al., 1979; Bruce et al., 1986; Lieberman et al., 1987b; Roache and Griffiths, 1987; Mattila et al. 1988; Loke, 1989; Yu et al., 1991; Bonnet and Arand, 1994a,b; Rush et al., 1994a,b) except when performed under suboptimal conditions, such as during the night (Nicholson et al., 1984; Borland et al., 1986; Rogers et al., 1989), when combating fatigue in the later part of test sessions (Rush et al., 1993), or in subjects who are not deprived of caffeine (Rees et al., 1999).

Caffeine facilitates performing substitution tests in suboptimal conditions, such as early in the morning or during the night, when fatigued, or during long tasks.

OTHER COGNITIVE TASKS

In two studies (Nicholson et al., 1984; Borland et al., 1986), caffeine improved the copying of symbols during the night in 20- to 26-year-old shift-workers and in 19- to 24-year-old females. Improvement was also found during the daytime in nondeprived young (20 to 25 years of age) and elderly (50 to 65 years of age) subjects (Rees et al., 1999), but this was not so in four caffeine-deprived men and five caffeine-deprived women 18 to 40 years of age (Bruce et al., 1986).

Only Anderson (1994) reported that caffeine improved performance on a verbal abilities task dose-dependently (1, 2, 3, and 4 mg/kg), but only in high impulsive subjects. For low impulsives, performance first improved and then declined with increasing dose. Erikson et al. (1985) found in a test of free recall of words that impulsivity did not play a role after a dose of 2 or 4 mg/kg. As for intelligence, caffeine (1, 2, 3, and 4 mg/kg) has been found to improve dose-dependently some measures of intelligence (Gupta, 1988a,b), but only in high impulsive or extroverted students. Still, high impulsives performed worse than low impulsives under caffeine in a word categorization task in which they had to sort words for their semantic meaning (Gupta, 1991) but performed better when they had to sort rhyming words.

One study has shown that a 200-mg dose of caffeine tended to improve the number of solved problems and the number of correct solutions on a concentration performance test (Dimpfel et al., 1993), whereas a 400-mg dose tended to impair performance. A similar performance impairment was found in a study by Marsden and Leach (2000) in a realistic daily life setting in which 12 male marines had to solve navigation problems by using information from charts. Weiss and Laties (1962) found no effect of 200 mg of caffeine on solving chess problems.

Other cognitive tasks that have been used in caffeine research include arithmetic (Weiss and Laties, 1962; Loke et al., 1985; Roache and Griffiths, 1987; Loke, 1990; Mitchell and Redman, 1992; Bonnet and Arand, 1994b; Wright et al., 1997) writing speed (Landrum et al., 1988), reading comprehension (Weiss and Laties, 1962; Landrum et al., 1988; Mitchell and Redman, 1992), reading speed (Weiss and Laties, 1962), sentence completion (Loke, 1990), solving anagrams (Smith et al., 1991b), classification of pictures (Swift and Tiplady, 1988), and card sorting (Loke et al., 1985; Loke, 1990). None of these cognitive or intellectual activities was found to be affected significantly by caffeine, except that one study found an improvement of addition performance (Uematsu et al., 1987), although the 600-mg dose reduced the number of completed additions (Loke, 1990).

LEARNING AND MEMORY

Learning

Paired-Associate Learning

In a paired-association learning paradigm, subjects typically are exposed to word pairs of a high or a low degree of semantic association (e.g., tree-apple; car-sea). The instruction is to learn the word pairs. Subjects are then given the first word of each pair as a stimulus for recall of the second. Thus, paired-association learning tasks involve cued recall, where the cue is provided by the experimenter rather than "created" by the subject self as in free recall.

Caffeine does not seem to affect paired-association learning performance, neither when recall is assessed immediately (Mattila et al., 1988; Smith et al., 1991b; Yu et al., 1991) nor when it is assessed after a delay, usually of 20 or 30 min (Smith et al., 1991b).

Serial Learning (Intentional Learning)

The effect of caffeine on learning has been examined by assessing changes in recall performance of words from lists (immediately or after a delay, usually 20 min) as a function of repeated presentation of information. Caffeine (200 and 100 mg) does not appear to influence the learning of words across two or six presentations (Landrum et al., 1988; Rees et al., 1999; Terry and Phifer, 1986), nor does a 3- or 6-mg/kg dose influence the learning of numbers across six presentations (Loke et al., 1985). Also, caffeine does not affect the learning of mental mazes (Bättig et al., 1984; King and Henry, 1992) or the learning of 10-response sequences using three buttons on a response panel (Rush et al., 1993, 1994a,b). Apparently, caffeine has no influence on intentional learning.

Incidental Learning

In contrast to the tasks discussed above, in which subjects are told that their memory is tested later (*intentional* learning), a few caffeine studies have used memory tasks where subjects are not told before that there will be a memory test (*incidental* learning). The basic idea behind the incidental learning paradigm is that more knowledge is gained over the subjects' information processing activities at the time of learning, while in intentional learning subjects may be inclined to perform additional processing activities (rehearsal, memory aids) in order to improve their performance.

Jarvis (1993) prompted subjects to fill in words of which only the first letter was depicted after having had two full presentations of complete words. During these full presentations they were instructed to focus on the word in the inner circle of a sheet full of words. The nondeprived subjects with a daily caffeine use ranging from one to greater than seven cups per day performed dose-dependently better.

Gupta (1991) also used an incidental learning paradigm in 20-year-old students to assess caffeine's effects (1, 2, 3, or 4 mg/kg) on different encoding processes involved in memory. The subjects first performed an acoustic and a semantic word categorization task and then were tested, unexpectedly, for the free recall of the words. It was assumed that the acoustic categorization task (words that rhyme) required shallow or nonsemantic processing of the verbal material, while the semantic task (using the meaning of words) demanded deep or semantic processing. The results showed that caffeine facilitated acquisition and recall in high impulsive subjects after rhyming acquisition but impaired it after semantic acquisition. Caffeine had no effect on the recall performance of low impulsives. In a subsequent study in 300 male students (19 to 24 years old) (Gupta, 1993), similar results were obtained with identical caffeine doses with respect to recognition performance. These findings indicate that for high impulsives, that is, subjects with *low* arousal levels, caffeine facilitates the encoding of the physical properties of verbal material (rhyming) but impairs the utilization of semantic information (i.e., meaning). These results indicate that high impulsives, extroverts or sensation seekers, who are assumed to have low arousal levels, may benefit from caffeine only in tasks that are relatively easy.

Only one other study (Loke, 1992) examined the effect of a 200-, 400-, or 600-mg caffeine dose on incidental learning in male and female students with an average age of 19 years. Caffeine did not affect the free recall of words following an incidental learning condition in which the subjects had to repeat the words. In sum, caffeine facilitates incidental learning in tasks in which information is presented passively; in tasks in which material is learned intentionally, caffeine has no effect.

MEMORY

Human memory can be divided into working memory (WM) and long-term memory (LTM). WM stores information over brief intervals of time during which further processing can be performed (e.g., recognition). Only limited amounts of information of which we are aware can be stored in this WM. LTM contains large amounts of information stored for considerable periods of time. We are not aware of this information until it is activated and becomes part of working memory.

LTM and WM can be deliberately accessed during task performance; they are then *explicit* memories. *Implicit* memories are memory representations, which cannot be directly accessed. An example of implicit memory is the association between a specific stimulus and a response. This association is formed after practice and can only be evoked by specific stimuli.

Free Recall

In an immediate free recall task with a supraspan word list, subjects are presented a list of unrelated words, exceeding their memory span, and then asked to recall as many words as possible in any order. The basic finding is that words occurring at the beginning and end of the list are recalled better than words in the middle of the list, producing so-called primacy and recency effects, respectively. This recall pattern is referred to as the *serial position effect*. It is usually hypothesized that the more recent words are retained in the primary, working memory, while the earlier items are retained in the secondary or "long-term" memory.

Caffeine was found either to exert no effects on free recall performance in 14 studies (Loke et al., 1985; Loke, 1988, 1992; Foreman et al., 1989; Smith et al., 1991a, 1992a, 1993a,b, 1994b; Smith, 1994; Riedel, 1995; Warburton, 1995; Wright et al., 1997; Rees et al., 1999), improve recall in six studies (Arnold et al., 1987; Barraclough and Foreman, 1994; Rogers and Dernoncourt, 1998; Schmitt, 2001a,b; Ryan et al., 2002), or to impair recall in three studies (Erikson et al., 1985; Terry and Phifer, 1986; Loke, 1992).

In one study (Arnold et al., 1987), improvements with 2 and 4 mg/kg of caffeine were seen only in 75 female students at the third level of practice, while decrements were seen in males at the second and third levels of practice with 2 mg/kg. Schmitt (2001a) found this improvement only in 50-year-old subjects, not in younger and older ones. In a study by Erikson et al. (1985), caffeine's effects (2 and 4 mg/kg) were detrimental under a slow rate of stimulus presentation only for the female students, while no effects were observed for males. Similar puzzling effects were found in high impulsive students (detrimental effects) but not in low impulsives (Smith et al., 1994b).

A few studies have evaluated caffeine's effects on recall more precisely by taking into account the position of the words in a list. Some authors concluded that neither the primacy nor the recency effect is affected specifically by caffeine (Erikson et al., 1985; Arnold et al., 1987; Smith et al., 1994b). Instead, it was observed (Barraclough and Foreman, 1994) that 2 and 4 mg of caffeine/kg, at least for males, exerted its largest effect on recall of the middle portion of the list. According to the authors, this selective caffeine effect may be the consequence of a general increase in CNS activity "leading to increased salience of stimuli that are normally of low recall priority." Other immediate free recall tasks include the recall of eight-digit numbers (Nicholson et al., 1984) and the recall of five- and six-letter nouns (Mitchell and Redman, 1992). In the former study, 300 mg of caffeine impaired recall, although not significantly, while in the latter study 4 mg/kg of caffeine showed no effects. In addition, caffeine has been observed to impair the immediate reproduction of numbers during the night (Nicholson et al., 1984), but in another study (Rogers et al., 1989), also measuring during the night, caffeine exerted no effect on number reproduction. Finally, 200 mg of caffeine has been reported (Linde, 1994) to impair, in 36 first-year male and female psychology students, the immediate reproduction of spatial relationships of verbal information after normal sleep but to improve it after a vigil, indicating a compensation of fatigue. In sum: caffeine does not seem to improve immediate free recall of words, letters, or digits consistently. Especially in situations of low arousal and with tasks of minimal complexity, caffeine may improve free recall.

Delayed Free Recall and Recognition

A few studies report beneficial effects of caffeine on delayed free recall (Warburton, 1995; Ryan et al., 2002) and delayed cognition (Riedel, 1995), using mostly a delay shorter than 30 min. In general, caffeine appears not to help the *delayed* free recall of word lists (Loke et al., 1985; Loke, 1988; Smith et al., 1992b, 1994b; Schmitt, 2001a,b) or the *delayed* recognition of words (Loke et al., 1985; Loke, 1988; Smith et al., 1994a; Warburton, 1995). In one study (Smith et al., 1994b), a 4-mg/kg caffeine dose impaired the delayed recognition of words, but this was observed only for high impulsive subjects; no caffeine effects were seen for low impulsives. Also, 200 mg of caffeine tended to impair the delayed recognition of pictures (Roache and Griffiths, 1987), while 400 mg induced better recognition than 600 mg; both higher doses elicited a better response than 200 mg. In general, caffeine does not affect delayed recall and recognition.

Task Load

In versions of focused and divided selective attention tasks, such as used by Lorist et al. (1994b, 1995, 1996), subjects are instructed to detect the possible appearance of a memory set item among the display items on a screen. Increased RTs with increasing task load, defined by the product of memory set size (memory load, ML) and display set size (display load, DL), suggest that subjects apply serial, limited-capacity searches, in which each memory set item is compared sequentially to all display items. An increase in the number of targets to be memorized (ML) or in the items presented on the display (DL) results in more extensive search processes of longer duration and a more pronounced negativity of the recorded brain activity.

Display Load

A positive effect of 3 mg of caffeine/kg on RT in the low-DL condition that was absent in the high-DL condition indicated that DL manipulations were influenced by caffeine (Lorist et al., 1996). The decrease of caffeine's effects with increasing DL corresponds to results indicating that caffeine facilitates performance only in simple or moderately complex tasks. In more complex tasks caffeine may have either no effect or may even impair performance (Weiss and Laties, 1962; Humphreys and Revelle, 1984).

The decreasing effect of caffeine on WM with increasing DL implies that with a *moderate* dose of caffeine, performance improves as long as energetic supplies increase up to a certain peak; beyond this optimum it deteriorates (Yerkes and Dodson, 1980). Because caffeine stimulates the availability of energetic resources, the positive effects of caffeine in the low-DL condition, and the absence of effects of caffeine in the high-DL condition, might be interpreted as reflecting the turning point from an optimal level to an overly high level of energetic supplies involved in WM processes. The conclusion is that caffeine facilitates performance on tasks that appeal to WM to a small degree but hinders performance on tasks presumed to appeal to WM heavily.

Memory Search

Sternberg's memory-search paradigm (Sternberg, 1969) is often used to evaluate caffeine's effects on the retrieval of information from memory. In this paradigm, the subject is presented a sequence of stimuli and on each trial has to decide, as fast as possible, whether the stimulus is a member of a small memorized set of stimuli. ML is manipulated by varying the size of the memory set. It appeared that 200 mg of caffeine increased sensimotor speed, but the scanning time of WM was either not affected or increased. The latter finding implies that caffeine slowed central memory scanning. The interaction between ML and caffeine on RT did not reach significance, suggesting that memory search was not affected by caffeine. The study by Kerr et al. (1991), using a constant memory set size of four items, reported that only 300 mg of caffeine speeded RT. In those task conditions in which task load was varied by varying the memory set size, no caffeine effects were observed on RTs. Apparently, memory load effects are not affected by caffeine; in other words, memory search is insensitive to caffeine.

Target Detection

In selective attention tasks, subjects have to decide whether a memory set item is present on the display. In other words, memory set letters are associated with being targets to which a response should be made.

From the brain potentials, the latency of the P3b component, representing stimulus evaluation or input processing activities, was on average shorter after caffeine than after placebo, and this effect depended on display load. The absence of an effect of caffeine on search-related negativity in selective attention tasks, which is a reflection of central processes, indicates that the effects of caffeine originate from an effect on input-related processes and *not* from an effect on central processes (Lorist et al., 1996).

Memory Span

In particular, the digit span task has been used to assess caffeine's effects on the capacity of WM. Span tasks involve the evaluation of the maximum number of unrelated items that can be recalled in the correct order immediately after presentation (immediate recall). In four out of five studies, caffeine in doses from 100 to 600 mg did not affect memory span performance (Borland et al., 1986; Roache and Griffiths, 1987; Davidson and Smith, 1989; Smith et al., 1993). In a study by Davidson and Smith (1989), 390 mg of caffeine was seen to impair backwards digit recall under a noise condition but to improve it under a no-noise condition. In a study by Humphreys and Revelle (1984), the effect of 100 mg of caffeine apparently depended on personality. Low impulsives performed better than high impulsives on delayed recall, but worse on immediate recall. Rees et al. (1999) found in caffeine young and elderly subjects not deprived of caffeine that 250 mg of caffeine somewhat improved performance on a taxing running digit span memory test. Caffeine also set off the decline in performance, as found in the placebo condition. The central information processing stage consists of a broad range of functionally different processes. Therefore, caffeine may affect serial comparisons and binary decisions as well as response selection processes. The safest conclusion about the effect of caffeine on central-related processes is that it is limited to WM.

ATTENTION

Attention, the *control* of information processing, may be categorized as focused attention, divided attention, and sustained attention.

FOCUSED ATTENTION

Focused, or selective, attention tasks generally are used to assess the subject's ability to select relevant information from a pool of both relevant and irrelevant information.

Stroop Test

Several caffeine studies have used variants of the well-known Stroop Color and Word Test. This test requires subjects to focus their attention on the color of printed words or on the number of identical digits series, while ignoring their meaning. Caffeine's effects are inconsistent; three studies showed positive effects (Hasenfratz and Bättig, 1992; Kenemans and Verbaten, 1998; Kenemans et al., 1999), three had null findings (Borland et al., 1986; Riedel, 1995; Edwards et al., 1996), and one had negative results (Foreman et al., 1989). Interestingly, in nondeprived subjects with an average age of 52 years, an improvement was followed by an impairment with time on task (Hameleers et al., 2000).

"Filter" or Selective Attention Tasks

Several authors (Smith et al., 1991a, 1992b; Smith, 1994; Riedel, 1995; Lorist et al., 1996; Ruijter et al., 2000a,b) used a visual focused-attention ("filter") task that required subjects to respond, as fast as possible, to a centrally presented letter that was sometimes surrounded by other, irrelevant letters. Another attention ("search") task was used, similar to the former except that the subject did not know at which of two possible locations a specified letter would appear. Basically, neither of these tasks was found to be affected by caffeine based on behavior parameters;

however, caffeine-induced changes in brain activity did show that the brain worked faster and more energetically.

Loke (1992) used two versions of a visual search task, utilizing stimulus frames of four items. One version required subjects to search for target digits among consonant distracters (easy automatic processing). The other version demanded search for target consonants among consonant distracters (controlled processing). The assumption was that the former task involved *automatic* target detection and the latter *subject-controlled* search. Processing (i.e., memory) load was manipulated by varying the number of targets (either two or four ML targets). The results showed that caffeine exhibited neither main nor interaction effects on performance, a result similar to that found by Lorist et al. (1996) (see above).

DIVIDED ATTENTION

Studies examining effects of caffeine on divided attention all used the dual-task paradigm. Mostly, subjects were required to perform a primary tracking task or a visual selective attention task simultaneously with a secondary visual or auditory vigilance task. In four studies (Borland et al., 1986; Rogers et al., 1989; Kerr et al., 1991; Kourtidou-Papadeli et al., 2002), caffeine improved dual-task performance, but in three other studies (Croxton et al., 1985; Zwyghuizen-Doorenbos et al., 1990; Rosenthal et al., 1991) caffeine induced no effects. Although the basis for the difference in outcomes is not clear, it should be noted that in two (Borland et al., 1986; Rogers et al., 1989) of the four research groups reporting benefits, the caffeine was given shortly before midnight, and the dual task was carried out during the night.

SUSTAINED ATTENTION

Considerable research efforts have been devoted to assess the effects of caffeine on sustained attention in different types of vigilance tasks. Important parameters of vigilance performance are the overall level of performance and the decrement in performance over time, called *time on task (TOT) effects*. Various vigilance tasks have been utilized to study caffeine's effects on sustained attention. Most often these were versions of the Auditory Vigilance Task (AVT) (Wilkinson, 1968) and the Bakan (1959) task; those of the latter type are also called *oddball tasks*.

In *auditory vigilance tasks* (Lieberman et al., 1987a,b; Fagan et al., 1988; Zwyghuizen-Doorenbos et al., 1990; Rosenthal et al., 1991), lasting either 60 or 40 min, subjects are asked to detect the occurrence of slightly deviant (e.g., of longer or shorter duration) tones occurring infrequently and randomly within a continuous series of standard tones. The tones are presented against a background of white noise at a rate of one every 2 sec. Caffeine was found to improve the overall number and/or speed of correct detections, while the number of false alarms was not significantly affected (Lieberman et al., 1987a,b; Fagan et al., 1988; Zwyghuizen-Doorenbos et al., 1990; Rosenthal et al., 1991) or tended to decrease (Kozena et al., 1986). Three studies (Nicholson et al., 1984; Fagan et al., 1988; Rosenthal et al., 1991) also provided information regarding changes in auditory vigilance performance over time. It appeared that caffeine reduced the vigilance decrement seen with placebo; in other words, it compensated for the TOT effects. Using a visual vigilance test, Lieberman et al. (2002) found that 200 and 300 mg of caffeine improved performance efficiency in sleep-deprived subjects. Both accuracy and speed measures showed a positive effect of caffeine. In subjects who were not sleep-deprived, similar effects of caffeine were also found (Fine et al., 1994).

Other studies have used versions of the Bakan task, the *Rapid Information Processing (RIP) task*, in which subjects usually are presented single digits on a visual screen and instructed to detect the occurrence of three successive odd or even digits. Stimulus presentation rate typically is fast (100 digits per minute), and task duration across studies ranges from 1 to 30 min (usually 20 to 30 min). In comparison with versions of the AVT, which may be identified as "sensory" types of vigilance tasks, the Bakan task versions may be better characterized as "cognitive" vigilance tasks (Davies and Parasuraman, 1982).

In 13 studies using visual or auditory versions of the Bakan task, caffeine was found to improve the overall number and/or speed of correct detections, whereas the number of false alarms (when reported) was reduced or not affected by caffeine (Borland et al., 1986; Eaton-Williams and Rusted in Rusted, 1994; Rapoport et al., 1981; Nicholson et al., 1984; Pons et al., 1988; Swift and Tiplady, 1988; Rogers et al., 1989; Lipschutz et al., 1990; Smith et al., 1990, 1992b; Frewer and Lader, 1991; Rosenthal et al., 1991; Wright et al., 1997). Improvements by caffeine were also found (Rapoport et al., 1981) with a 3- and 10-mg/kg caffeine dose in 11-year-old boys, but not in 22-year-old or 65- to 75-year-old men (Swift and Tiplady, 1988). Smith et al. (1991a) found that caffeine showed no effects on vigilance performance, but in their study the task used lasted only 1 min and had a lower presentation rate. It seems that performing vigilance tasks depends particularly on adequate arousal levels. Fatigue is regularly found to be compensated by caffeine. This finding is reported as the compensation for TOT effects (Smith et al., 1990; Frewer and Lader, 1991; Hasenfratz et al., 1991, 1994; Hasenfratz and Bättig, 1993, 1994; Bonnet and Arand, 1994a,b; Warburton, 1995; Smit and Rogers, 2000).

Additional information given by Kozena et al. (1986) on performance changes over time indicated that vigilance decrement in this type of task was not affected by caffeine. This finding has been confirmed in several studies (Hasenfratz et al., 1991, 1994; Hasenfratz and Bättig, 1993, 1994), using a visual version of the RIP task in which the digits were presented in a *subject-paced* manner rather than at a fixed rate. Surprisingly, contrary to these findings, Bättig and Buzzi (1986) found less performance decrement over time when caffeine was given.

In sum, these findings indicate that caffeine mostly improves the level of vigilance in AVT and Bakan types of vigilance tasks.

Although differences in the information-processing demands between tasks (e.g., sensory vs. cognitive) could play a role in the sometimes noted failure of caffeine to affect vigilance decrements in the Bakan task, other differences in task parameters may explain the inconsistency of findings, such as sensory modality, stimulus rate, and task duration.

In general, most studies examining caffeine's effects on vigilance performance while utilizing a diversity of visual and auditory vigilance tasks found that caffeine, to a greater or lesser extent, improved the overall level of vigilance. Caffeine exerted some compensating effect on the decrement in vigilance over time, although there are findings showing that caffeine failed to do so. Finally, Swift and Tiplady (1988) reported that caffeine increased the ratio of false alarms to hits, indicating a shift in the subject's response criterion (more speed, less accuracy) rather than a change in detection efficiency (less speed, more accuracy).

MENTAL FATIGUE

Mental fatigue often arises as a consequence of performing mentally demanding tasks for a prolonged period of time. An important factor in the fatiguing potential of cognitive tasks is task complexity. In general, tasks that show straightforward deterioration over time are mostly simple, monotonous tasks, such as Wilkinson vigilance-like tasks or simple reaction time tasks. In more complex and perhaps arousing, intrinsically interesting tasks, fatiguing effects are often less direct. As for the capacity of tasks to attract interest, complex tasks may be more interesting than monotonous simple tasks such as vigilance tasks, simple reaction time tasks, and so on.

The role of caffeine in performing cognitive tasks is often interpreted as changes in subjective energy. Indeed, Bruce and colleagues (Bruce et al., 1986) and Zwyghuizen-Doorenbos et al. (1990) did show that caffeine diminished feelings of tiredness and increased alertness, which could be interpreted as a caffeine-induced increased level of arousal. Another line of reasoning comes from theories stressing that the energetic states of subjects play a controlling role in information processing (Clubley et al., 1979; Sanders, 1983; Humphreys and Revelle, 1984). Support for this view comes from studies showing that caffeine increases EEG power. According to the Humphreys-Revelle model, the information transfer component of cognitive tasks should be facilitated by an increase of arousal; since caffeine is supposed to do so, numerous studies have been performed to verify this assumption. Although there are inconsistent findings, the position still holds that the effects of caffeine on cognitive function are mediated predominantly by arousal or the energetic state of the subject. It is obvious that many factors play a role in this state, such as time of day, personality, age, and so forth. It is beyond the scope of this discussion to ascertain the role of each factor. In an attempt to interpret information processing more satisfactorily and in more detail, multiple energetic resources in the brain have been assumed that supposedly perform a modulating role in cognitive operations (Pribram and McGuiness, 1975; Sanders, 1983; Hockey, 1986). As for the role of caffeine in these energetic resources, only recently have specific, technically sophisticated studies been performed (Lorist et al., 1994a,b, 1995; Ruijter et al., 1999, 2000a,b, 2001). These show, based on both behavioral and electrophysiological data (EEG-ERP), that caffeine has specific energy-enhancing effects on the input and output stages of information processing, but not on central cognitive, processing stages.

The more complex RIP task that has been used by Hasenfratz et al. (Hasenfratz et al., 1991, 1994; Hasenfratz and Bättig, 1993, 1994) produced significant fatigue-induced performance decrements across sessions. Interestingly, no interaction was found between arousal state and caffeine. Corresponding results were reported by Regina et al. (1974), Plath et al. (1991), and Lorist et al. (1994a,b) in well-rested and fatigued subjects, implying that caffeine may improve performance independent of state. Results of these studies suggest that caffeine might be able to let subjects invest effort even in a well-rested condition, when mentally or physically fatigued, and, possibly, also when physically exhausted.

CONCLUSIONS

This chapter considered the behavioral and electrophysiological results from studies on caffeine and mental performance. The results from different studies sometimes are at variance with each other. A major potential source for these varying results appears to relate to differences in caffeine dose, experimental design, protocol, and methods and procedures of testing. This great diversity across studies seems to reflect a lack of consensus on the appropriate methods to employ in coffee and caffeine research. Another potential source for the variable caffeine results lies in the nature of caffeine's effects. It seems that the influence of caffeine on performance typically is (1) of a modest size; (2) selective, in that some features of performance are more sensitive than others; (3) complex, perhaps representing *patterns* of behavioral facilitation and interference; and (4) not constant, in that it can be moderated by a wide variety of variables. Taking into account these considerations and to emphasize some consistency in results, the following conclusions can be made:

- Caffeine can apparently improve the performance of a wide variety of mental tasks directly and also indirectly by reducing decrements in performance under suboptimal conditions of alertness. This conclusion is based on the findings that caffeine can improve performance across various cognitive, learning, memory, and attention tasks.
- The efficacy of caffeine to reduce impairments in mental efficiency under states of reduced alertness is one of the most consistent findings in caffeine research. The results, however, are not exclusively limited to suboptimal conditions, since benefits of caffeine have also been observed under optimal alertness conditions (Lorist et al., 1994a,b).
- Caffeine may affect performance of sensory-perceptual tasks. These effects take the form of either performance facilitation or inhibition, possibly depending on dose, subject variables, and other unspecified variables. The available evidence is limited, however, and more research is clearly needed here.

- Caffeine often does not affect performance on purely cognitive tasks, as noted by Weiss and Laties (1962). This conclusion, however, should be qualified in that caffeine does seem to have the potential to improve cognitive performances that are timed, as assessed by RT, decision-making, or cancellation tasks. Also, cognitive tests requiring *speed* are more sensitive to caffeine's beneficial effects than tests involving intellectual power.
- Perceptual-motor task demands may be the principal determinants of caffeine's effects on speeded cognitive performance; effects are most often observed if these task demands are relatively high, for instance, if stimulus quality is low.
- Caffeine usually does not affect performance of learning and memory tasks. Although some studies occasionally have found caffeine to affect memory and learning performance, either facilitory or inhibitory, these effects typically emerged as complex interactions with dose, subject, and task variables. These caffeine effects may represent effects on the encoding, or the attention devoted to the information, rather than direct and specific effects on the storage or retrieval of information in short-term and working memory.
- Finally, the ingestion of caffeine is likely to improve general levels of performance in vigilance tasks. That is, its efficacy is evident throughout a period of vigilance, resulting in a steady, overall higher level of performance. This caffeine effect is rather robust and does not appear to depend on the type of vigilance task.

Two tentative general mechanisms or factors that may account for most of the observed caffeine effects emerge as particularly salient: (1) an indirect, nonspecific "alertness," "arousal," or "processing resources" factor, presumably accounting for why effects of caffeine generally are most pronounced when task performance is sustained or degraded under suboptimal conditions, and (2) a direct, and more specific, "perceptual-motor" speed or efficiency factor that may explain why under optimal conditions certain aspects of human performance and information processing (e.g., those related to sensation, perception, motor preparation, and execution) are more sensitive to caffeine's effects than other aspects (e.g., those related to cognition, memory, and learning).

On the basis of the evidence currently available, more definite and detailed conclusions and theoretical claims cannot be drawn.

REFERENCES

Anderson, K.J. (1994) Impulsivity, caffeine and task difficulty: a within-subject test of the Yerkes-Dodson Law. *Personal and Indivdual Differences*, 16, 813–829.

Arnold, M.E., Petros, T.V., Beckwith, B.F., Coons, G. and Gorman, N. (1987) The effect of caffeine, impulsivity, and sex on memory for word lists. *Physiology and Behavior*, 41, 25–30.

Bakan, P. (1959) Extraversion-introversion and improvement in an auditory vigilance task. *British Journal of Psychology*, 50, 325–332.

Barraclough, S. and Foreman, N. (1994) Factors influencing recall of supraspan word lists: caffeine dose and introversion. *Pharmacopsychoecologia*, 7, 229–236.

Bättig, K. and Buzzi, R. (1986) Effect of coffee on the speed of subject-paced information processing. *Neuropsychobiol*, 16, 126–130.

Bättig, K., Buzzi, R., Martin, J.R. and Feierabend, J.M. (1984) The effects of caffeine on physiological functions and mental performance. *Experientia*, 40, 1218–1223.

Böhme, M. and Böhme, H.R. (1985) Der einflusz hormonaler Kontrazeptiva und des Coffeins auf den Fransworth-Munsell-100 hue Test. *Zentralblat für Gynäkologie*, 107, 1300–1306.

Bonnet, M.H. and Arand, D.L. (1994a) Impact of naps and caffeine on extended nocturnal performance. *Physiology and Behavior*, 56, 103–109.

Bonnet, M.H. and Arand, D.L. (1994b) The use of prophylactic naps and caffeine to maintain performance during a continuous operation. *Ergonomics*, 37, 1009–1020.

Borland, R.G., Rogers, A.S., Nicholson, A.N., Pascoe, P.A. and Spencer, M.B. (1986) Performance overnight in shiftworkers operating a day-night schedule. *Aviation, Space and Environmental Medicine,* 57, 241–249.

Bruce, M., Scott, N., Lader, M. and Marks, V. (1986) The psychopharmacological and electrophysiological effects of single doses of caffeine in healthy human subjects. *British Journal of Clinical Pharmacology,* 22, 81–87.

Bullock, W.A. and Gilliland, K. (1993) Eysenck's arousal theory of introversion-extraversion: a converging measures investigation. *Journal of Personality and Society Psychology,* 64, 113–123.

Clubley, M., Bye, C.E., Henson, T.A., Peck, A.W. and Riddington, C.J. (1979) Effects of caffeine and cyclizine alone and in combination on human performance, subjective effects and EEG activity. *British Journal of Clinical Pharmacology,* 7, L157–L163.

Croxton, J.S., Putz-Anderson, V.R. and Setzer, J.V. (1985) Attributional consequences of chemical exposure and performance feedback. *Journal of Applied and Social Psychology,* 15, 313–329.

Davidson, R.A. and Smith, B.D. (1989) Arousal and habituation: differential effects of caffeine, sensation seeking and task difficulty. *Personality and Individual Differences,* 10, 111–119.

Davies, D.R. and Parasuraman, R. (1982) *The Psychology of Vigilance,* Academic Press, London.

Diamond, A.L. and Cole, R.E. (1979) Visual threshold as a function of test area and caffeine administration. *Psychonomic Science,* 20, 109–111.

Diamond, A.L. and Smith, E.M. (1974). The effects of caffeine on terminal dark adaptation, in *Sensation and Measurement,* Moskowitz, H.R. et al., Eds., D. Reidel Publishing. Co., Dordrecht, pp. 339–349.

Dimpfel, W., Schober, F. and Spüler, A. (1993) The influence of caffeine on human EEG under resting conditions and during mental loads. *Clinical Investigation,* 71, 197–207.

Edwards, S., Brice, C., Craig, C. and Penri-Jones, R. (1996) Effects of caffeine, practice and mode of presentation on Stroop task performance. *Pharmacology, Biochemistry and Behavior,* 54, 309–315.

Erikson, G.C., Hager, L.B., Houseworth, C., Dungan, J., Petros, T. and Beckwith, B.E. (1985) The effects of caffeine on memory for word lists. *Physiology and Behavior,* 35, 47–51.

Fagan, D., Paul, D.L., Tiplady, B. and Scott, D.B. (1994) A dose-response study of the effects of inhaled nitrous oxide on psychological performance and mood. *Psychopharmacology,* 116, 333–338.

Fagan, D., Swift, C.G. and Tiplady, B. (1988) Effects of caffeine on vigilance and other performance tests in normal subjects. *Journal of Psychopharmacology,* 2, 19–25.

Fine, B.J. and McCord, L. (1991) Oral contraceptives use, caffeine consumption, field dependence and the discrimination of colors. *Perceptual and Motor Skills,* 73, 931–941.

Fine, B.J., Kobrick, J.L., Lieberman, H.R., Marlowe, B., Riley, R.H. and Tharion, W.J. (1994) Effects of caffeine or diphenhydramine on visual vigilance. *Psychopharmacology,* 114, 233–238.

Foreman, N., Barraclough, S., Moore, C., Mehta, A. and Madon, M. (1989) High doses of caffeine impair performance of a numerical version of the Stroop task in men. *Pharmacology, Biochemistry and Behavior,* 32, 399–403.

Frewer, L.J. and Lader, M. (1991) The effects of caffeine on two computerized tests of attention and vigilance. *Human Psychopharmacology,* 6, 119–128.

Gupta, U. (1988a) Effects of impulsivity and caffeine on human cognitive performance. *Pharmacopsychoecologia,* 1, 33–41.

Gupta, U. (1988b) Personality, caffeine and human cognitive performance. *Pharmacopsychoecologia,* 1, 79–84.

Gupta, U. (1991) Differential effects of caffeine on free recall after semantic and rhyming tasks in high and low impulsives. *Psychopharmacology,* 105, 137–140.

Gupta, U. (1993) Effects of caffeine on recognition. *Pharmacology, Biochemistry and Behavior,* 44, 393–396.

Hameleers, P.A.H.M., Van Boxtel, M.P.J., Hogervorst, E., Riedel, W.J., Houx, P.J., Buntinx, F. and Jolles, J. (2000) Habitual caffeine consumption and its relation to memory, attention, planning capacity and psychomotor performance across multiple age groups. *Human Psychopharmacology and Clinical Exploration,* 15, 573–581.

Hasenfratz, M. and Bättig, K. (1992) No psychophysiological interactions between caffeine and stress? *Psychopharmacology,* 109, 283–290.

Hasenfratz, M. and Bättig, K. (1993) Dose-effect relationships between caffeine and various psychophysiological parameters, in *Proceedings of the 15th International Scientific Colloquium on Coffee,* Vol. II, pp. 476–482.

Hasenfratz, M. and Bättig, K. (1994) Acute dose-effect relationships of caffeine and mental performance, EEG, cardiovascular and subjective parameters. *Psychopharmacology*, 114, 281–287.

Hasenfratz, M., Buzzini, P., Cheda, P. and Bättig, K. (1994) Temporal relationships of the effects of caffeine and alcohol on rapid information processing. *Pharmacopsychoecologia*, 7, 87–97.

Hasenfratz, M., Jacquet, F., Aeschbach, D. and Bättig, K. (1991) Interactions of smoking and lunch with the effects of caffeine on cardiovascular functions and information processing. *Human Psychopharmacology*, 6, 277–284.

Hockey, G.R.J. (1986) Changes in operator efficiency as a function of environmental stress, fatigue, and circadian rhythms, in *Handbook of Perception and Human Performance*, Vol. 2, Boff, K.R., Kaufman, L. and Thomas, J.P., Eds., Wiley, New York, pp. 44-1 to 44-49.

Humphreys, M.S. and Revelle, W. (1984) Personality, motivation, and performance: a theory of the relationship between individual differences and information processing. *Psychological Reviews*, 91, 153–184.

Jacobson, B.H. and Edgley, B.M. (1987) Effects of caffeine in simple reaction time and movement time. *Aviation, Space and Environmental Medicine*, 58, 11533–11556.

Jarvis, M.J. (1993) Does caffeine intake enhance absolute levels of cognitive performance? *Psychopharmacology*, 110, 45–52.

Kenemans, J.L. and Verbaten, M.N. (1998) Caffeine and visuo-spatial attention. *Psychopharmacology*, 135, 353–360.

Kenemans, J.L., Wieleman, S.T., Zeegers, M. and Verbaten, M.N. (1999) Caffeine and Stroop interference. *Pharmacology, Biochemistry and Behavior*, 63, 589–598.

Kerr, J.S., Sherwood, N. and Hindmarch, I. (1991) Separate and combined effects of the social drugs on psychomotor performance. *Psychopharmacology*, 104, 113–119.

King, D.J. and Henry, G. (1992) The effect of neuroleptics on cognitive and psychomotor function a preliminary study in healthy volunteers. *British Journal of Psychiatry*, 160, 647–653.

Kohen, L. and Schneider, K. (1986) Der Einflusz von Theophyllin und Coffein auf die sensorische Netzhautfunktion des Menschen. *Fortschung in Ophtalmologie*, 83, 338–344.

Kourtidou-Papadeli, C., Papadelis, C., Louizos, A.-L. and Guiba-Tziampiri, O. (2002) Maximum cognitive performance and physiological time trend measurements after caffeine intake. *Cognitive Brain Research*, 12, 407–415.

Kozena, L., Frantik, E. and Horvath, M. (1986) The effect on vigilance of an analgesic combination (Ataralgin) and its components, guaiphenesine and caffeine. *Activitas Nervosa Superior (Praha)*, 28, 153–154.

Kuznicki, J.T. and Turner, L.S. (1986) The effects of caffeine on caffeine users and non-users. *Physiology and Behavior*, 37, 397–408.

Landrum, R.E., Meliska, C.J. and Loke, W.H. (1988) Effects of caffeine and task experience on task performance. *Psychologia: Annals of the International Journal of Psychology*, 31, 91–97.

Lieberman, H.R. (1992) Caffeine, in *Handbook of Human Performance: Health and Performance*, Vol. 2, Smith, A.P. and Jones, D.M., Eds., Academic Press, London, pp. 49–19.

Lieberman, H.R., Wurtman, R.J., Emde, G.G. and Coviella, I.L.G. (1987a) The effects of caffeine and aspirin on mood and performance. *Journal of Clinical Psychopharmacology*, 7, 3315–3320.

Lieberman, H.R., Wurtman, R.J., Emde, G.G., Roberts, C. et al. (1987b) The effects of low doses of caffeine on human performance and mood. *Psychopharmacology*, 92, 308–312.

Lieberman, H.R., Tharion, W.J., Shukitt Hale, B., Speckman, K.L. and Tulley, R. (2002) Effects of caffeine, sleep loss, and stress on cognitive performance and mood during U.S. Navy SEAL training. *Psychopharmacology*, 164, 250–261.

Linde, L. (1994) An auditory attention task: a note on the processing of verbal information. *Perceptual and Motor Skills*, 78, 563–570.

Lipschutz, L., Berman, S. and Spielman, A.J. (1990) Acute and chronic caffeine administration: alertness, cognition and diurnal variation. *Sleep Research*, 19, 74–80.

Loke, W.H. (1988) Effects of caffeine on mood and memory. *Physiology and Behavior*, 44, 367–372.

Loke, W.H. (1989) Effects of caffeine on task difficulty. *Psychologica Belgica*, 29, 51–62.

Loke, W.H. (1990) Effects of repeated caffeine administration on cognition and mood. *Human Psychopharmacology*, 5, 339–348.

Loke, W.H. (1992) The effects of caffeine and automaticity on a visual information processing task. *Human Psychopharmacology*, 7, 379–388.

Loke, W.H., Hinrichs, J.V. and Ghoneim, M.M. (1985) Caffeine and diazepam: separate and combined effects on mood, memory, and psychomotor performance. *Psychopharmacology,* 87, 344–350.

Lorist, J. (1998) Caffeine and information processing in man, in *Nicotine, Caffeine and Social Drinking: Behaviour and Brain Function,* Snel, J. and Lorist, M.M., Eds., Harwood Academic Publishers, London, pp. 185–200.

Lorist, M.M., Snel, J. and Kok, A. (1994a) Influence of caffeine on information processing stages in well rested and fatigued subjects. *Psychopharmacology,* 113, 411–421.

Lorist, M.M., Snel, J., Kok, A. and Mulder, G. (1994b) Influence of caffeine on selective attention in well rested and fatigued subjects. *Psychophysiology,* 31, 525–534.

Lorist, M.M., Snel, J., Kok, A. and Mulder, G. (1996) Acute effects of caffeine on selective attention and visual search processes. *Psychophysiology,* 33, 354–361.

Lorist, M.M., Snel, J., Mulder, G. and Kok, A. (1995) Aging, caffeine, and information processing: an event-related potential analysis. *Electroencephalography and Clinical Neurophysiology,* 96, 354–361.

Marsden, G. and Leach, J. (2000) Effects of alcohol and caffeine on maritime navigational skills. *Ergonomics,* 43, 17–26.

Mattila, M., Seppala, T. and Mattila, M.J. (1988) Anxiogenic effect of yohimbine in healthy subjects: comparison with caffeine and anatagonism by clonidine and diazepam. *International Psychopharmacology,* 3, 215–229.

Mitchell, P.J. and Redman, J.R. (1992) Effects of caffeine, time of day and user history on study-related performance. *Psychopharmacology,* 109, 121–126.

Münte, T.F., Heinze, H.J., Künkel, H. and Scholz, M. (1984) Personality traits influence the effects of diazepam and caffeine on CNV magnitude. *Neuropsychobiology,* 12, 60–67.

Nicholson, A.N., Stone, B.M. and Jones, S.J. (1984) Studies on the possible central effects in man of a neuropeptide (ACTH 4-9 analogue). *European Journal of Clinical Pharmacology,* 27, 561–565.

Plath, D., Roehrs, T.A., Zwyghuizen-Doorenbos, A., Sicklesteel, J., Wittig, R.M. and Roth, T. (1991) The alerting effects of caffeine after sleep restriction. *Sleep Research,* 20, 124–129.

Pons, L., Trenque, T., Bielecki, M., Moulin, M. and Potier, J.C. (1988) Attentional effects of caffeine in man: comparison with drugs acting upon performance. *Psychiatry Research,* 23, 329–333.

Pribram, K.H. and McGuiness, D. (1975) Arousal, activation and effort in the control of attention. *Psycholoical Reviews,* 82, 116–149.

Rapoport, J.L., Jensvold, M., Elkins, R., Buchsbaum, M.S., Weingartner, H., Ludlow, C. et al. (1981) Behavioral and cognitive effects of caffeine in boys and adult males. *Journal of Nervous and Mental Disease,* 169, 726–732.

Rees, K., Allen, D. and Lader, M. (1999) The influences of age and caffeine on psychomotor and cognitive function. *Psychopharmacology,* 145, 181–188.

Regina, E.G., Smith, G.M., Keiper, C.S. and McKelvey, R.K. (1974) Effects of caffeine on alertness in simulated automobile driving. *Journal of Applied Psychology,* 59, 483–489.

Riedel, W. (1995) Caffeine attenuates scopolamine-induced memory impairment in humans, unpublished manuscript, Rijks Universiteit Limburg, Maastricht.

Roache, J.D. and Griffiths, R.R. (1987) Interaction of diazepam and caffeine: behavioral and subjective dose effects in humans. *Pharmacology, Biochemistry and Behavior,* 26, 801–812.

Rogers, A.S., Spencer, M.B., Stone, B.M. and Nicholson, A.N. (1989) The influence of a 1 h nap on performance overnight. *Ergonomics,* 32, 1193–1205.

Rogers, P.J. and Dernoncourt, C. (1998) Regular caffeine consumption: a balance of adverse and beneficial effects for mood and psychomotor performance. *Pharmacology, Biochemistry and Behavior,* 59, 1039–1145.

Rosenthal, L., Roehrs, T., Zwyghuizen-Doorenbos, A., Plath, D. and Roth, T. (1991) Alerting effects of caffeine after normal and restricted sleep. *Neuropsychopharmacology,* 4, 103–108.

Ruijter, J., De Ruiter, M.B. and Snel, J. (2000a) The effects of caffeine on visual selective attention to colour: an ERP study. *Psychophysiology,* 37, 427–439.

Ruijter, J., De Ruiter, M.B., Snel, J. and Lorist, M.M. (2000b) The influence of caffeine on spatial-selective attention: an ERP study. *Clinical Neurophysiology,* 111, 2223–2233.

Ruijter, J., Lorist, M.M. and Snel, J. (1999) The influence of different doses of caffeine on visual task performance. *Journal of Psychophysiology,* 13, 37–48.

Ruijter, J., Lorist, M.M., Snel, J. and De Ruiter, M.B. (2001) The influence of caffeine on sustained attention: an ERP study. *Pharmacology, Biochemistry and Behavior,* 66, 29–38.

Rush, C.R., Higgins, S.T., Bickel, W.K. and Hughes, J.R. (1994a) Acute behavioral and cardiac effects of lorazepam and caffeine, alone and in combination, in humans. *Behavioural Pharmacology,* 5, 245–254.

Rush, C.R., Higgins, S.T., Hughes, J.R. and Bickel, W.K. (1994b) Acute behavioral and cardiac effects of triazolam and caffeine, alone and in combination, in humans. *Experimental and Clinical Psychopharmacology,* 2, 211–222.

Rush, C.R., Higgins, S.T., Hughes, J.R., Bickel, W.K. and Wiegner, M.S. (1993) Acute behavioral and cardiac effects of alcohol and caffeine, alone and in combination, in humans. *Behavioural Pharmacology,* 4, 562–572.

Rusted, J. (1994) Caffeine and cognitive performance: effects on mood or mental processing. *Pharmacopsychoecologia,* 7, 49–54.

Ryan, L., Hatfield, C. and Hofstetter, M. (2002) Caffeine reduces time-of-day effects on memory performance in older adults. *Psychological Science,* 13, 68–71.

Sanders, A.F. (1983) Towards a model of stress and human performance. *Acta Psychologica,* 53, 61–96.

Schicatano, E.J. and Blumenthal, T.D. (1994) Caffeine delays habituation of the human acoustic startle reflex. *Psychobiology,* 22, 117–122.

Schmitt, J.A.J. (2001a) Caffeine improves memory performance during distraction in middle-aged, but not in young or old subjects, unpublished manuscript, Universiteit Maastricht.

Schmitt, J.A.J. (2001b) Low dose of caffeine does not affect memory functions or focussed attention in middle-aged or elderly subjects, unpublished manuscript, Universiteit Maastricht.

Smit, H.J. and Rogers, P.J. (2000) Effects of low doses of caffeine on cognitive performance, mood and thirst in low and higher caffeine consumers. *Psychopharmacology,* 152, 167–173.

Smith, A., Kendrick, A., Maben, A. and Salmon, J. (1994a) Effects of breakfast and caffeine on cognitive performance, mood and cardiovascular functioning. *Appetite,* 22, 39–55.

Smith, A., Maben, A. and Brockman, P. (1994b) Effects of evening meals and caffeine on cognitive performance, mood and cardiovascular functioning. *Appetite,* 22, 57–65.

Smith, A.P. (1994) Caffeine, performance, mood and status of reduced alertness. *Pharmacopsychoecologia,* 7, 75–86.

Smith, A.P., Kendrick, A.M. and Maben, A.L. (1992a) Effects of breakfast and caffeine on performance and mood in the late morning and after lunch. *Neuropsychobiology,* 26, 198–204.

Smith, A.P., Kendrick, A. and Maben, A. (1992b) Effects of caffeine, lunch and alcohol on human performance, mood and cardiovascular function. *Proceedings of the Nutrition Society,* 51, 325–333.

Smith, A.P., Brockman, P., Flynn, R., Maben, A. and Thomas, M. (1993a) Investigation of the effects of coffee on alertness and performance during the day and night. *Neuropsychobiology,* 27, 217–223.

Smith, A.P., Maben, A.L. and Brockman, P. (1993b) The effects of caffeine and evening meals on sleep and performance, mood and cardiovascular functioning the following day. *Journal of Psychopharmacology,* 7, 203–206.

Smith, B.D., Davidson, R.A. and Green, R.L. (1993c) Effects of caffeine and gender on physiology and performance: further tests of a biobehavioral model. *Physiology and Behavior,* 54, 415–422.

Smith, A.P., Rusted, J.M., Eaton-Williams, P., Savory, M. and Leathwood, P. (1990) Effects of caffeine given before and after lunch on sustained attention. *Neuropsychobiology,* 23, 160–163.

Smith, A.P., Rusted, J.M., Savory, M., Williams, P.E. and Hall, S.R. (1991a) The effects of caffeine, impulsivity and time of day on performance, mood and cardiovascular function. *Journal of Psychopharmacology,* 5, 120–128.

Smith, B.D., Rafferty, J., Lindgren, K., Smith, D.A. and Nespor, A. (1991b) Effects of habitual use and acute ingestion: testing a biobehavioral model. *Physiology and Behavior,* 51, 131–137.

Smith, D.L., Tong, J.E. and Leigh, G. (1977) Combined effects of tobacco and caffeine on the components of choice reaction-time, heart rate, and hand steadiness. *Perceptual and Motor Skills,* 45, 635–639.

Sternberg, S. (1969) On the discovery of processing stages: some extension of Donders' method. *Acta Psychologica,* 30, 276–315.

Swift, C.G. and Tiplady, B. (1988) The effects of age on the response to caffeine. *Psychopharmacology,* 94, 29–31.

Terry, W.S. and Phifer, B. (1986) Caffeine and memory performance on the AVLT. *Journal of Clinical Psychology,* 42, 860–863.

Tharion, W.J., Kobrick, J.L., Lieberman, H.R. and Fine, B.J. (1993) Effects of caffeine and dyphenhydramine on auditory evoked cortical potentials. *Perceptual and Motor Skills,* 76, 707–715.

Uematsu, T., Mizuno, A., Itaya, T., Suzuki, Y., Kanamaru, M. and Nakashima, M. (1987) Psychomotor performance tests using a microcomputer: evaluation of effects of caffeine and chlorpheniramine in healthy human subjects. *Japanese Journal of Psychopharmacology,* 7, 427–432.

Warburton, D.M. (1995) The effects of caffeine on cognition and mood without caffeine abstinence. *Psychopharmacology,* 119, 66–70.

Weiss, B. and Laties, V.G. (1962) Enhancement of human performance by caffeine and the amphetamines. *Pharmacology Reviews,* 14, 1–36.

Wilkinson, R.T. (1968) Sleep deprivation: performance tests for partial and selective sleep deprivation. *Progress in Clinical Psychology,* 8, 28–43.

Wright, K.P., Badia, P., Myers, B.L. and Plenzer, S.C. (1997) Combination of bright light and caffeine as a countermeasure for impaired alertness and performance during extended sleep deprivation. *Journal of Sleep Research,* 6, 26–35.

Yerkes, R.M. and Dodson, J.D. (1980) The relation of strength of stimuli to rapidity of habit-formation. *Journal of Comparative Neurology and Psychology,* 18, 459–482.

Yu, G., Maskray, V., Jackson, S.H.D., Swift, C.G. and Tiplady, D. (1991) A comparison of the central nervous system effects of caffeine and theophylline in elderly subjects. *British Journal of Clinical Pharmacology,* 32, 341–345.

Zahn, T.P. and Rapoport, J.L. (1987) Autonomic nervous system effects of acute doses of caffeine in caffeine users and abstainers. *International Journal of Psychophysiology,* 5, 33–41.

Zwyghuizen-Doorenbos, A., Roehrs, T.A., Lipschutz, L., Timms, V. and Roth, T. (1990) Effects of caffeine on alertness. *Psychopharmacology,* 100, 36–39.

5 Effects of Coffee and Caffeine on Mood and Mood Disorders

Miguel Casas, Josep Antoni Ramos-Quiroga, Gemma Prat, and Adil Qureshi

CONTENTS

Introduction 73
Mood Disorders 74
    Clinical Issues 74
    Neurobiology of Depression 75
    Pharmacological Treatment 75
Caffeine, Mood, Mood Disorders, and Schizophrenia 76
    Caffeine and Mood 76
    Caffeine, Mood Disorders, and Schizophrenia 77
Caffeine and Psychotropic Medication 78
Conclusions 79
References 79

INTRODUCTION

Methylxanthine (MTX) consumption is one of the most common customs in the world, occurring in most every society and culture (James, 1991; Barone and Roberts, 1996). Methylxanthines, which include caffeine, theophylline, and theobromine, can be found in a variety of different forms, including coffee, tea, mate, cola beverages, and pharmaceutical and nutritional products, all of which contain different concentrations of distinct MTXs (James, 1991; Barone and Roberts, 1996).

In human subjects, MTX consumption is closely related to its gratifying (Griffiths and Mumford, 1995) and psychostimulant (French et al., 1994) effects. Consumption of low to moderate doses of MTXs increases arousal, vigilance, and motor activity; decreases the need to sleep; produces sensations of well-being and energy; and facilitates cognitive capacities. These stimulant effects seem to improve mood, and therefore coffee and tea are widely used breakfast beverages. Nevertheless, the ingestion of more elevated doses can result in sensations of discomfort, including anxiety, nervousness, and insomnia (James, 1991).

It remains to be seen whether the invigorating effects typical of daily caffeine consumption help to regulate mood and could be used in the treatment of depression or, conversely, whether they would complicate affective symptomatology. In this chapter we will review the effects of caffeine on mood and mood disorders, introducing first a brief review of the psychopathology of depressive states.

MOOD DISORDERS

Clinical Issues

The word *depression* refers both to a severe clinical condition and to the brief, mild, downward mood swings that are commonly experienced in daily life. A depressive disorder is not the same as a passing blue mood, and in the clinical context *depression* refers not simply to depressed mood, but also to a syndrome comprising mood disorder, psychomotor changes, and a variety of somatic and vegetative disturbances. A depressive disorder is a real illness that involves the body, mood, and thoughts. It can affect the way one eats and sleeps, feels about oneself, and thinks about things. Ethnicity can influence the experience and communication of symptoms of depression, and in some cultures depression may primarily be a somatic rather than an emotional experience (Kirmayer, 2001).

The Diagnostic and Statistical Manuals (DSM) are handbooks developed by the American Psychiatric Association that contain listings and descriptions of psychiatric diagnoses, analogous to the International Classification of Diseases Manuals. The DSM-IV-RT of the American Psychiatric Association (2000) includes different depressive disorders, but the three most common types are major depression, dysthymia, and bipolar disorder. The essential feature of a major depressive episode is a period of at least 2 weeks during which there is either depressed mood or the loss of interest or pleasure in nearly all activities. Such a disabling episode of depression may occur only once, but it more commonly occurs several times over the lifespan and is diagnosed as recurrent major depressive disorder. The reemergence of depression after recovery from an episode appears to contribute significantly to the burden of depressive disorders; as many as 75 to 80% of depressed patients experience a recurrence of depression at some point in their lives (Angst, 1992). Dysthymia is a less severe type of depression, characterized by at least 2 years of depressed mood for more days than not, accompanied by additional depressive symptoms that do no meet criteria for a major depressive episode. Another type of depression occurs in bipolar disorder, more popularly known as manic-depressive illness, which is characterized by cycling mood changes with severe highs (mania) and lows (depression). When in the depressed cycle, an individual with bipolar disorder can have any or all of the symptoms of a depressive disorder.

The U.S. National Comorbidity Survey (NCS) found that 17.3% of the general population had experienced an episode of major depression at some point during their lives (Kessler et al., 1994), and in any given 1-year period 9.5% of the population, or about 18 million American adults, suffer from a depressive illness (Regier et al., 1984). The global burden of mental illness is expected to increase in magnitude over the coming decades. The World Health Organization's (WHO) Global Burden of Disease Survey estimates that by the year 2020 major depression will be second only to ischemic heart disease in the amount of disability experienced by sufferers (World Health Organization, 1996).

Major depression and bipolar disorder are responsible for much of the suicide in the U.S. Two factors that characterize affective disorders are their episodic nature and suicidality. In the last 2 years suicide has been the 11th leading cause of death, surpassing HIV infection (Hoyert et al., 2001). In 1999, there were 29,199 self-inflicted fatalities in the U.S., a rate of one suicide every 18 minutes. Suicide is the third leading cause of death among young people in the U.S. (Hoyert et al., 2001). Depressive disorders are associated with about 80% of suicidal events. Suicide attempts have been reported in about 25% of bipolar patients, with suicide success rates of about 15%. A strong predictor of suicide is at least one previous suicide attempt in the history of the patient, which increases the mortality risk to 100 to 140 times that of the general population (Ahrens et al., 1995). Depression increases mortality significantly through suicide, but also by accidents and exacerbation of medical illness.

Neurobiology of Depression

Abnormalities in monoaminergic systems affecting norepinephrine, serotonin, and dopamine have been found in depressive states (Stahl, 2000). Research indicates reduced cerebrospinal fluid and urinary concentrations of metabolites, decreased plasma precursor concentrations, and clinical effectiveness of drugs that increase monoamine neurotransmission in depressed patients (Charney, 1998). Recently, it has been shown that a deficiency in the mesocorticolimbic dopamine transmission system is specifically related to the symptoms of depressive disorders, consisting of decreased experience of pleasure or interest in previously enjoyed activities (Naranjo et al., 2001; Cardenas et al., 2002).

Simplistic mechanistic interpretations based on monoamine deficiency are precluded by findings that show a significant delay of the onset of the action of antidepressant agents, despite initially increasing monoamine levels (Blier and de Montigny, 1998). In order to explain such a delay in antidepressant action, some authors have postulated the so-called neurotransmitter receptor hypothesis, which holds that an up-regulation of postsynaptic monoamine receptors is involved in depression, as a consequence of monoaminergic depletion (Blier and de Montigny, 1998). According to this hypothesis, therapeutic effects of antidepressants are delayed since some time is needed to reverse the up-regulation of monoamine receptors. Direct evidence for this theory is generally lacking, although modifications of receptor density have been found in depressive patients, mainly in postmortem studies (Mendelson, 2000).

Recently, this theory has been changed to a signal transduction pathway dysfunction of the monoamine receptors. A deficiency in the second and third messenger systems could lead to a deficient cellular response and, consequently, to a dysfunction of monoamine neurotransmission. The most relevant finding supporting this theory is that long-term antidepressant administration increases basal and stimulated adenylyl cyclase activity, increases c-AMP-dependent phosphorylation, and increases CREB levels (Duman et al., 1997; Chen et al., 1999; Young et al., 2002).

Other neurotransmitter systems have been implicated in the neurobiology of depression, such as GABA, acetylcholine, corticotropin-releasing factor (CRF), somatostatin, and neuropeptide Y (NPY). GABA is a major inhibitory neurotransmitter in the brain. Decreased GABAergic neurotransmission may constitute a component in the cascade of biochemical events associated with depressive disorders (Petty, 1995). Corticotropin-releasing factor is a peptide found both in the hypothalamus and in extrahypothalamic systems, such as forebrain limbic areas and brainstem nuclei. Nemeroff et al. (1991) found that depressed individuals and suicide victims exhibited elevated cerebrospinal fluid CRF levels compared with controls. Also, electroconvulsive therapy and chronic fluoxetine treatment decreased cerebrospinal fluid CRF levels.

Neuropeptide Y is a peptide widely distributed in the central nervous system and has been implicated in depression. Decreased cerebrospinal fluid NPY levels have been observed in depressed individuals compared with controls (Gjerris et al., 1992). Somatostatin, another peptide that is widely distributed in the brain, interacts with several neurotransmitter systems and has been implicated in depression through decreased cerebrospinal fluid levels (Pazzaglia et al., 1995).

Pharmacological Treatment

A wide range of antidepressants are now available, including tricyclic antidepressants (TCAs), monoamine oxidase inhibitors (MAOIs), reversible monoamine oxidase inhibitors (RIMAs), selective serotonin reuptake inhibitors (SSRIs), selective norepinephrine reuptake inhibitors (SNRI), serotonin and norepinephrine reuptake inhibitors (SNRIs), norepinephrine and dopamine reuptake inhibitor (bupropion), and atypical antidepressants such as mirtazapine and trazodone. New antidepressants are more selective than TCAs, are generally as effective as TCAs in the treatment of major depression, and have an improved side-effect profile (Stahl, 2000). Caffeine is not accepted as standard pharmacological treatment for mood disorders; nevertheless, it is probable that some individuals could use it as an antidepressant in the first stages of the illness. Unfortunately, no controlled clinical trials have been done testing the usefulness of MTX in depressive symptomatology.

CAFFEINE, MOOD, MOOD DISORDERS, AND SCHIZOPHRENIA

CAFFEINE AND MOOD

To understand the euthymic effect of caffeine and its possible beneficial effects in affective disorders, it will be helpful to review its mechanism of action (Chapter 1 of this book) and to take into account that many of these mechanisms overlap with the mechanism of action of current antidepressant medication.

Many biochemical actions of caffeine can reverse the abnormalities in the monoamine systems observed in depression. Caffeine's blocking of A_1 adenosine receptors may increase the levels of catecholamines and serotonin (Fredholm, 1995). In this respect, it is relevant that caffeine can increase serotonin release in limbic areas and dopamine release in the prefrontal cortex, an effect also obtained with antidepressants (Fredholm, 1995; Acquas et al., 2002).

Furthermore, caffeine is also able to inhibit phosphodiesterase, the enzyme responsible for the hydrolysis of the intracellular second messenger cyclic adenosine monophosphate (Dunlop et al., 1981), indicating that AMPc levels should be increased. Chronic antidepressant treatment has the same effects on this second messenger (Duman et al., 1997; Chen et al., 1999; Young et al., 2002).

Finally, a role for adenosine transmission has been found in depression. Adenosine receptor function in platelets is blunted in patients with major depression, suggesting that A_{2A} receptors are down-regulated due to an excess of adenosine neurotransmission (Berk et al., 2001). Adenosine antagonists such as caffeine are potentially effective in the treatment of depression due to the increased platelet aggregation found in this disorder (Musselman et al., 1996). Indeed, adenosine agonists have been effective in the treatment of depression in animal models (Williams, 1989; Sarges et al., 1990).

The gratifying and psychostimulant effects of caffeine are supposed to be responsible for its euphorigenic effects, and diverse clinical trials support these effects on healthy volunteers. Standardized instruments exist to evaluate effects on mood, such as the Visual Analog Mood Scale (VAMS), the Profile of Mood States (POMSs), The Nestlé Visual Analog Mood Scale (NVAMS), and the Stanford Sleepiness Scale (SSS), that are usually referenced in caffeine trials (Bättig and Welzl, 1993). The first studies were carried out by Goldstein's group, and their results indicated a dose-dependent effect of caffeine on mood, suggesting that the subjective effects of caffeine are a function of caffeine-consuming habits and plasma caffeine levels (Goldstein et al., 1965, 1969). Later, Lieberman et al. (1987) showed that low doses (64 mg) can have beneficial effects on mood. Other authors indicated that doses of 100 to 300 mg could improve mood or had positive effects on mood (Leatherwood and Pollet, 1983; Griffiths et al., 1989; Stern et al., 1989). In a double-blind study that examined the effects of caffeine on mood at breakfast time (patients were given either coffee with 4 mg/kg or decaffeinated coffee), the authors found that subjects given caffeine reported that they felt more alert, more contented, more attentive, more friendly, less bored, and more sociable than those subjects given decaffeinated coffee (Smith et al., 1992). The authors of that study observed that the effects of caffeine were present over a range of times and were not modified by prior consumption of breakfast or subsequent consumption of lunch. Penetar et al. (1993), in a double-blind study with 50 healthy males, showed that caffeine at different oral doses (150, 300, or 600 mg/70 kg) produced significant alerting and long-lasting beneficial mood effects in subjects deprived of sleep for 48 h compared with placebo.

Recently, Smith et al. (1999), in a randomized controlled trial with 144 subjects, found that the volunteers of a group that consumed breakfast cereal and caffeinated coffee had more positive mood and less fatigue than those in a group that ate no breakfast and consumed decaffeinated coffee.

Quinlan et al. (2000), in a randomized full crossover study, gave 17 subjects varying doses of caffeine after overnight or 3-h abstention. The subjects were given tea or coffee, water, or, in the case of the controls, no beverage. The study found that, relative to consumption of hot water, consumption of coffee or tea was associated with three indicators of improved mood: increases in energetic arousal and hedonic tone and a decrease in sedation. This article completes the previous results of the same group (Quinlan et al., 1997; Hindmarch et al., 1998) and agrees with previous studies (Lieberman et al., 1987; Warburton, 1995). Lieberman et al. (2002) in a randomized clinical trial with 68 U.S. Navy volunteers exposed to severe environmental stress and sleep deprivation suggested that moderate doses of caffeine can improve mood state even in the most adverse situations.

Nevertheless, the effect of caffeine withdrawal on mood responses to caffeine intake is also disputed, and some authors have suggested that reversal of caffeine withdrawal is a major component of the effects of caffeine on mood (James, 1998; Lane et al., 1998; Yeomans et al., 2002). Other authors disagree with this view and have found positive effects on mood in subjects that were not in caffeine withdrawal (Warburton et al., 2001). Brice and Smith (2002) compared a realistic drinking regimen (multiple small doses, 4×65 mg over a 5-h period) with a single large dose (200 mg) to evaluate the effects of caffeine doses on mood. Their results suggested that there are not differences between these two caffeine dosages. At low doses the effects of caffeine on mood do not appear to be strongly dose-dependent (Lieberman et al., 1987), but at high doses (400 to 500 mg) the mood benefits of caffeine can be reversed, leading to increases in tension and anxiety (Loke, 1988). Also, a few studies have shown that caffeine has no statistically significant effects on mood or that it has negative effects on mood (Svensson et al., 1980; Estler, 1982; Kuznicki and Turner, 1985; Loke et al., 1985; Loke, 1988; Herz, 1999).

Caffeine, like taurine in energy drinks, is progressively being introduced into a wide variety of new beverages. In a recent study, moderate doses of caffeine and taurine improved mood in subjects who were not in caffeine withdrawal (Warburton et al., 2001).

CAFFEINE, MOOD DISORDERS, AND SCHIZOPHRENIA

It is well established that psychiatric populations consume a large amount of caffeine and nicotine, as much as two to three times more than the general population (Worthington et al., 1996). This can be interpreted on the basis of the self-medication theory, which postulates that the consumption of some substances can be explained as self-treatment in accordance with previous psychopathological symptoms (Khantzian, 1985).

In this respect, caffeine is a weak reinforcer that could, to some extent, reduce the anhedonia present in depressive states (Naranjo et al., 2001; Cardenas et al., 2002). Nevertheless, although several authors have suggested that caffeine could act as a weak, temporary antidepressant drug (Neil et al., 1978; Leibenluft et al., 1993) its specific effects on depressive patients have yet to be clarified.

Caffeine consumption did not worsen the response to antidepressants in outpatients with major depressive disorders (Worthington et al., 1996). In a prospective study, the extent of caffeine consumption at baseline was a significantly positive predictor of improvement in somatic symptoms and hostility, as measured by the change in symptoms questionnaire scale scores after 8 weeks of treatment with 20 mg/d of fluoxetine (Worthington et al., 1996). Leibenluft et al. (1993) examined the relationship between depressive symptoms and the self-reported use of caffeine in 26 normal volunteers and four groups of psychiatric outpatients (35 subjects with major depression, 117 patients with seasonal affective disorder, 16 patients with alcohol dependence, and 24 patients with comorbid primary depression and secondary alcohol dependence). They found that patients of all diagnostic groups were more likely than normal volunteers to report using caffeine in response to depressive symptoms. Also, no difference in caffeine use between patient groups was observed. Other studies that have examined the relationship between caffeine consumption and depressive symptomatology have observed more frequent caffeine abuse among patients with comorbid obsessive-compulsive disorder (OCD) and bipolar disorder than among nonbipolar OCD patients (Perugi et al., 1997) and have found early-age use of caffeine among patients with substance-related disorders and dysthymia (Eames et al., 1998).

A few clinical case reports associate caffeine intake with an exacerbation of manic symptoms (Machado-Vieira et al., 2001) as has been found with antidepressants. Furthermore, two older studies indicate that a higher level of depression is directly related to the amount of caffeine intake among both college students (Gilliland and Andress, 1981) and psychiatric inpatients (Greden et al., 1978), groups with greater consumption than that of subjects that reported the highest depression scores. Recently, Lande and Labbate (1998) in a study that examined plasma caffeine concentrations in new psychiatric outpatients found that caffeine use was significantly correlated with Beck Depression Inventory (BDI) scores but not significantly correlated with State-Trait Anxiety Inventory (STAI) scores and that there were no significant correlations between plasma caffeine concentration and reported caffeine use and BDI or STAI scores. They suggest that caffeine may be responsible for anxious or depressive symptoms, although this contribution may be modest and only result from excessive use. Tanskanen et al. (2000) indicated that heavy coffee drinking (seven or more cups/day) may be associated with higher rates of suicide.

Kawachi et al. (1996) carried out a prospective 10-year follow-up study to examine the relationship of caffeine intake to risk of death from suicide. They found an inverse correlation between coffee drinking and suicide in a cohort of 86,626 female registered nurses, and the authors attributed their results to the mood-elevating effect of coffee. Also, a strong inverse association has been reported between daily coffee intake and risk of suicide in a prospective 8-year follow-up study of 128,934 subjects (Klatsky et al., 1993).

Caffeine consumption is particularly elevated among schizophrenic patients, of whom about 80% consume nicotine and 38% have been observed to consume more than 555 mg of coffee a day (the equivalent of about five to six cups per day) (Mayo et al., 1993; Hughes et al., 1998). The first studies involving the use of caffeine in schizophrenic patients focused on the possible negative effects on this population (De Freitas and Schwartz, 1979). However, Koczapski et al. (1989) found that caffeinated coffee did not impair the behavior of 33 schizophrenic inpatients compared with decaffeinated coffee. Mayo et al. (1993), in a double-blind crossover study of 26 long-stay schizophrenic patients, found that consumption of caffeine did not increase anxiety or depression levels. In that study, the investigators measured serum caffeine levels to confirm compliance when the wards changed to decaffeinated products, and no significant changes in level of anxiety and depression were observed. The elevated caffeine consumption among schizophrenic patients has also been correlated with an improvement of negative symptoms (Lucas et al., 1990). In this experimental study carried out with 13 schizophrenic subjects who had been caffeine-free for 6 weeks, caffeine was administered intravenously (10 mg per kg of body weight). Results show that caffeine improved negative symptoms, such as mood, but did not increase anxiety scores among these patients.

CAFFEINE AND PSYCHOTROPIC MEDICATION

A possible potential limitation of the therapeutic effects of caffeine in depressive patients is the metabolic interactions between caffeine and antidepressant drugs. The polycyclic aromatic hydrocarbon-inducible cytochrome P450, CYP1A2 participates in the metabolism of caffeine as well as in certain selective serotonin reuptake inhibitor drugs (Yoshimura et al., 2002). This metabolic enzyme is inhibited by the administration of fluvoxamine and the metabolism of caffeine is decreased, followed by the appearance of toxic effects such as seizures, delirium, and increases in heart rate (Carrillo and Benitez, 2000; Yoshimura et al., 2002). The clearance of caffeine decreased by 80% and its half-life increased by 500% during concomitant intake of fluvoxamine (Jeppesen et al., 1996). Spigset (1998) suggested that many of the adverse drug reactions traditionally attributed to fluvoxamine are in fact caffeine-related symptoms of toxicity as a consequence of inhibition by fluvoxamine of the metabolism of caffeine from dietary sources. Fluoxetine, paroxetine, and sertraline also inhibit CYP-1A2, but to a lesser degree than fluvoxamine (Carrillo and Benitez, 2000).

Antipsychotics are usually used to treat bipolar disorders and psychotic depressive episodes. Diverse interactions between antipsychotics and constituents of caffeinated beverages have been observed, indicating that caffeine interferes with the therapeutic effect of such drugs. One explanation is that consumption of phenothiazines (a chemical antipsychotic group) with coffee or tea causes the formation of precipitates that could inactivate oral doses of the drug, but the clinical significance of this finding remains unclear (Hirsch, 1979). Another explanation is the competitive pharmacokinetic interaction between antipsychotics and caffeine at the CYP1A2 enzyme level. Clozapine was the first marketed antipsychotic drug labeled as atypical. Caffeine seems to show competitive inhibition of clozapine metabolism and could alter plasma clozapine concentrations and then precipitate adverse effects (Carrillo and Benítez, 2000). Olanzapine, another atypical antipsychotic, may significantly impair the clearance of caffeine, mainly through CYP1A2 inhibition (Carrillo and Benítez, 2000). Nevertheless, a beneficial effect of MTX in psychotic patients taking classic neuroleptics has also been described; decreasing extrapyramidal side effects in these instances help justify the self-medication hypothesis (Casas et al., 1988, 1989a,b).

Lithium is the first-line treatment for the management of acute mania and the prophylaxis of bipolar disorder. Lithium also has the narrowest gap between therapeutic and toxic concentrations of any drug routinely prescribed in psychiatric medicine. Caffeine can modify renal lithium clearance and thus affect serum lithium concentrations (Carrillo and Benítez, 2000). It has been suggested that caffeine intake increases lithium excretion rates, and caffeine withdrawal has been documented to increase serum lithium concentrations by an average of 24% (Mester et al., 1995).

CONCLUSIONS

Although caffeine does not produce a clearly defined effect on mood, positive findings of significant mood effects are more numerous than negative findings. These contradictory results can be explained by different trial methodologies, different criteria for including or excluding subjects, acute or chronic caffeine intake, different caffeine dosages, and caffeine withdrawal effects. Further studies are necessary to clarify the potential antidepressant properties of caffeine; nevertheless, it is possible that caffeine and other methylxanthines could have clinical efficacy in some disorders that involve brain monoamine dysfunction. At present we can accept that habitual caffeine consumption can clearly improve mood and not worsen depressive disorder, but very high caffeine doses can impair mood disorders. Finally, it is necessary to point out that potential limitations of the therapeutic effects of caffeine in depressive patients are metabolic interactions between caffeine and antidepressant drugs.

REFERENCES

Acquas, E., Tanda, G. and Di Chiara, G. (2002) Differential effects of caffeine on dopamine and acetylcholine transmission in brain areas of drug-naïve and caffeine-pretreated rats. *Neuropsychopharmacology*, 27, 182–193.
Ahrens, B., Berghöfer, A., Wolf, T. and Müller-Oerlinghausen, B. (1995) Suicide attempts, age and duration of illness in recurrent affective disorders. *Journal of Affective Disorders*, 36, 43–49.
American Psychiatric Association (2000) Diagnostic and Statistical Manual of Mental Disorders: DSM-IV-TR. American Psychiatric Association, Washington, DC.
Angst, J. (1992) Epidemiology of depression. *Psychopharmacology*, 106, 71–74.
Barone, J.J. and Roberts, H.R. (1996) Caffeine consumption. *Food and Chemical Toxicology*, 34, 119–129.
Bättig, K. and Welzl, H. (1993) Psychopharmacological profile of caffeine, in *Caffeine, Coffee, and Health*, Garattini, S., Ed., Raven Press, New York, pp. 213–253.

Berk, M., Plein, H., Ferreira, D. and Jersky, B. (2001) Blunted adenosine A2a receptor function in platelets in patients with major depression. *European Neuropsychopharmacology*, 11, 183–186.

Blier, P. and de Montigny, C. (1998) Possible serotonergic mechanisms underlying the antidepressant and anti-obsessive-compulsive disorder responses. *Biological Psychiatry*, 44, 313–323.

Brice, C.F. and Smith, A.P. (2002) Effects of caffeine on mood and performance: a study of realistic consumption. *Psychopharmacology*, 164, 188–192.

Cardenas, L., Tremblay, L.K., Naranjo, C.A., Herrmann, N., Zack, M. and Busto, U.E. (2002) Brain reward system activity in major depression and comorbid nicotine dependence. *Journal of Pharmacology and Experimental Therapeutics*, 302, 1265–1271.

Carrillo, J.A. and Benítez, J. (2000) Clinically significant pharmacokinetic interactions between dietary caffeine and medications. *Clinical Pharmacokinetics*, 39, 127–153.

Casas, M., Ferré, S., Guix, T. and Jane, F. (1988). Theophylline reverses haloperidol-induced catalepsy in the rat: possible relevance to the pharmacological treatment of psychosis. *Biological Psychiatry*, 24, 642–648.

Casas, M., Ferré, S., Cobos, A., Grau, J.M. and Jane, F. (1989a) Relationship between rotational behavior induced by apomorphine and caffeine in rats with unilateral lesion of the nigrostriatal pathway. *Neuropharmacology*, 28, 407–409.

Casas, M., Torrens, M., Duro, P., Pinet, C., Alvarez, E. and Udina, C. (1989b) Utilisation des méthylxanthines, caféine et théophylline dans le traitement des troubles extrapyramidaux secondaires aux traitements par les neuroleptiques. *L'Encéphale*, 15, 177–180.

Charney, D.S. (1998) Monoamine dysfunction and the pathophysiology and treatment of depression. *Journal of Clinical Psychiatry*, 59, 11–14.

Chen, G., Hasanat, K.A., Bebchuk, J.M., Moore, G.J., Glitz, D. and Manji, H.K. (1999) Regulation of signal transduction pathways and gene expression by mood stabilizers and antidepressants. *Psychosomatic Medicine*, 61, 599–617.

De Freitas, B. and Schwartz, G. (1979) Effects of caffeine in chronic psychiatric patients. *American Journal of Psychiatry*, 136, 1337–1338.

Duman, R.S., Heninger, G.R. and Nestler, E.J. (1997) A molecular and cellular theory of depression. *Archives of General Psychiatry*, 54, 597–606.

Dunlop, M., Larkins, R.G. and Court, J.M. (1981) Methylxanthine effects on cyclic adenosine 3´ 5´–monophosphate phosphodiesterase activity in preparations of neonatal rat cerebellum: modification by trifluoperazine. *Biochemical and Biophysical Research Communications*, 98, 850–857.

Eames, S.L., Westermeyer, J. and Crosby, R.D. (1998) Substance use and abuse among patients with comorbid dysthymia and substance disorder. *American Journal of Drug and Alcohol Abuse*, 24, 541–550.

Estler, C.J. (1982) Caffeine, in *Psychotropic Agents, Part III, Alcohol and Psychotomimetics, Psychotropic Effects of Central Acting Drugs*, Hoffmeister, F. and Stille, G., Eds., Springer-Verlag, Berlin, pp. 369–389.

Fredholm, B.B. (1995) Adenosine, adenosine receptors and the actions of caffeine. *Pharmacology and Toxicology*, 76, 93–101.

French, J.A., Wainwright, C.J. and Booth, D.A. (1994) Caffeine and mood: individual differences in low-dose caffeine sensitivity. *Appetite*, 22, 277–279.

Gilliland, K. and Andress, D. (1981) Ad lib caffeine consumption, symptoms of caffeinism, and academic performance. *American Journal of Psychiatry*, 138, 512–514.

Gjerris, A., Widerlov, E., Werdelin, L. and Ekman, R. (1992) Cerebrospinal fluid concentrations of neuropeptide Y in depressed patients and in controls. *Journal of Psychiatry and Neuroscience*, 17, 23–27.

Goldstein, A., Kaizer, S. and Warren, R. (1965) Psychotropic effects of caffeine in man. II. Alertness, psychomotor coordination and mood. *Journal of Pharmacology and Experimental Therapeutics*, 150, 146–151.

Goldstein, A., Kaizer, S. and Whitby, O. (1969) Psychotropic effects of caffeine in man. IV. Quantitative and qualitative differences associated with habituation to coffee. *Clinical Pharmacology and Therapeutics*, 10, 489–497.

Greden, J.F., Fontaine, P., Lubetsky, M. and Chamberlin, K. (1978) Anxiety and depression associated with caffeinism among psychiatric inpatients. *American Journal of Psychiatry*, 135, 963–966.

Griffiths, R.R. and Mumford, G.K. (1995) Caffeine: a drug of abuse?, in *Psychopharmacology: The Fourth Generation of Progress*, Bloom, F.E. and Kupfer, D.J., Eds., Raven Press, New York, pp. 1699–1713.

Griffiths, R.R., Bigelow, G.E. and Liebson, I.A. (1989) Reinforcing effects of caffeine in coffee and capsules. *Journal of Experimental and Analytical Behavior*, 52, 127–140.

Herz, R.S. (1999) Caffeine effects on mood and memory. *Behavioural Research and Therapeutics*, 37, 869–879.

Hindmarch, I., Quinlan, P.T., Moore, K.L. and Parkin, C. (1998) The effects of black tea and other beverages on aspects of cognition and psychomotor performance. *Psychopharmacology*, 139, 230–238.

Hirsch, S.R. (1979) Precipitation of antipsychotic drugs in interaction with coffee or tea. *Lancet*, 2, 1130–1131.

Hoyert, D.L., Arias, E., Smith, B.L., Murphy, S.L. and Kochanek, K.D. (2001) Deaths: final data for 1999. *National Vital Statistics Report 49*, National Center for Health Statistics, Hyattsville, MD.

Hughes, J.R., McHugh, P. and Holtzman, S. (1998) Caffeine and schizophrenia. *Psychiatric Services*, 49, 1415–1417.

James, J.E. (1991) *Caffeine and Health*. Academic Press, London.

James, J.E. (1998) Acute and chronic effects of caffeine on performance, mood, headache and sleep. *Neuropsychobiology*, 38, 32–41.

Jeppesen, U., Loft, S., Poulsen, H.E. and Brsen, K. (1996) A fluvoxamine-caffeine interaction study. *Pharmacogenetics*, 6, 213–222.

Kawachi, I., Willett, W., Colditz, G., Stampfer, M. and Speizer, F. (1996) A prospective study of coffee drinking and suicide in women. *Archives of Internal Medicine*, 156, 521–525.

Kessler, R.C., McGonagle, K.A., Zhao, S., Nelson, C.B., Hughes, M., Eshleman, S. et al. (1994) Lifetime and 12-month prevalence of DSM-III-R psychiatric disorders in the United States: results from the National Comorbidity Survey. *Archives of General Psychiatry*, 51, 8–19.

Khantzian, E.J. (1985) The self-medication hypothesis of addictive disorders: focus on heroin and cocaine dependence. *American Journal of Psychiatry*, 142, 1259–1264.

Kirmayer, L.J. (2001) Cultural variations in the clinical presentation of depression and anxiety: implications for diagnosis and treatment. *Journal of Clinical Psychiatry*, 62, Suppl. 13, 22–28.

Klatsky, A.L., Armstrong, M.A. and Friedman, G.D. (1993) Coffee, tea, and mortality. *Annals of Epidemiology*, 3, 375–381.

Koczapski, A., Paredes, J., Kogan, C., Ledwidge, B. and Higenbottam, J. (1989) Effects of caffeine on behavior of schizophrenic inpatients. *Schizophrenia Bulletin*, 15, 339–344.

Kuznicki, J.T. and Turner, L.S. (1985) Effects of caffeine on users and non-users. *Physiology and Behavior*, 37, 397–408.

Lande, R.G. and Labbate, L.A. (1998) Caffeine use and plasma concentrations in psychiatric outpatients. *Depression and Anxiety*, 7, 130–133.

Lane, J.D. and Phillips-Bute, B.G. (1998) Caffeine deprivation affects vigilance performance and mood. *Physiology and Behavior*, 65, 171–175.

Leatherwood, P. and Pollet, P. (1983) Diet-induced mood changes in normal populations. *Journal of Psychiatric Research*, 17, 147–154.

Leibenluft, E., Fiero, P.L., Bartko, J.J., Moul, D.E. and Rosenthal, N.E. (1993) Depressive symptoms and the self-reported use of alcohol, caffeine, and carbohydrates in normal volunteers and four groups of psychiatric outpatients. *American Journal of Psychiatry*, 150, 294–301.

Lieberman, H.R., Tharion, W.J., Shukitt-Hale, B., Speckman, K.L. and Tulley, R. (2002) Effects of caffeine, sleep loss, and stress on cognitive performance and mood during U.S.Navy SEAL training. *Psychopharmacology*, 164, 250–261.

Lieberman, H.R., Wurtman, R.J., Emde, G.C. and Civiella, I.L.G. (1987) The effects of caffeine and aspirin on mood and performance. *Journal of Clinical Psychopharmacology*, 7, 315–320.

Loke, W.H. (1988) Effects of caffeine on mood and memory. *Physiology and Behavior*, 44, 367–372.

Loke, W.H., Hinrichs, J.V. and Ghoneim, M.M. (1985) Caffeine and diazepam: separate and combined effects on mood, memory, and psychomotor performance. *Psychopharmacology*, 87, 344–350.

Lucas, P.B., Pickar, D., Kelsoe, J., Rapaport, M., Pato, C. and Hommer, D. (1990) Effects of the acute administration of caffeine in patients with schizophrenia. *Biological Psychiatry*, 28, 35–40.

Machado-Vieira, R., Viale, C.I. and Kapczinski, F. (2001) Mania associated with an energy drink: the possible role of caffeine, taurine, and inositol. *Canadian Journal of Psychiatry*, 46, 454–455.

Mayo, K.M., Falkowski, W. and Jones, C.A. (1993) Caffeine: use and effects in long-stay psychiatric patients. *British Journal of Psychiatry*, 162, 543–545.

Mendelson, S.D. (2000) The current status of the platelet 5-HT(2A) receptor in depression. *Journal of Affective Disorders*, 57, 13–24.

Mester, R., Toren, P., Mizrachi, I., Wolmer, L., Karni, N. and Weizman, A. (1995) Caffeine withdrawal increases lithium blood levels. *Biological Psychiatry*, 37, 348–350.

Musselman, D.L., Tomer, A., Manatunga, A.K., Knight, B.T., Porter, M.R., Kasey, S. et al. (1996) Exaggerated platelet reactivity in major depression. *American Journal of Psychiatry*, 153, 1313–1317.

Naranjo, C.A., Tremblay, L.K. and Busto, U.E. (2001) The role of the brain reward system in depression. *Progress in Neuropsychopharmacology and Biological Psychiatry*, 25, 781–823.

Neil, J.F., Himmelhoch, J.M., Mallinger, A.G., Mallinger, J. and Hanin, I. (1978) Caffeinism complicating hypersomnic depressive episodes. *Comprehensive Psychiatry*, 19, 377–385.

Nemeroff, C.B., Bissette, G., Akil, H. and Fink, M. (1991) Neuropeptide concentrations in the cerebrospinal fluid of depressed patients treated with electroconvulsive therapy: corticotrophin-releasing factor, beta-endorphin and somatostatin. *British Journal of Psychiatry*, 158, 59–63.

Pazzaglia, P.J., George, M.S., Post, R.M., Rubinow, D.R. and Davis, C.L. (1995) Nimodipine increases CSF somatostatin in affectively ill patients. *Neuropsychopharmacology*, 13, 75–83.

Penetar, D., McCann, U., Thorne, D., Kamimori, G., Galinski, C., Sling, H. et al. (1993) Caffeine reversal of sleep deprivation effects on alertness and mood. *Psychopharmacology*, 113, 359–365.

Perugi, G., Akiskal, H.S., Pfanner, C., Presta, S., Gemignani, A., Milanfranchi, A. et al. (1997) The clinical impact of bipolar and unipolar affective comorbidity on obsessive-compulsive disorder. *Journal of Affective Disorders*, 46, 15–23.

Petty, F. (1995) GABA and mood disorders: a brief review and hypothesis. *Journal of Affective Disorders*, 34, 275–281.

Quinlan, P.T., Lane, J. and Aspinall, L. (1997) Effects of hot tea, coffee and water ingestion on physiological responses and mood: the role of caffeine, water and beverage type. *Psychopharmacology*, 134, 164–173.

Quinlan, P.T., Lane, J., Moore, K.L., Aspen, J., Rycroft, J.A. and O'Brien, D.C. (2000) The acute physiological and mood effects of tea and coffee: the role of caffeine level. *Pharmacology Biochemistry and Behavior*, 66, 19–28.

Regier, D.A., Myers, J.K., Kramer, M., Robins, L.N., Blazer, D.G., Hough, R.L. et al. (1984) The NIMH Epidemiologic Catchment Area program: historical context, major objectives, and study population characteristics. *Archives of General Psychiatry*, 41, 934–941.

Sarges, R., Howard, H.R., Browne, R.G., Lebel, L.A., Seymour, P.A. and Koe, B.K. (1990) 4-Amino[1,2,4]triazolo[4,3-a]quinoxalines: a novel class of potent adenosine receptor antagonists and potential rapid-onset antidepressants. *Journal of Medicinal Chemistry*, 33, 2240–2254.

Smith, A.P., Clark, R. and Gallagher, J. (1999) Breakfast cereal and caffeinated coffee: effects on working memory, attention, mood, and cardiovascular function. *Physiology and Behavior*, 67, 9–17.

Smith, A.P., Kendrick, A.M. and Maben, A.L. (1992) Effects of breakfast and caffeine on performance and mood in the late morning and after lunch. *Neuropsychobiology*, 26, 198–204.

Spigset, O. (1998) Are adverse drug reactions attributed to fluvoxamine caused by concomitant intake of caffeine? *European Journal of Clinical Pharmacology*, 54, 665–666.

Stahl, S.M. (2000) *Essential Psychopharmacology: Neuroscientific Basis and Practical Application*, 2nd ed., Cambridge University Press, Cambridge.

Stern, K.N., Chait, L.D. and Johanson, C.E. (1989) Reinforcing and subjective effects of caffeine in normal volunteers. *Psychopharmacology*, 98, 81–88.

Svensson, E., Persson, L.O. and Sjoberg, L. (1980) Mood effects of diazepam and caffeine. *Psychopharmacology*, 67, 73–80.

Tanskanen, A., Tuomilehto, J., Viinamäki, H., Vartiainen, E., Lehtonen, J. and Puska, P. (2000) Joint heavy use of alcohol, cigarettes and coffee and the risk of suicide. *Addiction*, 95, 1699–1704.

Warburton, D.M. (1995) Effects of caffeine on cognition and mood without caffeine abstinence. *Psychopharmacology*, 119, 66–70.

Warburton, D.M., Bersellini, E. and Sweeney, E. (2001) An evaluation of a caffeinated taurine drink on mood, memory and information processing in healthy volunteers without caffeine abstinence. *Psychopharmacology*, 158, 322–328.

Williams, M. (1989). Adenosine antagonists. *Medical Research Reviews*, 9, 219–243.

World Health Organization (1996) The global burden of disease, in *The Global Burden of Disease: A Comprehensive Assessment of Mortality and Disability from Diseases, Injuries, and Risk Factors in 1990 and Projected,* Murray, C.J.L. and Lopez, A.D., Eds., Harvard School of Public Health, Cambridge, MA.

Worthington, J., Fava, M., Agustin, C., Alpert, J., Nierenberg, A.A., Pava, J.A. et al. (1996) Consumption of alcohol, nicotine, and caffeine among depressed outpatients: relationship with response to treatment. *Psychosomatics*, 37, 518–522.

Yeomans, M.R., Ripley, T., Davies, L.H., Rusted, J.M. and Rogers, P.J. (2002) Effects of caffeine on performance and mood depend on the level of caffeine abstinence. *Psychopharmacology*, 164, 241–249.

Yoshimura, R., Ueda, N., Nakamura, J., Eto, S. and Matsushita, M. (2002) Interaction between fluvoxamine and cotinine or caffeine. *Neuropsychobiology*, 45, 32–35.

Young, L.T., Bakish, D. and Beaulieu, S. (2002) The neurobiology of treatment response to antidepressants and mood stabilizing medications. *Journal of Psychiatry and Neuroscience*, 27, 260–265.

6 Age-Related Changes in the Effects of Coffee on Memory and Cognitive Performance

Martin P. J. van Boxtel and Jeroen A. J. Schmitt

CONTENTS

Introduction 85
Cognitive Aging 86
Caffeine Consumption Habits and Age 87
Caffeine Intake and Age-Related Differences in Cognitive Function 88
Habitual Caffeine Intake Studies 88
Caffeine, Age, and Arousal 91
Acute Caffeine Effects in Older Persons 93
References 94

INTRODUCTION

The use of caffeine-containing drinks and food is a widespread habit in our modern culture. In Western societies people tend to start regular consumption of coffee or black tea before adulthood, and the total amount of intake remains relatively stable until old age (Hameleers et al., 2000). It is well established that caffeine, the main psychoactive ingredient present in these beverages, has a mild stimulating effect on the central nervous system (Smith, 2002a). Acute effects of caffeine have typically been documented in the domains of vigilance, (selective) attention, and information processing speed (Riedel and Jolles, 1996). The acute effects of caffeine are related to the adenosine-A_1 and A_{2A} antagonism of caffeine in the brain, which in turn stimulates the release and turnover of several central neurotransmitter substances, including acetylcholine and noradrenaline (Nehlig et al., 1992; Fredholm et al., 1999). The distribution of adenosine receptors is widespread in the human brain, but receptor sites are particularly abundant in some areas that are involved in higher-order processes (e.g., the hippocampus, a brain structure that is critical for memory formation).

Caffeine appears to be metabolized similarly in young and old individuals (Blanchard and Sawers, 1983). However, due to the lower lean body mass in older people, the bioavailability of caffeine in this group may be higher and may lead to higher blood and tissue concentrations. Although the metabolism of and the physiological response to caffeine is relatively independent of age, there may be age-related differences in the sensitivity to this compound in some organ systems, including the brain (Massey, 1998). Unfortunately, studies on the behavioral effects of caffeine have generally been performed in young to middle-aged, healthy individuals, which may complicate the interpretation of results in an aging context.

The purpose of this chapter is to summarize the literature on the age-dependency in the relationship between caffeine use and cognitive function, with special reference to caffeine's potential to prevent, postpone, or counteract age-related decline of such functions, as part of the aging process. To this end we will start with describing some of the hallmarks of usual cognitive aging and the way in which nutrition and health-related factors are involved in mediating the change of cognitive skills over time. Next, based on the available literature to date, we will evaluate the potential of habitual caffeine use to prevent age-related cognitive decline. Finally, the acute effects of coffee or caffeine use are discussed in an aging perspective.

COGNITIVE AGING

Age-associated cognitive decline (AACD) is generally considered an inevitable consequence of the normal aging process (Jolles et al., 1995b). If the efficiency in brain function is defined in terms of neurocognitive abilities, it is well established that calendar age almost linearly predicts reduced performance in virtually every cognitive domain (Schaie, 1994; Reischies, 1998). In the course of adult life the acquisition and processing of new information becomes less efficient and, in combination with a reduced capacity to retain information, this mechanism leads to lower levels of explicit memory function. In addition, processes related to attention and strategy use tend to deteriorate in older persons, albeit at a slower pace (Jolles et al., 1995b). An important common denominator in this observed decline within cognitive domains is a robust reduction in basic information processing speed (Salthouse, 1992). For example, age-related differences in working memory performance (a memory domain that is considered important for the ongoing buffering and manipulation of task-related information) are to a large extent accounted for by individual differences in basic information processing speed (Salthouse, 1994).

Apart from this gradual decline, another frequently observed phenomenon in cognitive aging studies is an increase in between-individual variance as a function of age. One of the core issues in cognitive gerontology is to identify factors that may account for these cognitive differences between age peers, particularly those that may be amenable to intervention strategies. On the other hand, it appears that in every random population sample, individuals can be identified within a specific age-decade who function at the cognitive level of young and healthy adults. Such individuals have been characterized as aging "successfully," as opposed to the term "usual" used for those who follow a trajectory of gradual decline, and "pathological" for those in whom loss of function has resulted in a progressive loss of independence and, ultimately, in a dementia syndrome (Rowe and Kahn, 1987).

The complex neurobiological mechanisms behind AACD and the age-related increase in performance variability are still far from elucidated, but genetic and health-related factors have been implicated to account for some of the individual differences. For example, vascular risk factors such as chronically elevated blood pressure, unfavorable fatty acid profile, or diabetes have consistently been associated with reduced efficiency of the central nervous system, particularly in older persons, both in cross-sectional and longitudinal population studies (e.g., Van Boxtel et al., 1998; Meyer et al., 2000). Indeed, the prevalence of disease in other organ systems rises sharply after the age of 40 has been reached (Van Boxtel et al., 1998). Many of these pathological conditions may add to the metabolic burden on the cognitive system in old age, including disorders of kidney, liver, thyroid or pulmonary function (Tarter et al., 1988). Apart from the effect of overt diseases, there now is good ground to assume that a healthy lifestyle, including adequate aerobic training (Buchner et al., 1992), abstinence from smoking, moderate drinking habits (Kalmijn et al., 2002), and a healthy diet rich in micronutrients and vitamins (Riedel and Jorissen, 1998) will promote better cognitive functioning in older individuals. Many of the identified health-related factors that affect the cognitive aging process are modulated by the presence of genetic factors, of which the apolipoprotein E (particularly Apo-E4) genotype has the most important impact (Haan et al., 1999). Apo-E4 is a "susceptibility gene" that proportionally increases the risk of early cognitive decline (mild cognitive impairment, or MCI, a borderline state between normal aging and dementia) and Alzheimer's disease in population studies (Smith, 2002b).

Thus, the neurobiological basis of age-related cognitive decline is a complex interplay between genes and environment. It has been suggested that there may be ways to intervene in the health behavior of individuals in order to promote a better "offset" in cognitive reserve capacity. Due to the psychoactive properties of caffeine, in many acute-dose studies this compound has been found to be a potential cognitive enhancer that may ameliorate age-related cognitive decline to some degree (Riedel and Jolles, 1996). Cognitive enhancers are a pharmacologically heterogeneous group of compounds (e.g., neuropeptides and cholinergic and monoaminergic agents) that can improve performance in at least one domain of cognitive performance. It is clear from the previous discussion that the beneficial impact of such compounds on the cognitive aging process is largest when they improve basic neurobiological processes that underpin the efficiency of performance in a wide array of cognitive domains. From this theoretical perspective, compounds that improve attentional processes or basic speed of information processing are the most promising in the prevention of AACD in usual aging (Jolles et al., 1995b).

As stated, caffeine is a mild central nervous system stimulant and its use is widespread, which makes the study of the effects of habitual caffeine use on cognitive aging an interesting public health issue.

CAFFEINE CONSUMPTION HABITS AND AGE

Before turning to the literature on caffeine intake and cognitive function, we will first discuss some aspects of the age-dependency in coffee and caffeine use. There is evidence from population studies that the intake of caffeine is higher in younger to middle-aged adults (Jarvis, 1993; Hameleers et al., 2000), but age trends may vary among different countries. In a recent population-based Dutch study, an estimation was made of the daily caffeine intake as a function of caffeine source and calendar age (Figure 6.1) (Van Boxtel et al., 2003). An age-stratified group of 928 individuals reported their caffeine intake habits in a dedicated caffeine intake questionnaire. Average intake levels were computed based on the caffeine content, expressed in standard units of coffee, (black) tea, cola and energy drink (Van Boxtel et al., 2002). The mean overall intake ranged between 580 mg in people aged 40 (±1) years and 310 mg in 75-year-old individuals. The figure indicates that in this study the intake of caffeine in tea was relatively stable over age groups (50 to 70 mg/day). Caffeine intake in soft drink (cola) added only marginally to the overall caffeine intake and was

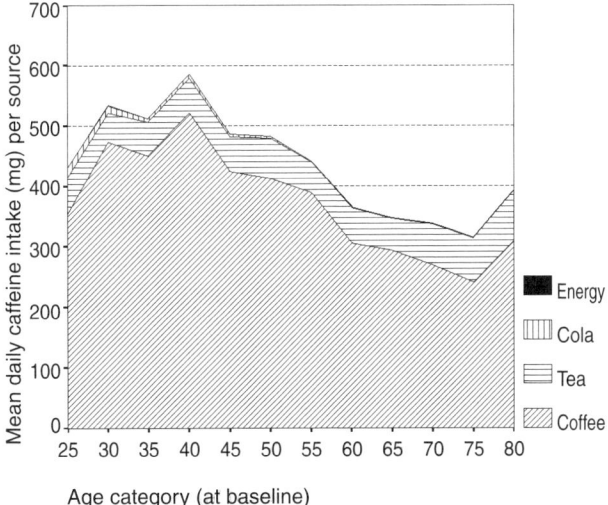

FIGURE 6.1 Total caffeine intake as a function of age and of beverage type (N = 928). "Energy" refers to energy drinks, as a contemporary source of caffeine.

substantial only in the younger age categories. Furthermore, the overall contribution of caffeine contained in modern beverages such as energy drinks was almost negligible. In this study the intake of coffee, as the major source of caffeine in the diet, was relatively stable over the follow-up period of 6 years. The proportion of individuals who did not change their coffee intake ranged between 68% (30 to 47 years) and 83% (65 to 87 years). The reasons that 15 to 18% of the respondents (depending on their age group) cut coffee consumption in the past 6 years were related to general health (43%), cardiovascular complaints (8%), sleeping complaints (10%), or other concerns (48%). There were no clear age trends in the reasons to reduce intake. The majority of this population (61%) started drinking coffee before the age of 15, and the reported peak intake of coffee was most often between 21 and 30 years.

In a large sample taken from a U.K. population (N = 7414), a strong inverse relationship was found between the average intake of coffee and tea (Jarvis, 1993). In that study, a higher intake of coffee was not only associated with younger age, but also with perceived better health, higher educational level or social class, higher consumption of cigarettes and alcohol, and lower use of tranquilizers. These findings indicate that careful control for possible confounding variables may be necessary when habitual caffeine use is studied in relation to cognitive performance variables.

CAFFEINE INTAKE AND AGE-RELATED DIFFERENCES IN COGNITIVE FUNCTION

The majority of experimental studies have focused on the effects of caffeine use on cognition in terms of performance measures. The literature on how individuals perceive the psychotropic effects of caffeine, or coffee, still is limited. In a small, placebo-controlled acute-dose study in six younger (18 to 37 years) and six older (65 to 75 years) individuals, the younger group reported higher levels of alertness, calmness, interest, and steadiness on visual analog scales after the ingestion of 200 mg of caffeine (Swift and Tiplady, 1988). The authors of that study hypothesized in their discussion that older persons may be less able to report the subjective effects of psychoactive compounds. In one aging study (Van Boxtel et al., 2002) participants in different age groups were asked their opinion on several statements about the effects of coffee on behavior and cognition (Table 6.1). The analyses revealed that there was no strong agreement in this sample on statements about a positive effect of coffee consumption on memory, attention, or general performance (range 5 to 9%). More specifically, there were no significant trends in this opinion over the three age groups that were addressed. A larger proportion of participants agreed that coffee can be used to "wake up" (15%); this opinion was more prevalent in the younger group (30 to 47 years). Still, the majority of the respondents disagreed with the statements about positive effects of coffee use on cognitive functioning. These observations, therefore, make it unlikely that a substantial proportion of coffee users drink coffee in order to boost their individual performance.

Furthermore, there is no strong evidence to suggest that there are substantial differences over age groups in the awareness of the effects of habitual coffee or caffeine use on behavior.

HABITUAL CAFFEINE INTAKE STUDIES

Epidemiological evidence for a protective effect of habitual caffeine use on functional brain integrity comes from a recent study into factors associated with a reduced chance of Alzheimer's disease (AD) (Maia and De Mendonca, 2002). A group of 54 patients with probable AD were matched for age and sex with cognitively intact controls. The average daily consumption in the 20 years preceding the diagnosis of AD was 94 mg in the AD patient group and 199 mg in a comparable life episode in the control individuals. The odds ratio in a logistic regression analysis was 0.40 (CI 0.25 to 0.67), indicating a lower risk of AD in conjunction with higher caffeine intake levels. However, retrospective studies are often confounded and identified differences in coffee consump-

TABLE 6.1
Reported Opinions about Statements Regarding the Behavioral Effects of Coffee (%) by Levels of Age and Sex (N = 919)

Statement	Category	Missing	Strongly Agree	Agree	Neither Agree Nor Disagree	Disagree	Strongly Disagree
I drink coffee to wake up							
	30–47 yr	0	7	13	12	22	47
	50–62 yr	1	5	10	12	27	44
	65–87 yr[a]	5	3	6	14	24	47
	Male	2	3	10	11	22	52
	Female	3	7	9	14	27	39
	All	2	5	10	13	24	46
Coffee is bad for my health							
	30–47 yr	0	2	16	43	28	11
	50–62 yr	1	3	13	42	33	7
	65–87 yr	5	2	9	39	31	15
	Male	1	2	12	44	29	11
	Female	3	2	13	38	33	11
	All	2	2	13	41	31	11
Coffee improves my memory							
	30–47 yr	1	1	3	42	26	28
	50–62 yr	2	1	2	42	33	19
	65–87 yr	5	2	3	34	31	24
	Male	2	1	2	45	27	23
	Female	4	2	4	33	33	25
	All	3	2	3	39	30	24
I perform worse without coffee							
	30–47 yr	1	1	7	18	29	44
	50–62 yr	2	2	7	19	36	34
	65–87 yr	5	4	6	22	33	31
	Male	2	2	6	20	32	38
	Female	3	2	8	19	33	34
	All	3	2	7	20	32	36
I can concentrate better with coffee							
	30–47 yr	1	2	5	29	27	37
	50–62 yr	2	1	10	23	35	29
	65–87 yr	4	3	6	26	35	26
	Male	1	2	6	27	31	32
	Female	3	2	7	25	33	30
	All	2	2	7	26	32	31

[a] $p < .05$ (chi^2 test for trend).

tion may have been the result of report bias, or a behavioral adaptation to the cognitive decline that patients were confronted with. Two population-based studies have specifically addressed the relationship between habitual caffeine intake and cognitive performance in a cross-sectional setting (Jarvis, 1993; Hameleers et al., 2000); the group of Hameleers et al. (2000) recently extended their earlier findings with a longitudinal follow-up of their participants after 6 years (Van Boxtel et al., 2003). Both studies were based on the premise that caffeine's effects on cognitive performance do not wear off in long-term users and that older caffeine consumers benefit more from regular caffeine intake than younger individuals. The first study was performed in the U.K. as part of the Health and Lifestyle Survey in 7414 adults. All participants were interviewed about their average daily intake of coffee and tea. This information was translated into the average number of standard caffeine units per day, a value ranging from 0 to 8, in which the weight of one tea unit was half that of one coffee unit. Four cognitive tests were administered by a research nurse during a home visit: simple reaction time, choice reaction time, (delayed) incidental verbal memory, and visuospatial reasoning. All subsequent analyses were controlled for several background variables that were related to caffeine use, cognitive measures, or both: age; sex; social class; housing tenure; educational level; the use of alcohol, cigarettes, or tranquilizers; retirement; disablement; and perceived health. After controlling for these variables and, in addition, for the caffeine consumed in tea, a significant trend was found between better performance on all four cognitive measures and a higher coffee consumption. The same trends were found in models of simple reaction time and visuospatial reasoning when tea was used as a primary predictor of performance while controlling for coffee use and the other background characteristics. Again, all dose-related trends were apparent in all neurocognitive measures when an overall estimate of caffeine intake was used as predictor of performance. In addition, significant age by caffeine intake interactions were identified in all but one (visuospatial reasoning) cognitive measure, characterized by a stronger positive association between caffeine intake and performance in the oldest group (55+) than in both younger groups (16 to 34 and 35 to 54, respectively). These results were interpreted as being indicative that tolerance to the performance-enhancing effects of caffeine had not developed. The dose-effect relationship was considered the result of a general performance boost, mediated by an improvement of arousal or vigilance. It was suggested that the age by intake interactions could be the result of older people performing more below their own maximum level of performance than the younger participants, so they were able to gain more from a pharmacological agent that promotes their alertness. Although this study was the first to demonstrate the positive relationship between habitual caffeine intake and cognitive performance, it has also been criticized for the fact that the habitual users in the study may have been performing below their usual level due to caffeine withdrawal effects (James, 1994; Rogers and Dernoncourt, 1998).

A second study devoted to habitual caffeine use and cognition (Hameleers et al., 2000) was part of a larger Dutch research program into determinants of cognitive aging, the Maastricht Aging Study (or MAAS; Jolles et al., 1995a). It was designed to replicate the findings of Jarvis (1993), using a more rigorous control of testing conditions: all tests were administered in a behavioral laboratory. In addition, effects of caffeine withdrawal were controlled for by allowing participants to drink coffee ad libitum. A group of 1875 individuals stratified for age (24 to 81 years), sex, and level of occupational achievement were questioned about their usual coffee and tea intake and took part in an extensive neurocognitive test battery. Control variables were taken from the Jarvis (1993) study: education, sex, (actual) smoking, alcohol use, perceived health, housing tenure, and occupation (blue/white collar). The positive association between estimates of motor choice reaction time (simple and complex conditions; movement times) and verbal memory performance (delayed recall) on the one hand, and the daily consumed number of caffeine units on the other, was again apparent in this data set but was smaller than that reported by Jarvis (1993). In addition, a dose-effect relationship was found between movement time in the choice reaction time task and caffeine intake level. However, no linear effects were found on other cognitive measures, including those related to planning (verbal fluency) and cognitive flexibility (concept shifting test, Stroop Color and Word interference Test). Furthermore, the authors could not demonstrate interactions between age and caffeine intake, which suggested that caffeine intake effects were not differential over age groups. Thus, in both population-based studies, small associations between habitual caffeine intake and motor performance were recorded. It may be argued, however, that acute effects of caffeine (either in the acute phase of action or during decreasing blood levels, in this case withdrawal) are, at least to some extent, responsible for the associations that were found on both occasions. This issue was recently addressed by the latter research group (Van Boxtel et al., 2003). They focused on the longitudinal effects of habitual caffeine intake levels at baseline on cognitive performance in the group of 1376 participants who were available for a follow-up measurement after 6 years. Apart from the control variables that were identical to those used in the cross-sectional analyses, additional control for baseline performance was made in the regression models for cognitive performance at follow-up. Again, the movement times of both the simple and complex conditions of a choice reaction time task were related to caffeine intake, accounting for an additional explained variance of less than 1%. No associations between baseline intake and all other performance measures (including verbal memory) that were used in the cross-sectional study reached statistical significance.

Both discussed population studies may have been hampered by the fact that the estimation of habitual caffeine intake was, to some extent, inaccurate. No distinction was made for the use of decaffeinated coffee, or herbal tea, which both contain substantially lower amounts of caffeine. Furthermore, the close interplay between caffeine intake habits and other sociodemographic or lifestyle characteristics can in part attenuate the strength of the found associations. In either case, these associations will be an underestimation of the true relationship between habitual caffeine intake and cognitive performance measures. However, the relationship found so far is weak, and from a public health standpoint it seems unwarranted to promote the use of caffeine-containing beverages in order to prevent or postpone age-related cognitive decline.

CAFFEINE, AGE, AND AROUSAL

An alternative role for caffeine in combating age-related cognitive decline may lie in the *restoration*, rather than the prevention, of diminished performance through its acute effects on cognition. The effects of acute caffeine intake on mental performance have been documented in numerous research papers (for an extensive overview see Smith, 2002a), and although the effects may vary depending on the conditions of the subjects (e.g., arousal level), task characteristics (e.g., complexity, duration), and experimental design (e.g., caffeine dosage), caffeine intake predominantly has beneficial effects on cognitive performance. Caffeine is a mild stimulant and hence its cognition-enhancing effects are most pronounced when performance is somehow degraded due to a lowering of energetic resources (arousal), for example, due to fatigue, sleep deprivation, prolonged mental activity, or use of sedating psychoactive compounds.

Energetic resources also diminish with increasing age (Salthouse, 1988), as well as the speed of information processing (Cerella, 1990; Salthouse, 1992, 1994). These mechanisms are thought to underlie at least part of the cognitive impairments seen in the aging individual, as pointed out above. Given caffeine's well-known stimulant effects (Smith et al., 1999), an interaction between caffeine and age may be mediated by arousal-related mechanisms and, as such, caffeine could serve as a nutritional tool to counteract age-related cognitive decline.

The relationship between level of arousal and cognitive performance is not a linear one. Performance of a certain task is optimal at a specific arousal level, and deviation from this point may lead to either under- or overarousal and consequently to suboptimal performance (see Watters et al., 1997; Anderson et al., 1989). The relationship between arousal and task performance thus follows an inverted, U-shaped curve (Figure 6.2), which is described by the Yerkes-Dodson Law (Yerkes and Dodson, 1908). Furthermore, the optimal arousal level is thought to vary between different cognitive tasks. In short, the optimal level is considered to be higher for "easier" tasks, whereas the optimal level of arousal is low for more complex cognitive tasks. In other words, the U-shaped curve shifts towards the left with increasing task complexity (Anderson et al., 1989; Watters et al., 1997).

The arousal–task performance theory is supported by experiments showing that caffeine's effects on a specific task depend on the caffeine dosage and thus on the level of arousal increment. For example, Hasenfratz and Bättig (1994) have demonstrated that the performance of young volunteers on a rapid information processing task improved with a caffeine dose of 1.5 mg/kg,

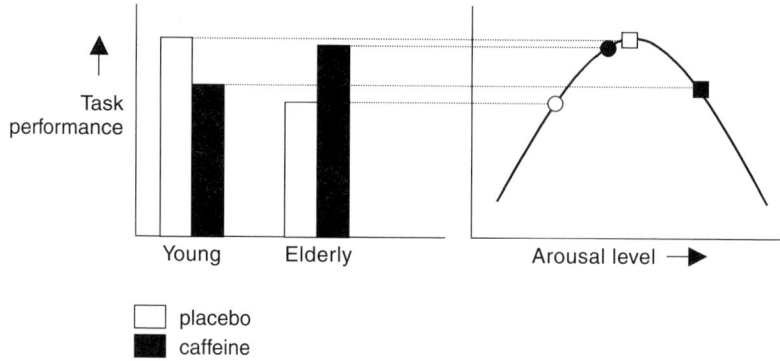

FIGURE 6.2 Illustration of the relationship between task performance and arousal level in young and elderly persons and the effect of arousal enhancement by caffeine intake.

whereas the effect progressively wore off with higher dosages of 3 and 6 mg/kg. A similar pattern was found for a continuous attention task in elderly subjects (Bryant et al., 1998). In that study, maximum improvement was achieved at the lowest dose (mean peak plasma concentration of 4.4 mg/l, average dose of 116 mg caffeine) and performance tended to revert to placebo scores with a higher (approximately double) dose. Loke (1988) showed that memory span increased with low (162 mg) but not with high (325 mg) caffeine doses. In a series of experiments Watters et al. (1997) confirmed the inverted-U hypothesis in a dose range of 0 to 600 mg of caffeine for a number of tasks requiring numerical and alphabetical manipulation. For these tasks, performance was optimal after the administration of 400 mg of caffeine. The task difficulty hypothesis has been proven to be more difficult to confirm, not least because it is quite difficult to justify the labeling of a task as "difficult" or "easy" (Anderson et al., 1989; Watters et al., 1997). Nevertheless, in accordance with the task difficulty hypothesis, performance of "simpler" tasks, such as reaction time tests, is more likely to benefit from the activating effects of caffeine, whereas caffeine may diminish performance in more complex tasks due to overarousal.

As energetic resources decrease with age, elderly may find themselves in a state of underarousal in an increasing number of instances of cognitive demand. In this respect, aging can be compared to other states of low arousal, for example, due to fatigue or sleep deprivation. Under such conditions, caffeine may increase arousal to a level closer to the optimum, and thus improve performance. From this premise, two mediating factors are important to consider. The first is the baseline arousal level in relation to the optimal level for a specific task. Since this optimum may vary depending upon task complexity, individuals may invest more or less of their available energetic resources to meet specific task demands. Thus, in case of declining arousal the level of performance may be maintained by the investment of spare energetic resources. In other words, the task requires more effort. The second mediating factor relates to the extent of the arousal enhancement by caffeine. The latter is mostly dependent upon caffeine dosage, although it is known that many factors may influence the pharmacokinetic properties of caffeine, and thus the brain availability. For example, cigarette smoking, use of oral contraceptives, and pregnancy may significantly prolong caffeine clearance (Fredholm et al., 1999). Importantly, the bioavailability of caffeine appears to be largely unaffected by age and the pharmacokinetic characteristics of caffeine do not differ between adult age groups (Blanchard and Sawers, 1983; Massey, 1998).

The baseline arousal level combined with the caffeine-induced shift in arousal ultimately determines the position on the U-curve. Young subjects are more susceptible to caffeine-induced overactivation because they are already operating close to an optimal arousal level. As is illustrated in Figure 6.2, a similar increase in arousal may produce impairment in young subjects (due to overactivation), but it may improve performance in the elderly. Furthermore, following this line of reasoning, it may be predicted that young subjects are more likely to experience cognitive improvement with lower caffeine dosages, whereas elderly subjects may benefit predominantly from higher caffeine dosages.

ACUTE CAFFEINE EFFECTS IN OLDER PERSONS

The following section will discuss the experimental findings relating to the cognitive effects of acute caffeine intake in the elderly with special regard to the possible relationship between dosage and age. The vast majority of our knowledge on the acute behavioral effects of caffeine is based on experiments with young or, at best, middle-aged subjects. Surprisingly few studies have investigated the acute cognitive effects of caffeine in an elderly population, and only a handful have actually compared effects across various age groups.

The majority of studies that have compared age groups have indeed shown that caffeine's effects may differ across the adult life span. Swift and Tiplady (1988) compared the cognitive effects of 200 mg of caffeine in a group of young (18 to 37 years) and elderly (65 to 75 years) subjects. In young subjects caffeine improved only simple movement speed (tapping task), whereas the elderly benefited from caffeine administration on more complex tasks (i.e., choice reaction time and continuous attention) for which performance was initially reduced due to aging. Lorist et al. (1995) examined the effects of 250 mg of caffeine on event related potentials (ERPs) in young (18 to 23 years) and older (60 to 72 years) volunteers. Their ERP data showed that the caffeine-induced increase in the availability of energetic resources for task performance was similar in young and elderly persons. This may be interpreted as evidence that the central pharmacodynamic (stimulatory) effects of caffeine remain stable across age groups. Interestingly, on a behavioral level, caffeine was able to counteract an age-related decline in stimulus evaluation during an attention/working memory task.

Rees et al. (1999) also found differential acute effects of 250 mg of caffeine in elderly (50 to 65 years) vs. young (20 to 25 years) subjects. Young subjects improved their performance predominantly on psychomotor parameters, such as tapping, simple reaction time, and response speed, while in the elderly subjects more complex cognitive functions were also improved by caffeine, including focused attention, symbol copying, and learning. Of interest is that the caffeine-induced improvements in the older participants often reflected a reversal of a performance decrement during the test days. In other words, performance deteriorated from the first (baseline) to the second test session on the day the elderly subjects received placebo, but this decline was not seen when caffeine was administered. Apparently, the elderly were more prone to fatigue due to test procedures and caffeine was able to attenuate this fatigue. This may at least partially explain why caffeine can exert more pronounced effects in the elderly.

A possible relationship between caffeine dosage, arousal, and age is illustrated by the results from a study by Hogervorst et al. (1998). Three age groups, young (26 to 34 years), middle-aged (46 to 54 years), and elderly (66 to 74 years), received 225 mg of caffeine and performed a number of cognitive tests. It was found that caffeine administration improved short-term memory performance (one trial word learning, short-term memory scanning) in the middle-aged group but not in the young or elderly groups. In the young group, memory performance was even impaired by caffeine. The authors suggested that 250 mg of caffeine may have shifted the arousal level beyond the optimum for these tasks in young subjects but towards the optimum in middle-aged subjects. Presumably, in older participants the arousal manipulation was not potent enough to induce a detectable change in performance. This could imply that elderly would have benefited more from higher caffeine dosages, but this remains to be established. In line with the notion of a dose-dependent relationship between age and caffeine's cognitive effects are data showing that 100 mg of caffeine was ineffective in producing performance changes in middle-aged and elderly subjects, using tasks similar to those of Hogervorst et al. (1998) (Schmitt et al., 2003).

One study (Amendola et al., 1998) failed to detect age-related differences in the effect of various dosages of caffeine on a number of neuropsychological tests measuring sustained and selective attention, reaction time, learning, and long-term memory. In this study 12 young (18 to 30 years) and 12 older (above 60 years) volunteers were tested after receiving 64, 128, or 256 mg of caffeine and placebo. Caffeine at the two highest doses improved performance on a vigilance task in both age groups, but no other cognitive effects were found. However, remarkably, none of the cognitive tests showed a typical age-related decline in performance, suggesting that the older group may have inadvertently consisted of a special subgroup of so-called successful aging individuals, or the study may have lacked sufficient power.

The limited amount of experimental data on possible age-related differences in the effect of caffeine does not allow definitive conclusions. However, several tentative mechanisms underlying such an effect may be identified. First, aging is associated with a decrease in baseline arousal, which provides an opportunity for a stimulating agent such as caffeine to restore arousal levels and hence facilitate cognitive functioning. This effect is even more pronounced in situations of prolonged mental activity in which the energetic resources are further depleted. In older persons, less compensatory capacity (spare energetic resources) is available, resulting in an accelerated drop in cognitive performance over time due to mental fatigue, which can be attenuated by caffeine. At present there is some, albeit very limited, evidence suggesting that older persons may benefit from relatively higher dosages of caffeine. However, the issue is far from being resolved and requires much further investigation. Ideally, this would involve testing the effects of multiple dosages in various age groups, using tasks with increasing cognitive complexity. If confirmed, from a cognition point of view it may be warranted to stimulate caffeine intake in the elderly to boost performance in task-specific situations. As pointed out earlier, caffeine intake typically increases in early adulthood, peaks during middle age, and subsequently declines with old age. Therefore, those individuals who may benefit most from the cognitive effects of caffeine may be the ones that use it the least.

REFERENCES

Amendola, C.A., Gabrieli, J.D.E. and Leberman, H.R. (1998) Caffeine's effects on performance and mood are independent of age and gender. *Nutritional Neuroscience,* 1, 269–280.

Anderson, K.J., Revelle, W. and Lynch, M.J. (1989) Caffeine, impulsivity, and memory scanning: a comparison of two explanations for the Yerkes-Dodson effect. *Motivation and Emotion,* 8, 614–624.

Blanchard, J. and Sawers, S.J.A. (1983) Comparative pharmacokinetics of caffeine in young and elderly men. *Journal of Pharmacokinetics and Biopharmacology,* 11, 109–126.

Bryant, C.A., Farmer, A., Tiplady, B., Keating, J., Sherwood, R., Swift, C.G. and Jackson, S.H. (1998) Psychomotor performance: investigating the dose-response relationship for caffeine and theophylline in elderly volunteers. *European Journal of Clinical Pharmacology,* 54, 309–313.

Buchner, D.M., Beresford, S.A.A., Larson, E.B., LaCroix, A.Z. and Wagner, E.H. (1992) Effects of physical activity on health status in older adults. II: Intervention studies. *Annual Review of Public Health,* 13, 469–488.

Cerella, J. (1990) Aging and information processing rate, in *Handbook of the Psychology of Aging,* Birren, J. and Schaie, K., Eds., Academic Press, San Diego, CA.

Fredholm, B.B., Bättig, K., Holmén, J., Nehlig, A. and Zvartau, E.E. (1999) Actions of caffeine in the brain with special reference to factors that contribute to its widespread use. *Pharmacological Reviews,* 51, 83–133.

Haan, M.N., Shemanski, L., Jagust, W.J., Manolio, T.A. and Kuller, L. (1999) The role of APOE epsilon4 in modulating effects of other risk factors for cognitive decline in elderly persons. *Journal of the American Medical Association,* 282, 40–46.

Hameleers, P.A.H.M., van Boxtel, M.P.J., Hogervorst, E., Houx, P.J., Buntinx, F., Riedel, W.J. and Jolles, J. (2000) Habitual caffeine consumption and its relation to memory, attention, planning capacity and psychomotor performance across multiple age groups. *Human Psychopharmacology,* 15, 573–581.

Hasenfratz, M. and Bättig, K. (1994) Acute dose-effect relationships of caffeine and mental performance, EEG, cardiovascular and subjective parameters. *Psychopharmacology,* 114, 281–287.

Hogervorst, E., Riedel, W., Schmitt, J.A.J. and Jolles, J. (1998) Caffeine improves memory performance during distraction in middle-aged, but not in young or old subjects. *Human Psychopharmacology,* 13, 277–284.

James, J.E. (1994) Does caffeine enhance or merely restore degraded psychomotor performance? *Neuropsychobiology,* 30, 124–125.

Jarvis, M. (1993) Does caffeine intake enhance absolute levels of cognitive performance? *Psychopharmacology,* 110, 45–52.

Jolles, J., Houx, P.J., van Boxtel, M.P.J. and Ponds, R.W.H.M., Eds. (1995a) *Maastricht Aging Study: Determinants of Cognitive Aging.* Neuropsychology Publishers, Maastricht, the Netherlands.

Jolles, J., Verhey, F.R.J., Riedel, W.J. and Houx, P.J. (1995b) Cognitive impairment in elderly people: predisposing factors and implications for experimental drug studies. *Drugs and Aging,* 7, 459–479.

Kalmijn, S., van Boxtel, M.P.J., Verschuren, M.W.M., Jolles, J. and Launer, L.J. (2002) Cigarette smoking and alcohol consumption in relation to cognitive performance in middle age. *American Journal of Epidemiology,* 156, 936–944.

Loke, W.H. (1988) Effects of caffeine on mood and memory. *Physiology and Behavior,* 44, 367–372.

Lorist, M.M., Snel, J., Mulder, G. and Kok, A. (1995) Aging, caffeine, and information processing: an event-related potential analysis. *Electroencephalography and Clinical Neurophysiology, Evoked Potentials,* 96, 453–467.

Maia, L. and De Mendonca, A. (2002) Does caffeine intake protect from Alzheimer's disease? *European Journal of Neurology,* 9, 377–382.

Massey, L.K. (1998) Caffeine and the elderly. *Clinical Pharmacology,* 13, 43–50.

Meyer, J.S., Rauch, G., Rauch, R.A. and Haque, A. (2000) Risk factors for cerebral hypoperfusion, mild cognitive impairment, and dementia. *Neurobiology of Aging,* 21, 161–169.

Nehlig, A., Daval, J.L. and Debry, G. (1992) Caffeine and the central nervous system: mechanisms of action, biochemical, metabolic and psychostimulant effects. *Brain Research Reviews,* 17, 139–170.

Rees, K., Allen, D. and Lader, M. (1999) The influences of age and caffeine on psychomotor and cognitive function. *Psychopharmacology,* 145, 181–188.

Reischies, F.M. (1998) Age-related cognitive decline and the dementia threshold, in *Handbook of Aging and Mental Health: An Integrative Approach,* Lomranz, J., Ed., Plenum Press, New York, pp. 435–448.

Riedel, W.J. and Jolles, J. (1996) Cognition enhancers in age-related cognitive decline. *Drugs and Aging,* 8, 245–274.

Riedel, W.J. and Jorissen, B.L. (1998) Nutrients, age and cognitive function. *Current Opinion in Clinical Nutrition and Metabolic Care,* 1, 579–585.

Rogers, P.J. and Dernoncourt, C. (1998) Regular caffeine consumption: a balance of adverse and beneficial effects for mood and psychomotor performance. *Pharmacology, Biochemistry and Behavior,* 59, 1039–1045.

Rowe, J.W. and Kahn, R.L. (1987) Human aging: usual and successful. *Science,* 237, 143–149.

Salthouse, T.A. (1988) Resource-reduction interpretations of cognitive aging. *Developmental Reviews,* 8, 238–272.

Salthouse, T.A. (1992) *Mechanisms of Age-Cognition Relations in Adulthood,* Lawrence Erlbaum Associates, Hillsdale, NJ.

Salthouse, T.A. (1994) The aging of working memory. *Neuropsychology,* 8, 535–543.

Schaie, K.W. (1994) The course of adult intellectual development. *American Psychologist,* 49, 304–313.

Schmitt, J.A.J., Hogervorst, E., Vuurman, E.F.P.M., Jolles, J. and Riedel, W.J. (2003) Memory and attention functions in middle-aged and elderly subjects are unaffected by a low, acute dose of caffeine. *Journal of Nutrition Health and Aging,* 7, 301–303.

Smith, A. (2002a) Effects of caffeine on human behavior. *Food and Chemical Toxicology,* 40, 1243–1255.

Smith, B.D., Tola, K. and Mann, M. (1999) In *Caffeine and Behavior: Current Views and Research Trends,* Gupta, B.S. and Gupta, U., Eds., CRC Press, Boca Raton, FL, pp. 87–135.

Smith, J.D. (2002b) Apolipoproteins and aging: emerging mechanisms. *Ageing Research Reviews,* 1, 345–365.

Swift, C.G. and Tiplady, B. (1988) The effects of age on the response to caffeine. *Psychopharmacology,* 94, 29–31.

Tarter, R.E., van Thiel, D.H. and Edwards, K.L. (1988) *Medical Neuropsychology: The Impact of Disease on Behavior.* Plenum Press, New York.

Van Boxtel, M.P.J., Buntinx, F., Houx, P.J., Metsemakers, J.F.M., Knottnerus, J.A. and Jolles, J. (1998) The relation between morbidity and cognitive performance in a normal aging population. *Journal of Gerontology,* 53A, M146–M154.

Van Boxtel, M.P.J., Bosma, H. and Jolles, J. (2002) The effects of habitual caffeine use on cognitive change: a longitudinal perspective, research report, Universiteit Maastricht, Maastricht, the Netherlands.

Van Boxtel, M.P.J., Schmitt, J.A.J., Bosma, H. and Jolles, J. (2003) The effects of habitual caffeine use on cognitive change: a longitudinal perspective, *Pharmacology, Biochemistry and Behavior,* 75, 921–927.

Watters, P.A., Martin, F. and Schreter, Z. (1997) Caffeine and cognitive performance: the nonlinear Yerkes-Dodson Law. *Human Psychopharmacology,* 12, 249–257.

Yerkes, R.M. and Dodson, J.D. (1908) The relation of strength of stimuli to rapidity of habituation. *Journal of Comparative Neurology and Psychology,* 18, 459–482.

7 Neurodevelopmental Consequences of Coffee/Caffeine Exposure

Tetsuo Nakamoto

CONTENTS

Abstract 97
Introduction 98
Caffeine Exposure to Animals and Its Implications for Humans 98
Caffeine's Effects during Pregnancy on Brain Development 100
Caffeine's Effects during Lactation on Brain Development 102
Caffeine's Effects during Gestation and Lactation on Brain Development 104
Nutritional Status and Caffeine's Effects on Brain Development 105
Caffeine and Fetal Origins of Adult Disease 106
Conclusion 107
Acknowledgment 108
References 108

ABSTRACT

Caffeine is one of the most commonly consumed drugs in our society today. In spite of the fact that consumption of caffeine far exceeds our consumption of alcohol and tobacco, a basic understanding of caffeine's effects during gestation and lactation is still not clear. The effects of routine maternal caffeine consumption on fetal and neonatal neurodevelopment are controversial. Although pregnant women are advised to avoid caffeine-containing drinks, many continue to consume them. Unfortunately, this is particularly true for the less educated. Furthermore, many women continue to consume caffeine-containing beverages after delivery. The results of various animal studies on caffeine cannot be extrapolated to humans due in part to the different methods of caffeine administration. When they are, they offer confusing information. Some studies also use an unrealistic amount of caffeine. In addition, for unknown reasons, certain animals might be more susceptible to caffeine's effects than others, as permanent behavior changes may indicate. Nutritional factors also might modify caffeine's effects. It is conceivable that caffeine exposure before fertilization could also affect neurodevelopment; thus, chronic daily caffeine intake might be deleterious for those planning to have a family. It might be safe to state that overall evidence from animal studies suggests that routine caffeine intake during critical periods of growth could exert certain detrimental effects on fetal and neonatal neurodevelopment and be one of the causes of diseases in later life.

INTRODUCTION

Caffeine has recently been added to certain foods, such as ice cream, frozen yogurt, orange juice, and even water, and it is the most popular drug of the general public, surpassing nicotine and alcohol. Many women aged 18 or older consume the caffeine equivalent of two cups (Graham, 1978; Diamond, 1983) to four cups (Weidner and Istvan, 1985) of coffee per day. An estimated 70 to 95% of pregnant women receive caffeine from various sources each day, consuming an average of 2 to 2.5 cups of coffee per day (Graham, 1978; Martin and Bracken, 1987), and heavy caffeine consumption is particularly noted among pregnant women with fewer years of formal education (Martin and Bracken, 1987). Epidemiological studies have defined a heavy coffee user as one who daily consumes more than three cups (Martin and Bracken, 1987; Febsterm et al., 1991), four cups (Mills et al., 1993), or five cups of coffee (Furuhashi et al., 1985). It is still controversial, however, whether commonly consumed doses of caffeine by women during pregnancy and the early neonatal period are harmful to their fetuses and infants, even though most organs, including the brain, are rapidly growing during these critical periods and the results of animal studies indicate that caffeine can affect these organs.

At the postnatal age of 12 to 13 days, the rat's neocortex is considered to be a suitable model for studying the human neocortex around birth (Romijn et al., 1991). The ratio of embryo to maternal blood caffeine concentrations was approximately 1, indicating the free transfer of caffeine to the embryo (Kimmel et al., 1984). Caffeine also penetrates the blastocyst (Fabro and Siever, 1969) and accumulates in the fetal brain (Galli et al., 1975; Tanaka et al., 1987). In addition, caffeine diffuses readily into human and animal milk (Tyrala and Dodson, 1979; Gullberg et al., 1986) and was found in the growing brain of newborn rats (Nakamoto et al., 1988). Thus, it would not be surprising to learn that caffeine exerts presently unknown adverse effects on the normal development of the brain. The slightest changes that maternal caffeine consumption may cause during gestational and early neonatal periods of growth, which are critical periods, might result in irreversible effects in later life. In fact, the concept of fetal origins of adult disease has been discussed (Barker, 1995; Morley and Dwyer, 2001; Cooper et al., 2002), and the possibility of caffeine's role in the fetal origins of adult disease deserves some consideration.

A recent study indicated that heavy maternal caffeine consumption is associated with sudden infant death syndrome (SIDS) (Ford et al., 1998), although others have disputed this finding (Alm et al., 1999). When caffeine (3 mg or 6 mg/100 g BW) is administered during gestation, the caffeine-exposed pups grew more slowly, resulting in smaller adults (Tye et al., 1993); therefore, the authors suggested a link between human infants with apnea of prematurity when it occurs after the first week and an increased risk for later apnea and SIDS. Long-term maternal caffeine intake during gestation increases the pontine inhibition of the brain stem, and respiratory rhythm is more pronounced (Herlenius et al., 2002). These and other studies indicate that caffeine can have immediate and long-lasting effects on neurodevelopment. Therefore, it is essential for us to understand the effects of caffeine on brain development during gestation and the early neonatal period.

CAFFEINE EXPOSURE TO ANIMALS AND ITS IMPLICATIONS FOR HUMANS

Caffeine literature has put forth confusing information on the various methods of caffeine administration and on the extrapolations of the effects of caffeine. It is also important for investigations on caffeine intake by animal models to take into account the difference between the half-life ($t^1/_2$) of caffeine in humans and animals (Massey, 1991). In humans and animals, caffeine's half-life in one who is pregnant differs from that in one not pregnant, and the half-life in the neonate is also different from that in the adult. Caffeine concentration in the tissue of a pregnant woman rises three times if the half-life is increased from 4 to 12 h (Knutti et al., 1981). The half-life of plasma caffeine in the rat is shorter than that in the human. Caffeine's half-life in the adult human has been shown to be 3.1 h (Knutti et al., 1981), 5.2 h (Bonati and Garattini, 1984), or 4 to 6 h (Oser and Ford, 1981). In contrast, in the adult rat it is 0.88 h (Bonati and Garattini, 1984) or about 1.5 to 2 h (Oser and Ford, 1981), and in the adolescent 40-day-old rat, it is 2 h (Latini et al., 1980). Our estimation of the half-life in the adult female rat is about 2 h (unpublished observation).

The half-life of plasma caffeine during pregnancy, the neonatal period, and infancy is longer. In the human, the half-life in a nonpregnant woman is 3.1 h (Knutti et al., 1981). During pregnancy it increases from 4.7 to 10.5 h as pregnancy progresses, and at the end of pregnancy the half-life becomes the longest, at 10.5 h (Knutti et al., 1981) or even about 15 h (Brazier et al., 1983). In contrast, in pregnant rats, it is 3 to 4 h (Kitts et al., 1986), 5 h (Nakazawa et al., 1985), or 6.9 h (Leal et al., 1990). The half-life of the plasma caffeine of fetal rats is 9.9 h (Leal et al., 1990). A woman during late pregnancy consuming a cup of coffee is comparable to a nonpregnant woman consuming only one-third of a cup of coffee. Therefore, caffeine's effects on the developing fetus should be a serious concern. Maternal caffeine consumption may also have significant implications for the newborn. The half-life in human infants of 1.5 months and 3 to 4.5 months of age is 41 and 14 h, respectively (Aranda et al., 1979). No information is available on the half-life in newborn rats.

Theoretically, 2 mg/100 g of body weight (BW) caffeine exposure in animals is equivalent to an adult human drinking 10 cups of coffee (Gilbert and Pistey, 1973; Concannon et al., 1983), if one assumes that one cup of coffee contains an average of 100 mg of caffeine and that 50 kg is the average BW of an adult human female. However, directly equating the rat's caffeine intake with the human consumption of caffeine on a kilogram basis is misleading. It is reasonable to assume that the half-life of plasma caffeine in the rat can range from approximately one third to one sixth of that in the human, as these studies have shown. One must therefore consider the difference in caffeine's half-life in the rat and the human for a comparison of caffeine's exposure to both. Although the difference of the half-life between species is the simplest parameter, one may need to consider other factors, such as physiological and pharmacokinetic parameters and drug metabolism rates (Bonati et al., 1984).

Based on the caffeine intake in these animal studies, one can calculate the values of proportional human caffeine intake using the formula of metabolic body weight ($kg^{3/4}$) (Kleiber, 1961; Yeh et al., 1986). The values obtained as a result of the difference of the half-life between animals and humans and the calculated values obtained as a result of the difference in the metabolic BW between animals and humans are close enough for a valid comparison. Therefore, 1, 2, 4, and 6 mg of caffeine/100 g BW exposure to animals is approximately comparable to the daily human consumption of slightly more than one, two, four, and six cups of coffee, respectively, with the caffeine content per cup of coffee and body weight described above. For example, for the human, the average daily caffeine intake from all sources is about 5 mg/kg (Grossman, 1984). However, 5 mg/100 g BW for the animals is not a pharmacological dose.

Furthermore, one must use caution in comparing animals and humans when one uses metabolic BW ($kg^{3/4}$) (Kleiber, 1961). For example, a dose of caffeine of 2 mg/100 g BW given to dams is equivalent to the human intake of slightly more than two cups of coffee a day based on metabolic BW, if one assumes that one cup of coffee contains an average of 100 mg of caffeine and that 50 kg is the average BW of an adult human female. Rats weigh between 200 and 300 g, not in the range of kilograms; therefore, it is critical to calculate caffeine intake based upon the 100 g BW basis of rats, not simply to multiply 10 times (in this case, 20 mg/kg) to apply the results of the animal study to humans (Leon et al., 2002). This calculation is critical because metabolic BW on a 100-g BW basis in rats or per kilogram basis in rats is entirely different (Kleiber, 1961). Thus, comparison of the rat to the human for caffeine intake on a 100-g basis or kilogram basis in rats will result in different values of coffee consumption in humans, even if the assumption stated above is the same.

One must be particularly careful with the interpretation of data obtained from animal studies that use gavage (Jacombs et al., 1999) or the intraperitoneal (2.5 mg or 5 mg/100 g BW) (Sahir et al., 2000) or subcutaneous administration of caffeine because the effects of caffeine (3 mg/100 g BW) introduced to a subject by these methods are not the same as the effects of caffeine introduced to a subject in the daily diet or drinking water. Gavage, which is also known as intragastric feeding, and intraperitoneal and subcutaneous administration introduce the total amount of caffeine all at once, and in a practical sense results of these studies cannot be applied to humans. Needless to say, a person does not drink five cups of coffee all at once, but rather consumes five cups of coffee throughout the day. The divided dose and the single dose have different effects on fetuses (Jiritano et al., 1985), suggesting that if caffeine were administered through the diet or drinking water, the data obtained (Wilkinson and Pollard, 1994; Jacombs et al., 1999) would differ from data obtained with gavage. Although 2.5 to 5 mg/100 g BW of caffeine introduced in the diet or drinking water is not a pharmacological dose, the immediate introduction of this amount by gavage or intraperitoneal or subcutaneous injection could make this amount a pharmacological dose and be detrimental to animals, and therefore it is difficult to extrapolate results of these studies to human consumption. Many animal studies have used gavage, and I will therefore discuss these studies even though their results may have limited relevance for human caffeine consumption.

Some studies indicated the amount of caffeine that was given but did not indicate the BW basis. Therefore, in order to make a valid comparison of the studies, a calculation of the amount of caffeine exposure was based upon the BW presented. It is critical to pay attention to BW during pregnancy because the dam's weight increases gradually toward the end of pregnancy. Thus, the amount of caffeine in the diet has to be adjusted when it is based upon BW.

CAFFEINE'S EFFECTS DURING PREGNANCY ON BRAIN DEVELOPMENT

Needless to say, proper fetal and neonatal brain development and function are essential to the good health of the individual, and many environmental and congenital factors can interfere with the opportunity for this good health. Caffeine consumption seems to interfere in a number of significant ways, including decreased fertility (Wilcox et al., 1988).

A caffeine diet (0.5, 1, or 2 mg/100 g BW) was fed to different groups of dams from day 10 of gestation until day 22, and the fetuses were then removed surgically at day 22 (Yazdani et al., 1990). The 0.5 mg caffeine group had more fetal brain DNA content than the controls, but the 2 mg caffeine group had less. Protein content in the 2 mg caffeine group was higher in the noncaffeine controls, but the cholesterol content was lower. In another study, the caffeine diet (2 mg/100 g BW) was fed from day 3 of gestation, and the fetal brain was removed surgically at day 22. The fetal brain weight of the caffeine group was heavier than that of the noncaffeine controls (Yazdani et al., 1992), whereas DNA, protein, and cholesterol contents in the caffeine group tended to be less than in the noncaffeine control.

Maternal caffeine intake (6 mg/100 g BW) from fertilization to gestational day 20 was associated with a significant reduction in fetal cerebral weight and placental weight, and DNA and protein contents in the caffeine group were in general lower than in the control group (Tanaka et al., 1983). Maternal caffeine ingestion may be associated with a decreased volume of caffeine in maternal plasma and a high caffeine content in the fetal cerebrum (Tanaka et al., 1984). These data suggest that the starting time of caffeine consumption as well as the amount of maternal caffeine consumption during pregnancy might have important implications for compositional changes of the brain.

Different amounts of maternal caffeine exposure during gestation exerted different effects on DNA, protein, and cholesterol contents of the growing brain (Yazdani et al., 1990). However, the dam's brain showed only a minimal change from the control, indicating that the growing fetal brain is much more sensitive to caffeine exposure than the adult brain. It is well known that nutritional stresses that do not affect the developed adult brain easily affect the growing brain (Dobbing, 1968), in spite of apparent changes in the BW of the adult.

Three approximate amounts of caffeine (1.2, 2.4, or 9 mg/100 g BW) were fed with the diet to three different groups of dams during pregnancy. Their female offspring were then raised with a noncaffeine diet until adulthood (Enslen et al., 1980). The dopamine contents of the locus coeruleus of the offspring of two caffeine groups (2.4 and 9 mg groups) were significantly decreased. This decrease of dopamine contents may impair developing dopaminergic neurons due to caffeine exposure during the critical time, and caffeine's effects may be lasting.

One group of dams consumed 5 mg/100 g BW of caffeine daily from fertilization to gestational day 21, when fetuses were removed by Caesarean section. Another group consumed 2.5 mg/100 g BW caffeine daily under the same conditions. The BW and the weight of the fetal cerebrum for the 5 mg group were decreased, but for the 2.5 mg group, only the weight of the cerebrum was decreased (Tanaka et al., 1987). In addition, in another study, the brain weights of newborn rats whose dams received coffee (12.2 mg caffeine/100 g BW) during pregnancy were lower than those of the controls upon birth, but the brain weights of newborn rats whose dams received decaffeinated coffee (0.45 mg/100 g BW) were not (Groisser et al., 1982).

Caffeine's effects seem to be far-reaching, even beginning before fertilization. Premating caffeine ingestion (2.5 mg/100 g BW) for 130 days and during pregnancy caused an additional decrease of fetal cerebrum weight as well as decreased placental weight compared to a group that received caffeine during pregnancy alone (Tanaka et al., 1987), suggesting that habitual caffeine intake before pregnancy might influence the future health of fetuses.

In humans, maternal ingestion of two cups of coffee during the last trimester decreases placental blood supply (Gressens et al., 2001), and perinatologic risks may be present (Kirkinen et al., 1983). Because the human brain growth spurt occurs around the time of birth (Dobbing and Sands, 1973), the decrease of placental blood supply due to the daily consumption of two cups of coffee might impair the developing brain and lead to certain latent symptoms of which we are currently unaware. It is also important to keep in mind that the half-life of caffeine is three times longer in the later part of pregnancy (Knutti et al., 1981).

Caffeine ranging from 0.45 mg to 2 mg/100 g BW (Groisser et al., 1982; Yazdani et al., 1990), which is comparable to one half to two cups of coffee, does not seem to affect brain weight, yet some biochemical parameters (Yazdani et al., 1990) are affected. This indicates the need for more research in this area. However, a large dose of caffeine, comparable to 2.5 to 12 cups of coffee, apparently affects brain weight (Groisser et al., 1982; Tanaka et al., 1983, 1987).

Caffeine (2.5 mg/100 g BW) was gavaged to dams whose fetuses were at embryonic day (E) 8 and 9 and killed at E10. The regions of open fetal neural tube were proportionately much higher in the caffeine group than in the control. Because no neural-tube defects were observed at birth, the effect of caffeine may be growth retardation rather than a specific developmental defect (Wilkinson and Pollard, 1994). Note again that the 2.5 mg of caffeine was gavaged to the dams.

On the other hand, caffeine (1.5 mg or 3 mg/100 g BW) was gavaged to dams from E2 to E11, and at both dosage levels somite number and the extent of neural-tube closure were significantly reduced at E12 (Jacombs et al., 1999). Furthermore, in the 3 mg caffeine group the forebrain cavity was significantly enlarged and bounded by a reduced, irregularly aligned neuroepithelium (Jacombs et al., 1999). The higher caffeine group (3 mg) received an approximate daily intake of three cups of coffee, although some epidemiological studies of humans consider three cups of coffee to be a high amount of caffeine (Martin and Bracken, 1987; Febsterm et al., 1991). Note that both caffeine groups received caffeine via gavage.

Pregnant mice were injected intraperitoneally with caffeine (1.25, 2.5, or 5 mg/100 g BW) once a day between E8 and E10, and the fetuses were examined at E9, E10, E13, and E17 (Sahir et al., 2000). In the experimental group, accelerated primitive neuroepithelium evagination into telencephalic vesicles occurred. However, the dose-dependent effect seemed to be reversible during subsequent neuronal migration if caffeine exposure was discontinued (Sahir et al., 2000). Caffeine-induced inhibition of cAMP-dependent protein kinase (PKA) plays a role in early telencephalic evagination (Sahir et al., 2001). In an *in vitro* study, premature evagination of telencephalic vesicles was present in 50% of caffeine-treated embryos in which caffeine concentration in the medium was adjusted to an amount comparable to heavy caffeine consumption (Marret et al., 1997).

In spite of limited caffeine exposure during pregnancy, acceleration of primitive neuroepithelium evagination into telencephalic vesicles (Sahir et al., 2000) and gene modulation in postimplantation embryos occurred (Sahir et al., 2001). In addition, caffeine regionally modified the schedule and/or rate of neural-cell proliferation (Marret et al., 1997). Whether these animals at a later age will exhibit abnormal behavior or abnormal physiological function is currently unknown, although these anatomical changes were reversible upon the discontinuation of caffeine. These studies seem to indicate that persistent minimal histologic defects and/or brain-function impairment are real possibilities in heavy caffeine users (Marret et al., 1997; Sahir et al., 2000, 2001), but keep in mind that two studies (Sahir et al., 2000, 2001) administered caffeine intraperitoneally.

Daily prenatal exposure to caffeine through drinking water (4.4 mg/100 g BW) showed decreased locomotor activity at postnatal day 73, 117, and 171 (Hughes and Beveridge, 1987), whereas the locomotion of offspring whose dams were fed coffee containing caffeine (0.45 or 12.2 mg/100 g BW) showed an increase at day 30 (Groisser et al., 1982), in spite of the lack of caffeine exposure after birth in both studies. The early and widespread expression of A_1-adenosine receptors mRNA in the normal development of the fetal brain has been reported (Weaver, 1996). Chronic exposure of the fetal brain to adenosine antagonists such as caffeine during the critical time could influence and permanently alter postnatal behavior. The motor stimulant effect of caffeine was correlated with its affinity for adenosine receptors in the brain (Snyder et al., 1981).

In humans the long-term consequences of prenatal caffeine intake of approximately two cups of coffee during early pregnancy to one and a half cups at midpregnancy were shown to be null at the age of 7 years (Barr and Streissguth, 1991). However, the validity of this conclusion for life beyond 7 years of age for these children has to be carefully determined in a future study because animal studies have shown latent behavioral modification in adolescents (Sobotka et al., 1979; Guillet and Dunham, 1995).

Caffeine, not its metabolites, is responsible for teratogenic effects on fetuses (Jiritano et al., 1985). Examination of the disposition of caffeine and its metabolites, theophylline, theobromine, and paraxantine, in the 20-day-old fetal brain following a single maternal dose of 0.5 or 2.5 mg of caffeine/100 g BW showed that the fetal, but not the adult, brain accumulates theophylline, theobromine, and paraxanthine. However, the specific effects of these metabolites on brain development are unknown (Wilkinson and Pollard, 1993).

CAFFEINE'S EFFECTS DURING LACTATION ON BRAIN DEVELOPMENT

Different methods of caffeine exposure also have different effects on suckling pups during lactation. Suckling pups receive a much greater amount of caffeine through gavage or subcutaneous injection than through caffeine in the diet or caffeine dissolved in the drinking water of lactating dams because the amount of caffeine in the former is based upon the pups' BW and the pups directly receive caffeine. In the latter, pups only receive caffeine indirectly through the maternal milk, and the amount of caffeine provided to the dam is based upon the dam's BW and is fed to the lactating dams. As a result, suckling pups receive caffeine throughout the day from maternal milk except when dams are out of the nest, whereas caffeine delivered to the pups by gavage is usually administered once a day.

In some instances, early studies introduced large doses of caffeine by gavage. In an *in vivo* study, caffeine (4 or 8 mg/100 g BW) was administered by gavage from postnatal days 2 through 20. Although myelin protein synthesis at days 21 to 24 decreased, it was recovered on days 27 to 28 (Fuller and Wiggins, 1981). Caffeine (2, 4, or 8 mg/100 g BW) was administered daily from birth to postnatal day 17 to pups by gastric intubation and then stopped. Pups were killed at days 17, 23, 30, or 70 (Fuller et al., 1982). A dose-dependent lag in brain weight was evident in the 4 and 8 mg group at day 30 and in the 8 mg group at day 70. Although there was no impairment in myelination in the 2 mg group, myelin recovery was significantly decreased in a dose-dependent manner at day 30; however, no deficit in myelin recovery was seen at day 70. With the advances in technology that have occurred since these studies were performed, the effects of caffeine on myelin formation merit further consideration. However, more importantly in these cases, 4 or 8 mg/100 g BW caffeine administration at once is too much for one to observe the typical effects of caffeine. A single daily administration of 10 mg/100 g BW of caffeine can cause different effects than four divided doses given at 3-h intervals throughout the day (Smith et al., 1987).

It is not surprising that different methods of caffeine administration have different effects on the development of the newborn rat's brain (Quinby et al., 1985; Nakamoto et al., 1988). Caffeine's effect on the developing brains of suckling pups in general seems to be greater when caffeine is delivered through maternal milk rather than by gavage, possibly because a higher concentration of caffeine in general stays in the body longer than it does with the single-dose administration of caffeine. When a single dose is given, the plasma caffeine level surges but gradually decreases (Jiritano et al., 1985). When dams are fed diets or drinking water supplemented with caffeine, pups are indirectly exposed to caffeine through the maternal milk and exposed to caffeine throughout the day.

The brains of newborn rats that received caffeine from dams that were fed a caffeine (1 mg/100 g BW)-supplemented diet showed a decrease of protein and cholesterol contents at day 15 (Nakamoto et al., 1988), whereas intragastric feeding of caffeine (1 mg/100 g BW) to the pups resulted in an increase of protein content of the brain at day 15 (Quinby et al., 1985). In an *in vitro* caffeine exposure study, cholesterol synthesis decreased in C-6 glia cells (Volpe, 1981). Because *de novo* synthesis of cholesterol is the primary brain source and cholesterol synthesis is a critical process in the developing brain, the effect of caffeine on decreased cholesterol contents may have important consequences for the development of the brain.

A caffeine diet (1 mg/100 g BW) was fed to lactating dams, and weaned male offspring were continuously fed this diet until day 43. The total brain weight as well as the weight of various parts of the brain (cerebellum, medulla oblongata, hypothalamus, striatum, cortex-midbrain, and hippocampus) all showed a decrease, as did total DNA content (Yazdani et al., 1988a). These various parts of the brain showed either an increase or decrease in DNA and protein concentrations (mg/g tissue). Because different parts of the brain grow at different rates, the effects of caffeine could influence to some degree different parts of the brain in different ways.

Rat pups that received caffeine (0.1 or 0.9 mg/100 g BW) through gavage from birth to day 6 showed no difference in brain weight in adulthood but did show persistent behavioral deficits into adulthood (Zimmerberg et al., 1991). In other studies, newborn rats were gavaged with caffeine, 2 mg/100 g BW of caffeine on day 2 and 1.5 mg/100 g BW on day 3 to 6 (Guillet and Kellogg, 1991a,b; Etzel and Guillet, 1994; Guillet and Dunham, 1995). Although their brain weights were not affected by these doses, neonatal exposure to caffeine had an effect on central nervous system (CNS) excitability that persisted into adulthood (Guillet and Dunham, 1995). In the cortex, cerebellum, and hippocampus, there was up-regulation of the adenosine A_1 receptor that persisted into young adulthood (14 to 90 days of age) in rats that had only limited exposure to caffeine in the early neonatal period (Guillet and Kellogg, 1991a). There is a marked increase of adenosine receptors in the developing brain in the first weeks of extrauterine life (Johansson et al., 1997), and limited neonatal caffeine exposure altered (Guillet and Kellogg, 1991b) the development of adenosine receptors in the thalamus and cerebellum of rats 14 to 31 days old (Etzel and Guillet, 1994).

These studies indicate that even a limited exposure to caffeine in the early neonatal period may influence later behavior. Caffeine exposure may also exert a critical influence on the developing human brain, a possibility that may be of particular importance since caffeine is used to treat the premature human neonate for apnea (Fisher and Guillet, 1997; Lee et al., 1997).

An *in vitro* study has shown that toxic levels of caffeine (50 μg/ml) could have a prejudicial effect on the number of proliferating glial cells and on the increase of hyaluronan secretion per cell, which could affect myelination onset (Marret et al., 1993).

CAFFEINE'S EFFECTS DURING GESTATION AND LACTATION ON BRAIN DEVELOPMENT

A caffeine diet (1 mg/100 g BW) was fed to dams beginning on day 13 of gestation and during lactation. The weaned male offspring at day 22 were then fed a noncaffeine diet until days 57 and 58 (Nakamoto et al., 1986). The weights of the medulla oblongata and striatum were significantly less than those of the noncaffeine controls, and the weights of other parts of the brain in the caffeine group also tended to be lower than in the controls. Furthermore, DNA contents of striatum, protein contents of medulla oblongata, and cholesterol contents of medulla oblongata, striatum, cortex-midbrain, and hippocampus in the caffeine group were significantly lower than in the controls, indicating that the amount of caffeine comparable to the daily human consumption of one cup of coffee during the gestational and lactational periods affects the developing brain. It is possible that the impairment of the CNS that would occur at this time would not appear until later in the lives of these offspring.

Maternal coffee intake also contributes to maternal and infant iron deficiency anemia (Munoz et al., 1988), and coffee drinking inhibits iron absorption (Morck et al., 1983); it is therefore not surprising that caffeine affects DNA content of the developing brain since iron is required for DNA and protein synthesis (Hironishi et al., 1999).

Caffeine (about 6 mg/100 g BW) was given to dams during gestation and lactation in their drinking water (Tanaka and Nakazawa, 1990). Although the weight of the cerebrum in the caffeine group was less than in the controls at day 1, no significant difference between the groups was observed. However, when rats received caffeine before mating, the weight of their brains was significantly decreased compared to the controls at day 1, although this weight difference disappeared with the continuous administration of caffeine to dams at postnatal days 5 and 10. Since caffeine consumption can be a daily habit, routine consumption prior to pregnancy could influence fetal growth if one becomes pregnant.

The amount of caffeine that is comparable to a human consuming up to three cups of coffee a day was given to dams in their drinking water during gestation and lactation (Aden et al., 2000). Their brains were periodically examined during gestation and lactation and minimal changes in A_1, A_{2A} receptors in the cortex, hippocampus, striatum, and cerebellum were found.

Caffeine (2.6 or 4.5 mg/100 g BW) was given to dams in their drinking water during gestation, and caffeine (2.5 or 3.5 mg/100 g BW) was also given during lactation (Hughes and Beveridge, 1991), and their offspring were tested at 1, 2, 4, and 6 months after birth. The effects of gestational and lactational exposures to caffeine were additive in their modification of the developing brain and were reflected in decreased motor activity in either dose combination of caffeine, suggesting that caffeine continuously affects behavior long after caffeine exposure ends.

Diazepam is used in the treatment of infantile seizures, and caffeine could interfere with diazepam as an adenosine antagonist. When caffeine (0.5 mg/100 g BW) was injected daily during dams' gestational and lactational periods, the percentage of cerebral-bound diazepam in pups dramatically fell at postnatal days 5 and 15 (Daval and Vert, 1986). However, it was recovered at day 25. On the other hand, caffeine (2 mg/100 g BW) was administered during the gestational, lactational, and growing periods for up to 93 days, and the caffeine diet was then changed to a noncaffeine diet until day 388. Only a certain group of caffeine-fed rats showed hyperactive behavior at day 128, and this behavior continued until day 388, when the experiment was terminated (Nakamoto et al., 1991). It should be noted that not all of the animals showed hyperactive behavior, demonstrating that chronic caffeine exposure did not affect all of the caffeine-fed groups to the same extent. Some animals in the caffeine-fed groups showed no effect. Thus, in humans, it could be possible that some may be affected more than others by caffeine exposure during the critical period of growth.

NUTRITIONAL STATUS AND CAFFEINE'S EFFECT ON BRAIN DEVELOPMENT

Protein malnutrition is a worldwide problem. Its incidence in industrialized society is not uncommon (Chase et al., 1980; Listernick et al., 1985), and the low calorie intake of women during pregnancy and lactation in developing countries is also widespread (Kusin et al., 1993). Nutritional deprivation, especially protein-calorie malnutrition, during pregnancy adversely affects neurological development. In addition, protein deficiency can alter the metabolism of drugs and influence their effectiveness and the sensitivity of tissues to them (Varma, 1981). Caffeine consumption increases in relation to the less formal education that a population has (Martin and Bracken, 1987), and it is widely known that those with little formal education are often at the low end of the socioeconomic ladder. A diminished economic level of the individual is also related to an increase in one's nutritional deficiency. The combination of increased caffeine consumption and significant incidences of protein and calorie inadequacy could have much stronger effects on growing offspring than on the offspring of normally nourished adults (Osofsky, 1975). Thus, the interaction between caffeine's effects and the nutritional status of a woman during pregnancy might require further attention.

Normally nourished dams (20% protein) and malnourished dams (8% protein) were subdivided, and the experimental groups received a caffeine (2 mg/100 g BW)-supplemented diet from day 10 of gestation to day 22 just before birth (Yazdani et al., 1988b). The fetuses were then removed surgically and weighed. The BW and DNA concentrations of the 20% protein group supplemented with caffeine were decreased compared to those of the noncaffeine control group of the same nutritional status. On the other hand, only the DNA concentrations of the 8% protein group with caffeine were decreased compared to those of the same nutritional controls. DNA synthesis at day 20 and 22 of gestation in the 20% protein group with caffeine was 74 and 13% of that of the noncaffeine controls, respectively, whereas that of the 8% protein group with caffeine at day 20 and 22 was 214 and 43% of that of the noncaffeine controls, respectively, suggesting that nutritional status during pregnancy plays an important role in DNA synthesis. Protein concentrations of the brain increased in the caffeine-fed 20 and 8% protein groups compared to the respective noncaffeine controls.

Normally nourished (20% protein) and malnourished (6% protein) pregnant dams were each subdivided, and the experimental groups received the caffeine diet (2 mg/100 g BW) from day 13 of gestation to delivery (Mori et al., 1984). Although the brain weights of the 6% protein-supplemented group given caffeine showed a significant increase, brain weight/BW showed no significant difference, whereas that of the 20% protein-supplemented group given caffeine showed a significant decrease compared to that of the noncaffeine controls of the same nutritional status. Although the amount of caffeine administered was the same in these studies (Mori et al., 1984; Yazdani et al., 1988b), the nutritional status (8% protein vs. 6% protein) and period of caffeine supplementation during pregnancy were different. Thus, it is not surprising that certain parameters of these studies are different. Nevertheless, it is clear that maternal nutrition influences the effects of caffeine on the developing brain.

Caffeine diets (2 mg/100 g BW) were fed to pregnant dams that received different nutrition (20, 12, or 6% protein diets) beginning on day 7 of gestation. On day 18 of gestation, prenatal fetal behavior was recorded, and postnatal nipple attachment and general motor activity of the newborn rats were studied (Yoshino et al., 1994). The findings indicate that prenatal caffeine consumption may produce lasting functional alterations in the nervous system affecting the emergence of suckling behavior and motor activity. In the normally nourished group (20% protein diet), nipple-attachment latencies of 2-day-old pups tended to shorten, whereas in the severe malnutrition group, latencies tended to increase. Nipple-attachment latencies in 1-day-old pups increased as the caffeine amount was increased through subcutaneous injection (Holloway, 1982). However, the amount of caffeine increased as much as 8 mg/100 g BW, which is not an amount that is practical for an extrapolation of its effects to humans.

The peak of neurogenesis in the rat trigeminal motor nucleus was delayed in the caffeine-supplemented groups in both the normally nourished groups and those receiving different degrees of malnutrition (Saito et al., 1995). Caffeine intake in combination with protein-energy malnutrition produced effects on the trigeminal nuclear center with various changes in DNA and protein contents.

One group of dams received a normally nourished diet (20% protein) and another group received a malnourished diet (6% protein) starting at delivery. Pups were randomly assigned with a constant number of eight to each dam. Half of the dams given the 20 or 6% protein diet were fed supplementary caffeine (2 mg/100 g BW). At day 15, pups were killed and their brains were removed (Nakamoto et al., 1989). The BW and protein concentrations of the brains in the 20% protein-with-caffeine group were significantly increased compared to those of the noncaffeine control group, but zinc concentrations and alkaline phosphatase activity of the caffeine group were significantly decreased. DNA and cholesterol concentrations of the 6% protein-with-caffeine group were significantly increased compared to the noncaffeine control group.

When caffeine (1 mg/100 g BW) was administered orally by gavage to the newborn pups whose dams were normally nourished (20% protein) or malnourished (6% protein), protein concentrations of the 20% protein-with-caffeine group were significantly increased compared to those of the noncaffeine controls, whereas DNA concentrations of the 6% protein-with-caffeine group were significantly increased compared to the noncaffeine control of the same nutritional status (Quinby et al., 1985). Although the newborn rats that received caffeine by gavage received a higher amount of caffeine than the newborns that received caffeine through maternal milk (Nakamoto et al., 1989), the data seem to indicate that caffeine given through maternal milk had a greater influence on the various parameters of the brain in newborn rats. Furthermore, the effects of caffeine in both groups were modified by their nutritional status.

When a caffeine-supplemented diet (2 mg/100 g BW) was continuously fed to two malnourished groups (12 or 6% protein diets) of lactating dams, the brains of the suckling offspring showed increased concentrations in the caffeine group of cyclo (His-Pro), a neuropeptide ubiquitous throughout the CNS (Mori et al., 1983). This neuropeptide was significantly increased as the degree of malnutrition increased, suggesting again that the nutritional status of the dam modifies caffeine's effects.

When a caffeine-supplemented diet (2 mg/100 g BW) was continuously fed to dams from day 9 of gestation until postnatal day 15, zinc contents of the brains were decreased, but this decrease returned to normal when zinc was supplemented with caffeine in the diet (Nakamoto and Joseph, 1991). Apparently, this decrease of zinc could impair the developing brain, since zinc is an essential metal of growth and development (Prasad, 1988). When caffeine (2 mg/100 g BW) was supplemented to the maternal diet from day 3 of gestation to day 22 just before birth, zinc decreased in the fetal brain. However, in this instance, the addition of zinc to the maternal diet did not return concentrations of zinc in the fetal brain to their original levels (Yazdani et al., 1992). These studies could indicate that caffeine exposure may influence not only zinc but also other minerals in the developing brain.

CAFFEINE AND FETAL ORIGINS OF ADULT DISEASE

Proper maternal nutrition during pregnancy is critical for the fetus to grow and develop physically and mentally to its full potential (Anderson, 2001), and fetal nutrition is widely known to play an important role in the future good health of the individual. However, exposure to an adverse environment *in utero* can lead to an increased risk of adult disease (Morley and Dwyer, 2001; Cooper et al., 2002), and "programmed" changes in physiology and metabolism can lead to a number of diseases in later life (Barker, 1995). Evidence indicates that caffeine consumption during critical periods may lead to disease in later life. For example, offspring from dams that received caffeine (3.5 mg/100 g BW) in drinking water during pregnancy developed significantly more frequent and more severe gastric lesions than did offspring from the control group at 200 days of age (Glavin and Krueger, 1985). Recently, we have shown that caffeine exposure during pregnancy caused a change in angiotensin II type 2 receptor gene expression in the placenta of pregnant rats that were fed a diet supplemented with caffeine, suggesting that caffeine intake alters gene expression of the developing organ in the early stages of life. This alteration may cause certain diseases that we are not aware of at the present time in the later life of the offspring (Tanuma et al., 2003). Chronic caffeine exposure for human fetal neurodevelopment during the critical growth period could result in disease and modified behavior in later life, as it does in animal studies (Hughes and Beveridge, 1987, 1990; Nakamoto et al., 1991). Thus, it may be critical to assess the early effect of caffeine in later life.

Certain changes in the brain have occurred in animal studies (Yazdani et al., 1990) due to an amount of caffeine comparable to that in two cups of coffee. Gestation is a critical period of growth for the CNS (Rodier, 1980). Therefore, it is not unimaginable that the injury that a nutritional factor such as caffeine consumption can cause during this critical period could result in far-reaching effects in the later years, and even that many adult diseases may have a fetal origin (Barker, 1995). However, as of now, we may not be aware of these effects because we have thus far not studied this area in depth.

Not many decades ago, we did not imagine how cigarette smoking (Wakschlag et al., 2002) or alcohol consumption (West et al., 1994; Bookstein et al., 2001) during pregnancy or the early neonatal periods could influence growing offspring and their behavioral development, and caffeine consumption is much more common than cigarette smoking or alcohol consumption in the general population. As animal data show, it is quite likely that caffeine consumption during the critical periods could affect human fetuses and lead to disease in later years.

CONCLUSION

Although the application of animal studies to humans in general has to be done carefully (Bonati et al., 1984), the results of animal studies strongly indicate that routine daily caffeine consumption during the gestational and early neonatal periods can have significant effects. Some believe that moderate caffeine consumption poses no measurable consequences for the fetus and newborn infant (Nehlig and Debry, 1994a,b), but the studies discussed in this chapter may suggest that caffeine exposure presents certain dangers to neurodevelopment during the critical growth period.

Although the deleterious effects of caffeine exposure during the critical growth period seem to be obvious, the long-term effects of exposure during this period of human growth are currently not known. Because the public has unlimited access to caffeine-containing foods, drinks, and over-the-counter drugs, they should be more informed about the danger of caffeine's effects during the gestational and early neonatal periods, just as they are informed about the dangers of alcohol and tobacco, two substances that they consume less of than caffeine. Unfortunately, many assume caffeine to be a relatively safe substance.

I propose that serious consideration be given to the concept of the interrelationship of caffeine consumption, fetal and neonatal development, and possible disease development in the later years. Many diseases in the later years of human life are believed to originate in early fetal life (Barker, 1995; Morley and Dwyer, 2001; Cooper et al., 2002). If this is so, it would not be surprising to learn that caffeine exposure in early life is responsible for various diseases and/or behaviors we are currently not aware of. In fact, the increased susceptibility to gastric lesions in later life has been demonstrated in animals exposed to caffeine during pregnancy (Glavin and Krueger, 1985), and SIDS could possibly be related to heavy maternal caffeine consumption (Ford et al., 1998). In general, numerous animal studies have shown permanent behavioral changes long after caffeine exposure had ended.

The Food and Drug Administration (FDA) has advised pregnant women to avoid caffeine-containing drinks (Goyan, 1980). Given evidence of the adverse effects of caffeine consumption, the FDA on an ongoing basis should strongly discourage the use of all caffeine-containing substances by pregnant or lactating women and by those who plan to have children until caffeine's effects on humans during these critical periods become further clarified.

ACKNOWLEDGMENT

The author wishes to acknowledge the assistance provided by M. Higgins, editorial consultant, E. Strother, librarian, and S. Ryan, who typed the manuscript.

REFERENCES

Aden, U., Herlenius, E., Tang, L.Q. and Fredholm, B.B. (2000) Maternal caffeine intake has minor effects on adenosine receptor ontogeny in the rat brain. *Pediatric Research*, 48, 177–183.

Alm, B., Wennergren, G., Norvenius, G., Skjaerven, R., Oyen, N., Helweg-Larsen, K. et al. (1999) Caffeine and alcohol as risk factors for sudden infant death syndrome. *Archives of Disease in Childhood*, 81, 107–111.

Anderson, A.S. (2001) Pregnancy as a time for dietary change? *Proceedings of the Nutrition Society*, 60, 497–504.

Aranda, J.V., Collinge, J.M., Zinman, R. and Watts, G. (1979) Maturation of caffeine elimination in infancy. *Archives of Disease in Childhood*, 54, 946–949.

Barker, D.J.P. (1995) The Welcome Foundation lecture, 1994. The fetal origins of adult disease. *Proceedings of the Royal Society of London Series B containing papers of a Biological Character*, 262, 37–43.

Barr, H.M. and Streissguth, A.P. (1991) Caffeine use during pregnancy and child outcome: a 7-year prospective study. *Neurotoxicology and Teratology*, 13, 441–448.

Bonati, M. and Garattini, S. (1984) Interspecies comparison of caffeine disposition, in *Caffeine*, Davis, P.B., Ed., Springer-Verlag, New York, p.50.

Bonati, M., Latini, R., Tognoni, G., Young, J.F. and Garattini, S. (1984) Interspecies comparison of *in vivo* caffeine pharmacokinetics in man, monkey, rabbit, rat, and mouse. *Drug Metabolism Reviews*, 15, 1355–1383.

Bookstein, F.L., Sampson, P.D., Streissguth, A.P. and Connor, P.D. (2001) Geometric morphometrics of corpus callosum and subcortical structures in the fetal-alcohol-affected brain. *Teratology*, 64, 4–32.

Brazier, J.L., Ritter, J., Berland, M., Kheufer, D. and Faucon, G. (1983) Pharmacokinetics of caffeine during and after pregnancy. *Developmental Pharmacology and Therapeutics*, 6, 315–322.

Chase, H.P., Kumar, V., Caldwell, R.T. and O'Brien, D. (1980) Kwashiorkor in the United States. *Pediatrics*, 66, 972–976.

Concannon, J.T., Braughler, J.M. and Schechter, M.D. (1983) Pre- and postnatal effects of caffeine on brain biogenic amines, cyclic nucleotides and behavior in developing rats. *The Journal of Pharmacology and Experimental Therapeutics*, 226, 673–679.

Cooper, C., Javaid, M.K., Taylor, P., Walker-Bone, K., Dennison, E. and Arden, N. (2002) The fetal origins of osteoporotic fracture. *Calcified Tissue International*, 70, 391–394.

Daval, J.L. and Vert, P. (1986) Effect of chronic exposure to methylxanthines on diazepam cerebral binding in female rats and their offsprings. *Developmental Brain Research*, 27, 175–180.

Diamond, J.P. (1983) Coffee drinking and U.S. lifestyles. *Tea & Coffee Trade Journal*, 155, 30–56.

Dobbing, J. (1968) Effects of experimental undernutrition on development of the nervous system, in *Malnutrition, Learning, and Behavior*, Scrimshaw, N.S. and Gordon, J.E., Eds., The MIT Press, Cambridge, MA, pp. 181–202.

Dobbing, J. and Sands, J. (1973) The quantitative growth and development of the human brain. *Archives of Disease in Childhood*, 48, 756–767.

Enslen, M., Milton, H. and Wurzner, H.P. (1980) Brain catecholamines and sleep states in offspring of caffeine-treated rats. *Experientia*, 36, 1105–1106.

Etzel, B.A. and Guillet, R. (1994) Effects of neonatal exposure to caffeine on adenosine A_1 receptor ontogeny using autoradiography. *Developmental Brain Research*, 82, 223–230.

Fabro, S. and Siever, S.M. (1969) Caffeine and nicotine penetrate the pre-implantation blastocyst. *Nature*, 233, 410–411.

Febsterm, K., Eskenazi, B., Wundham, G.C. and Swan, S.H. (1991) Caffeine consumption during pregnancy and fetal growth. *American Journal of Public Health*, 81, 458–461.

Fisher, S. and Guillet, R. (1997) Neonatal caffeine alters passive avoidance retention in rats in an age- and gender-related manner. *Developmental Brain Research*, 98, 145–149.

Ford, R.P.K., Schluter, P.J., Mitchell, E.A., Taylor, B.J., Scragg, R., Stewart, A.W. et al. (1998) Heavy caffeine intake in pregnancy and sudden infant death syndrome. *Archives of Disease in Childhood*, 78, 9–13.

Fuller, G.N. and Wiggins, R.C. (1981) A possible effect of the methylxanthines caffeine, theophylline and aminophylline on postnatal myelination of the rat brain. *Brain Research*, 213, 476–480.

Fuller, G.N., Divakaran, P. and Wiggins, R.C. (1982) The effect of postnatal caffeine administration on brain myelination. *Brain Research*, 249, 189–191.

Furuhashi, N., Sato, S., Suzuki, M., Hiruta, M., Tanaka, M. and Takahashi, T. (1985) Effects of caffeine ingestion during pregnancy. *Gynecologic and Obstetric Investigation*, 19, 187–191.

Galli, C., Spano, P.F. and Szyszka, K. (1975) Accumulation of caffeine and its metabolites in rat fetal brain and liver. *Pharmacological Research Communications*, 7, 217–221.

Gilbert, E.F. and Pistey, W.R. (1973) Effect on the offspring of repeated caffeine administration to pregnant rats. *Journal of Reproduction and Fertility*, 34, 495–499.

Gilbert, R.M. (1984) Caffeine consumption, in *The Methylxanthine Beverages and Foods: Chemistry, Consumption and Health Effects*, Spiller, G.A., Ed., Alan R. Liss, New York, pp. 185–213.

Glavin, G.B. and Krueger, H. (1985) Effects of prenatal caffeine administration on offspring mortality, open-field behavior and adult gastric ulcer susceptibility. *Neurobehavioral Toxicology and Teratology*, 7, 29–32.

Goyan, J.E. (1980) Food and Drug Administration, news release No. P80-36. Food and Drug Administration, Washington, DC.

Graham, D.M. (1978) Caffeine: its identity, dietary source, intake and biological effects. *Nutrition Reviews*, 36, 97–102.

Gressens, P., Mesples, B., Sahir, N., Marret, S. and Sola, A. (2001) Environmental factors and disturbances of brain development. *Seminars of Neonatology*, 6, 185–194.

Groisser, D.S., Rosso, P. and Winick, M. (1982) Coffee consumption during pregnancy: subsequent behavioral abnormalities of the offspring. *Journal of Nutrition*, 112, 829–832.

Grossman, E.M. (1984) Some methodological issues in the conduct of caffeine research. *Food and Chemical Toxicology*, 22, 245–249.

Guillet, R. and Dunham, L. (1995) Neonatal caffeine exposure and seizure susceptibility in adult rats. *Epilepsia*, 36, 743–749.

Guillet, R. and Kellogg, C. (1991a) Neonatal exposure to therapeutic caffeine alters the ontogeny of adenosine A1 receptors in brain of rats. *Neuropharmacology*, 30, 489–496.

Guillet, R. and Kellogg, C.K. (1991b) Neonatal caffeine exposure alters developmental sensitivity to adenosine receptor ligands. *Pharmacology, Biochemistry and Behavior*, 40, 811–817.

Gullberg, E.I., Ferrell, F. and Christensen, H.D. (1986) Effects of postnatal caffeine exposure through dam's milk upon weanling rats. *Pharmacology, Biochemistry and Behavior*, 24, 1695–1701.

Herlenius, E., Aden, U., Tang, L.Q. and Lagercrantz, H. (2002) Perinatal respiratory control and its modulation by adenosine and caffeine in the rat. *Pediatric Research*, 51, 4–12.

Hironishi, M., Ueyama, E. and Senba, E. (1999) Systematic expression of immediate early genes and intensive astrocyte activation induced by intrastriatal ferrous iron injection. *Brain Research*, 828, 145–153.

Holloway, W.R., Jr. (1982) Caffeine: effects of acute and chronic exposure on the behavior of neonatal rats. *Neurobehavioral Toxicology and Teratology*, 4, 21–32.

Hughes, R.N. and Beveridge, I.J. (1987) Effects of prenatal exposure to chronic caffeine on locomotor and emotional behavior. *Psychobiology*, 15, 179–185.

Hughes, R.N. and Beveridge, I.J. (1990) Sex- and age-dependent effects of prenatal exposure to caffeine on open-field behavior, emergence latency and adrenal weights in rats. *Life Sciences*, 47, 2075–2088.

Hughes, R.N. and Beveridge, I.J. (1991) Behavioral effects of exposure to caffeine during gestation, lactation or both. *Neurotoxicology and Teratology*, 13, 641–647.

Jacombs, A., Ryan, J., Loupis, A. and Pollard, I. (1999) Maternal caffeine consumption during pregnancy does not affect preimplantation development but delays early postimplantation growth in rat embryos. *Reproduction, Fertility and Development*, 11, 211–218.

Jiritano, L., Bortolotti, A., Gaspari, F. and Bonati, M. (1985) Caffeine disposition after oral administration to pregnant rats. *Xenobiotica*, 15, 1045–1051.

Johansson, B., Georgiev, V. and Fredholm, B.B. (1997) Distribution and postnatal ontogeny of adenosine A_{2A} receptors in rat brain: comparison with dopamine receptors. *Neuroscience*, 80, 1187–1207.

Kimmel, C.A., Kimmel, G.L., White, C.G., Grafton, T.F., Young, J.F. and Nelson, C.J. (1984) Blood flow changes and conceptual development in pregnant rats in response to caffeine. *Fundamental and Applied Toxicology*, 4, 240–247.

Kirkinen, P., Jouppila, P., Koivula, A., Vuori, J. and Puukka, K. (1983) The effect of caffeine on placental and fetal blood flow in human pregnancy. *American Journal of Obstetrics and Gynecology*, 147, 939–942.

Kitts, D.D., Scaman, C.H. and Shekhtman, K. (1986) Caffeine metabolism and its disposition in the pregnant rat and fetus. *Canadian Institute of Food Science and Technology Journal*, 19 (Abstr.), XL.

Kleiber, M. (1961) Body size and metabolic rate, in *The Fire of Life: An Introduction to Animal Energetics*, John Wiley & Sons, New York, pp. 177–216.

Knutti, R., Rothweiler, H. and Schlatter, C. (1981) Effects of pregnancy on the pharmacokinetics of caffeine. *European Journal of Clinical Pharmacology*, 21, 121–126.

Kusin, J.A., Kardjati, S. and Renqvist, U.H. (1993) Chronic undernutrition in pregnancy and lactation. *Proceedings of the Nutrition Society*, 52, 19–28.

Latini, R., Bonati, M., Marzi, E., Tacconi, M.T., Sadurska, B. and Bizzi, A. (1980) Caffeine disposition and effects in young and one-year-old rats. *Journal of Pharmacology*, 32, 596–599.

Leal, M., Barletta, M. and Carson, S. (1990) Maternal-fetal electrocardiographic effects and pharmacokinetics after an acute IV administration of caffeine to the pregnant rat. *Reproductive Toxicology*, 4, 105–112.

Lee, T.C., Charles, B., Steer, P., Flenady, V. and Shearman, A. (1997) Population pharmacokinetics of intravenous caffeine in neonates with apnea of prematurity. *Clinical Pharmacology and Therapeutics*, 61, 628–640.

Leon, D., Albasanz, J.L., Ruiz, M.A., Fernandez, M. and Martin, M. (2002) Adenosine A_1 receptor down-regulation in mothers and fetal brain after caffeine and theophylline treatments to pregnant rats. *Journal of Neurochemistry*, 82, 625–634.

Listernick, R., Christoffel, K., Pace, J. and Chiaramonte, J. (1985) Severe primary malnutrition in US children. *The American Journal of Diseases of Children*, 139, 1157–1160.

Marret, S., Delpech, B., Girard, N., Leroy, A., Maingonnat, C., Menard, J-F. et al. (1993) Caffeine decreases glial cell number and increases hyaluronan secretion in newborn rat brain cultures. *Pediatric Research*, 34, 716–719.

Marret, S., Gressens, P., Van-Maele-Fabry, G., Picard, J. and Evrard, P. (1997) Caffeine-induced disturbances of early neurogenesis in whole mouse embryo cultures. *Brain Research*, 773, 213–216.

Martin, T.R. and Bracken, M.B. (1987) The association between low birth weight and caffeine consumption during pregnancy. *American Journal of Epidemiology*, 120, 813–821.

Massey, L.K. (1991) Caffeine and bone: directions in research. *Journal of Bone and Mineral Research*, 6, 1149–1151.

Mills, J.L., Holmes, L.B., Aarons, J.H., Simpson, J.L., Brown, Z.A., Jovanovic-Peterson, L.G. et al. (1993) Moderate caffeine use and the risk of spontaneous abortion and intrauterine growth retardation. *Journal of American Medical Association*, 269, 593–597.

Morck, T.A., Lynch, S.R. and Cook, J.D. (1983) Inhibition of food iron absorption by coffee. *The American Journal of Clinical Nutrition*, 37, 416–420.

Mori, M., Wilber, J.F. and Nakamoto, T. (1983) Influences of maternal caffeine on the neonatal rat brains vary with the nutritional states. *Life Sciences*, 33, 2091–2095.

Mori, M., Wilber, J.F. and Nakamoto, T. (1984) Protein-energy malnutrition during pregnancy alters caffeine's effect on brain tissue of neonate rats. *Life Sciences*, 35, 2553–2560.

Morley, R. and Dwyer, T. (2001) Fetal origins of adult disease? *Clinical and Experimental Pharmacology and Physiology*, 28, 962–966.

Munoz, L.M., Lonnerdal, B., Keen, C.L. and Dewy, K.G. (1988) Coffee consumption as a factor in iron deficiency anemia among pregnant women and their infants in Costa Rica. *The American Journal of Clinical Nutrition*, 48, 645–651.

Nakamoto, T. and Joseph, F., Jr. (1991) Interaction between caffeine and zinc on brain in newborn rats. *Biology of the Neonate*, 60, 118–126.

Nakamoto, T., Hartman, A.D., Miller, H.I., Temples, T.E. and Quinby, G.E. (1986) Chronic caffeine intake by rat dams during gestation and lactation affects various parts of the neonatal brain. *Biology of the Neonate*, 49, 277–283.

Nakamoto, T., Joseph, F., Jr., Yazdani, M. and Hartman, A.D. (1988) The effects of different levels of caffeine supplemented to the maternal diet on the brains of newborn rats and their dams. *Toxicology Letters*, 44, 167–175.

Nakamoto, T., Hartman, A.D. and Joseph, F., Jr. (1989) Interaction between caffeine intake and nutritional status on growing brains in newborn rats. *Annals of Nutrition and Metabolism*, 33, 92–99.

Nakamoto, T., Roy, G., Gottschalk, S.B., Yazdani, M. and Rossowska, M. (1991) Lasting effects of early chronic caffeine feeding on rats' behavior and brain in later life. *Physiology and Behavior*, 49, 721–727.

Nakazawa, K., Tanaka, H. and Arima, M. (1985) The effect of caffeine ingestion on pharmacokinetics of caffeine and its metabolites after a single administration in pregnant rats. *Journal of Pharmacobio-Dynamics*, 8, 151–160.

Nehlig, A. and Debry, G. (1994a) Consequences on the newborn of chronic maternal consumption of coffee during gestation and lactation: a review. *Journal of the American College of Nutrition*, 13, 6–21.

Nehlig, A. and Debry, G. (1994b) Potential teratogenic and neurodevelopmental consequences of coffee and caffeine exposure: a review on human and animal data. *Neurotoxicology and Teratology*, 16, 531–543.

Oser, B.L. and Ford, R.A. (1981) Caffeine: an update. *Drug and Chemical Toxicology*, 4, 311–329.

Osofsky, H.J. (1975) Relationship between nutrition during pregnancy and subsequent infant and child development. *Obstetrical and Gynecological Survey*, 30, 227–241.

Prasad, A.S. (1988) Zinc in growth and development and spectrum of human zinc deficiency. *Journal of American College of Nutrition*, 7, 377–384.

Quinby, G.E., Batirbaygil, Y., Hartman, A.D. and Nakamoto, T. (1985) Effects of orally administered caffeine on cellular response in protein-energy malnourished neonatal rat brain. *Pediatric Research*, 19, 71–74.

Rodier, P.M. (1980) Chronology of neuron development: animal studies and their clinical implications. *Developmental Medicine and Child Neurology*, 22, 525–545.

Romijn, H.J., Hofman, M.A. and Gramsbergen, A. (1991) At what age is the developing cerebral cortex of the rat comparable to that of the full-term newborn human baby? *Early Human Development*, 26, 61–67.

Sahir, N., Bahi, N., Evrard, P. and Gressens, P. (2000) Caffeine induces *in vivo* premature appearance of telencephalic vesicles. *Developmental Brain Research*, 121, 213–217.

Sahir, N., Mas, C., Bourgeois, F., Simonneau, M., Evrard, P. and Gressens, P. (2001) Caffeine-induced telencephalic vesicle evagination in early post-implantation mouse embryos involves cAMP-dependent protein kinase (PKA) inhibition. *Cerebral Cortex*, 11, 343–349.

Saito, T., Narayanan, C.H., Joseph, F., Jr., Yoshino, S. and Nakamoto, T. (1995) Combined effects of caffeine and malnutrition on the development of the trigeminal nuclear center: autoradiographic and biochemical studies. *Physiology and Behavior*, 58, 769–774.

Smith, S.E., McElhatton, P.R. and Sullivan, F.M. (1987) Effects of administering caffeine to pregnant rats either as a single daily dose or as divided doses four times a day. *Food and Chemical Toxicology*, 25, 125–133.

Snyder, S.H., Katims, J.J., Annau, Z., Bruns, R.F. and Daly, J.W. (1981) Adenosine receptors and the behavioral actions of methylxanthines. *Proceedings of the National Academy of Sciences of the U.S.A.* 78, 3260–3264.

Sobotka, T.J., Spaid, S.L. and Brodie, R.E. (1979) Neurobehavioral teratology of caffeine exposure in rats. *Neurotoxicology*, 1, 403–416.

Tanaka, H., Nakazawa, K. and Arima, M. (1983) Adverse effect of maternal caffeine ingestion on fetal cerebrum in rat. *Brain and Development*, 5, 397–406.

Tanaka, H., Nakazawa, K., Arima, M. and Iwasaki, S. (1984) Caffeine and its dimethylxanthines and fetal cerebral development in rat. *Brain and Development*, 6, 355–361.

Tanaka, T. and Nakazawa, K. (1990) Maternal caffeine ingestion increases the tyrosine level in neonatal rat cerebrum. *Biology of the Neonate*, 57, 133–139.

Tanaka, T., Nakazawa, K. and Arima, M. (1987) Effects of maternal caffeine ingestion on the perinatal cerebrum. *Biology of the Neonate*, 51, 332–339.

Tanuma, A., Saito, S., Ide, I., Sasahara, H., Yazdani, M., Gottschalk, S. et al. (2003) Caffeine enhances the expression of the angiotensin II type 2 receptor mRNA in BeWo cell culture and in the rat placenta. *Placenta*, 24, 638–647.

Tye, K., Pollard, I., Karlsson, L., Scheibner, V. and Tye, G. (1993) Caffeine exposure in utero increases the incidence of apnea in adult rats. *Reproductive Toxicology*, 7, 449–452.

Tyrala, E.E. and Dodson, W.E. (1979) Caffeine secretion into breast milk. *Archives of Disease in Childhood*, 54, 787–789.

Varma, D.R. (1981) Protein deficiency and drug interactions: a Review. *Drug Development Research*, 1, 183–198.

Volpe, J.J. (1981) Effects of methylxanthines on lipid synthesis in developing neural systems. *Seminars in Perinatology*, 5, 395–405.

Wakschlag, L.S., Pickett, K.E., Cook, E., Benewitz, N.L. and Leventhal, B.L. (2002) Maternal smoking during pregnancy and severe antisocial behavior in offspring: a review. *American Journal of Public Health*, 92, 966–974.

Weaver, D.R. (1996) A_1-adenosine receptor gene expression in fetal rat brain. *Developmental Brain Research*, 94, 205–223.

Weidner, G. and Istvan, J. (1985) Dietary source of caffeine. *New England Journal of Medicine*, 313, 1421.

West, J.R., Chen, W.-J.A. and Pantazis, N.J. (1994) Fetal alcohol syndrome: the vulnerability of the developing brain and possible mechanisms of damage. *Metabolic Brain Disease*, 9, 291–322.

Wilcox, A., Weinberg, C. and Baird, D. (1988) Caffeinated beverages and decreased fertility. *Lancet*, December 24/31, 1453–1455.

Wilkinson, J.M. and Pollard, I. (1993) Accumulation of theophylline, theobromine and paraxanthine in the fetal rat brain following a single oral dose of caffeine. *Developmental Brain Research*, 75, 193–199.

Wilkinson, J.M. and Pollard, I. (1994) In utero exposure to caffeine causes delayed neural tube closure in rat embryos. *Teratogenesis, Carcinogenesis, and Mutagenesis*, 14, 205–211.

Yazdani, M., Fontenot, F., Gottschalk, S.B., Kanemaru, Y., Joseph, F., Jr. and Nakamoto, T. (1992) Relationship of prenatal caffeine exposure and zinc supplementation on fetal brain development. *Developmental Pharmacology and Therapeutics*, 18, 108–115.

Yazdani, M., Joseph, F., Jr., Grant, S., Hartman, A.D. and Nakamoto, T. (1990) Various levels of maternal caffeine ingestion during gestation affect biochemical parameters of fetal rat brain differently. *Developmental Pharmacology and Therapeutics*, 14, 52–61.

Yazdani, M., Hartman, A.D., Miller, H.I., Temples, T.E. and Nakamoto, T. (1988a) Chronic caffeine intake alters the composition of various parts of the brain in young growing rats. *Developmental Pharmacology and Therapeutics*, 11, 102–108.

Yazdani, M., Tran, T.H., Conley, P.M., Laurent, J. and Nakamoto, T. (1988b) The effect of protein malnutrition and maternal caffeine intake on the growth of fetal rat brain. *Biology of the Neonate*, 52, 86–92.

Yeh, J.K., Aloia, J.F., Semla, H.M. and Chen, S.Y. (1986) Influence of injected caffeine on the metabolism of calcium and the retention and excretion of sodium, potassium, magnesium, zinc and copper in rats. *Journal of Nutrition*, 116, 273–280.

Yoshino, S., Narayanan, C.H., Joseph, F., Jr., Saito, T. and Nakamoto, T. (1994) Combined effects of caffeine and malnutrition during pregnancy on suckling behavior in newborn rats. *Physiology and Behavior*, 56, 31–37.

Zimmerberg, B., Carr, K.L., Scott, A., Lee, H.H. and Weider, J.M. (1991) The effects of postnatal caffeine exposure on growth, activity and learning in rats. *Pharmacology, Biochemistry and Behavior*, 39, 883–888.

8 Caffeine's Effects on the Human Stress Axis

Mustafa al'Absi and William R. Lovallo

CONTENTS

Introduction 113
Caffeine Acts by Blocking Adenosine Receptors 114
    Effects of Caffeine on the HPA Axis during Rest 114
Caffeine and Catecholamines 116
Caffeine's Cardiovascular Effects during Rest 117
    Caffeine's Effects on Blood Pressure 117
    Caffeine's Effects on the Heart 118
Caffeine and HPA Responses to Stress 119
Caffeine's Effects on Cardiovascular Responses during Stress 119
Mental and Cognitive Stressors 120
Exercise, Psychomotor Stress, and Cold Pressor Challenge 120
Workplace and Related Settings outside the Laboratory 121
Individual Differences in the Effects of Caffeine 122
Other Individual Differences 124
Concluding Remarks and Future Directions 124
References 124

INTRODUCTION

Caffeine is one of the world's most widely used drugs. Surveys in the U.S. have shown that approximately 80% of adults regularly consume caffeine, averaging two to three cups of coffee a day (Bonham and Leaverton, 1979). Caffeine activates the central nervous system (CNS) (Rall, 1980; Nehlig et al., 1992), leading to behavioral, autonomic, and endocrine responses. Caffeine's effects on peripheral functions are widespread and are mediated by direct tissue effects along with hormonal and autonomic outputs. Caffeine increases circulating catecholamines and free fatty acids (Robertson et al., 1978; Pincomb et al., 1988). It also increases blood pressure (BP), both at rest and during behavioral stress, and in the lab and the workplace (Lane and Williams, 1985; Sung et al., 1990; James, 1993). The combined BP effect of caffeine plus stress is usually additive (Lane and Williams, 1985; Lovallo et al., 1991; Shepard et al., 2000). Limited evidence from animal models suggests that caffeine may have pathogenic effects when combined with stress (Henry and Stephens, 1980). Such evidence in humans is lacking, although evidence suggests that persons at risk for hypertension may have greater cardiovascular responses and stress endocrine changes when exposed to stress following ingestion of caffeine, both in the lab and in daily life (Shepard et al., 2000). Caffeine interacts with CNS functions, it elevates stress hormone secretion, and it has direct effects on the heart and blood vessels. These points of action suggest that caffeine may be capable of influencing stress responses. In this chapter, we review evidence of caffeine's effects on adrenal and sympathetic functions during rest and in response to acute stressful challenges.

CAFFEINE ACTS BY BLOCKING ADENOSINE RECEPTORS

Normal dietary intake of caffeine ranges from approximately 50 to 1500 mg per day, corresponding to consumption of one caffeinated soft drink up to 15 cups of coffee per day. In the U.S., the average reported adult intake is 250 mg per day, or about two to three cups of coffee. At these levels of intake, caffeine's physiological actions are due to competitive blockade of adenosine receptors (Fredholm, 1980, 1995; Smits et al., 1990; Fredholm et al., 1999). Adenosine is produced by all tissues as a function of the breakdown of adenosine triphosphate during cellular metabolism and neurotransmission (Hoyle, 1992; Takiyyuddin et al., 1994; Johnson et al., 2001), and all cells have receptors for adenosine. The widespread effects of adenosine therefore account for caffeine's broad spectrum of effects. In the CNS, adenosine is a putative physiological sleep factor that mediates the somnogenic effects of prior wakefulness. The duration and depth of sleep after periods of wakefulness appear to be profoundly modulated by elevated concentrations of adenosine in areas responsible for generalized arousal, and its gradual disappearance may account for the restorative effects of sleep (Porkka-Heiskanen et al., 1997).

Two actions of adenosine underlie caffeine's cardiovascular and endocrine effects in humans. First, adenosine acts on potassium channels to hyperpolarize cell membranes of neurons, vascular smooth muscle, and cardiac muscle (Belardinelli, et al., 1989, 1995; Suzuki et al., 2001). This membrane effect of adenosine causes reduced rates of neuronal transmission and lowered responses of the heart and blood vessels. Second, adenosine acts presynaptically to decrease rates of transmitter release in the central and autonomic nervous systems (Fredholm and Dunwiddie, 1988). This reduces sympathetic outflow to the heart, blood vessels, and adrenal medulla (Shinozuka et al., 2002). These effects lead to the general conclusion that adenosine acts to modulate the rate of neuronal firing and activation of target tissues.

Caffeine's competitive antagonism of adenosine is therefore responsible for its ability to increase activation of the central and autonomic nervous systems. This suggests that caffeine can have effects both at rest and during periods of stress. Indeed, caffeine is consumed in greater quantities by employees during times of increased work stress (Conway et al., 1979), raising the possibility that caffeine consumption often accompanies periods of mental stress in daily life. We know that persons vary in their responses to stress (al'Absi et al., 1997), and the possibility presents itself that caffeine has greater effects in some persons than in others.

Effects of Caffeine on the HPA Axis during Rest

Among the most important of caffeine's acute effects is its ability to alter the activity of the hypothalamic-pituitary-adrenocortical (HPA) axis. The HPA axis affects all tissues and participates in functions occurring at rest and during stress. The HPA cascade is initiated at the hypothalamic paraventricular nucleus (PVN), resulting in release of corticotropin-releasing hormone (CRH) by terminals at the median eminence. CRH enters the portal circulation to the anterior pituitary, where it stimulates cleavage of pro-opiomelanocortin (POMC), causing release of adrenocorticotropic hormone (ACTH) and beta-endorphin into the general circulation. Arginine vasopressin (AVP) is also secreted at the median eminence, especially during stress, and together with CRH markedly potentiates ACTH secretion. ACTH increases the synthesis and release of cortisol by the adrenal cortex. Cortisol in turn exerts negative feedback on the HPA at the level of the pituitary and hypothalamus (Lovallo and Thomas, 2000).

Cortisol exerts numerous central and peripheral effects, including liberation of glucose by the liver and increasing plasma volume by causing a shift of fluid from intracellular to extracellular compartments and aiding in nutrient transport (Wilson and Foster, 1992; Vander et al., 1994). Cortisol participates indirectly in sympathetic nervous system or SNS function by increasing rates of catecholamine synthesis and production of adrenergic receptors. Cortisol also crosses the blood-brain barrier to reach the CNS, where it exerts negative feedback effects on the HPA axis. It also acts at sites where CRH receptors are found, including the anterior cingulate gyrus, prefrontal cortex, hippocampus, and amygdala (Fink et al., 1988; Swanson and Simmons, 1989; Diorio et al., 1993).

Experiments in rats show that caffeine stimulates production of cortisol/corticosterone, ACTH, and beta-endorphin (Vernikos-Danellis and Harris, 1968; Arnold et al., 1982; Spindel et al., 1984). In humans, we and others have noted that caffeine may elevate cortisol and ACTH production at rest (Arnold et al., 1982; Lane et al., 1990; al'Absi et al., 1995; Lovallo et al., 1996). These results suggest that caffeine's adrenocortical actions in humans originate at the CNS. The effects of caffeine on ACTH and cortisol during rest were found to be comparable to that produced by acute stress (Lovallo et al., 1996). This suggests that caffeine alone may evoke an adrenocortical stress response during rest and in the absence of explicit stressful challenge. Such effects may counteract the effects of medications that normally suppress pituitary secretion. For example, caffeine increases the rate of recovery of adrenocortical functions after the administration of prednisolone, which suppresses corticosterone secretion in rats (Marzouk et al., 1991). Similarly, human studies have shown that ACTH and cortisol changes following caffeine ingestion may produce a false positive response to the dexamethasone suppression test, leading to escape in about a third of healthy persons (Uhde et al., 1985).

Most studies that have examined effects of caffeine in humans have used a dose range from 150 to 500 mg of caffeine (e.g., Rall, 1980; Spindel et al., 1984; al'Absi et al., 1998; Lane et al., 1990; Lovallo et al., 1996), or approximately 2 to 8 mg/kg. Low doses of caffeine tend to produce less consistent adrenocortical effects (see Lane et al., 1990; al'Absi et al., 1995). Early studies have shown that an oral dose of 250 mg of caffeine increased serum and urinary cortisol metabolites (Bellet et al., 1969; Avogardo et al., 1973), but other studies failed to show an effect of this dose during rest (e.g., Daubresse et al., 1973; Oberman et al., 1975; al'Absi et al., 1995). Similarly, caffeine increased catecholamine levels in some studies (Levi, 1967; Bellet et al., 1969; Robertson et al., 1978), but not in others (Jung et al., 1981). These inconsistencies have been attributed to the possibility that a low dose of caffeine would not be potent enough to produce consistent adrenocortical effects.

It should be noted here that the dose required to increase adrenocortical changes in rats tended to be high (approximately 20 mg/kg). This dose roughly equates to a 70-kg man drinking 10 to 15 cups of coffee in one sitting. Studies in humans that have used doses greater than 250 mg have provoked more consistent results. Cortisol levels were increased following ingestion of 500 mg of caffeine, but not after 250 mg of caffeine (Spindel et al., 1984). A dose of 500 mg of caffeine also increased beta-endorphin within 60 min of injestion, but no change in beta-endorphin levels were found after 250 mg. The beta-endorphin change also indicates the activation of the anterior pituitary and the cosecretion of ACTH (Guillemin et al., 1977).

Mechanisms that mediate caffeine's stimulatory effects on the HPA axis are not yet fully delineated. It has been suggested that these effects may be due to blockade of the adenosine receptor and to interference with cyclic adenosine monophosphate (AMP) phosphodiesterase at the hypothalamus (Rall, 1980; Nehlig et al., 1992), as discussed elsewhere in this chapter. Each of these can result in increased availability of cAMP, which stimulates expression of the corticotropin-releasing factor (CRF) gene (Snyder et al., 1981). Secretion of CRF at the median eminence of the hypothalamus in turn stimulates ACTH release by the pituitary, resulting in increased cortisol production by the adrenal cortex (Petrusz and Merchenthaler, 1992). Caffeine may therefore increase CRF, ACTH, and cortisol through cAMP accumulation in the median eminence or pituitary, as has been shown in rats (Marzouk et al., 1991). Caffeine may also contribute indirectly to increased ACTH release through stimulation of epinephrine secretion (Neville and O'Hare, 1984), which can enhance cAMP production in the pituitary.

TABLE 8.1
Caffeine's Effects on Catecholamine Release

Findings	Studies
Caffeine increased both epinephrine (E) and norepinephrine (NE)	Bellet et al. (1969); placebo-controlled, randomized open administration of coffee vs. hot water, caffeine: 220 mg, n = 13, E and NE increased to caffeine ($p < .01$)
	Robertson et al. (1978); caffeine (250 mg) or placebo to subjects free of caffeine for 3 wk; increased plasma NE by 75% and E by 207%
	Robertson et al. (1981); repeated dosing ($\times$ 3/d), E and NE increased on d 1 and 2 with tolerance developing by d 4 for both plasma and urine catecholamines
	Smits et al. (1985); placebo-controlled administration of 2 cups regular vs. decaffeinated coffee; E and NE increased in plasma ($p < .05$) relative to levels after placebo
Caffeine increased epinephrine (E)	Debrah et al. (1995); caffeine (250 mg) vs. placebo; plasma E increased ($p < .01$)
	Izzo et al. (1983); caffeine (250 mg) vs. placebo, 3-h increase in plasma E ($p < .05$)
	Lane (1994); caffeine (300 mg) vs. placebo; 37% elevation in urinary E during work
	Lane et al. (2002); caffeine (500 mg: 250 mg $\times$ 2) vs. placebo; 32% elevation in urinary E during work day and evening
	Laurent et al. (2000); caffeine (6 mg/kg) vs. placebo at baseline and at 90 min before cycling exercise test; E increased ($p < .01$)
	Levi (1967); caffeine (225 mg coffee vs. hot water in separate control group); urinary E increased 80% relative to baseline in caffeine group only; NE increased in both caffeine and placebo groups
	Patwardhan et al. (1980); caffeine (250 mg); urinary E increased ($p < .05$)
	Smits et al. (1983); "regular" coffee (2 cups) vs. placebo; plasma E increased (+150%)
	Smits et al. (1985); "regular" vs. "strong" coffee (2 cups) vs. decaf., water, and no intervention; E increased (+86%) to regular and (+140%) strong coffee vs. decaf ($p < .05$)
	Smits et al. (1986); "regular" coffee (2 cups); E increased in normotensives (+257%) and hypertensives (+115%) ($p < .01$)
	Smits et al. (1987); caffeine (250 mg i.v.); plasma E increased (+114%)
	Smits et al. (1989); caffeine (250 mg) vs. placebo; E increased (+92%) ($p < .01$)
	Takiyyuddin et al. (1994); oral caffeine; E release stimulated
Caffeine increased norepinephrine (NE)	Lane et al. (1990); caffeine (3.5 mg/kg) vs. placebo; plasma NE increased ($p < .01$)

CAFFEINE AND CATECHOLAMINES

The first reported study of caffeine's catecholamine effects was on dogs following intravenous administration, with the result that total catecholamine secretion increased (Deschaepdryver, 1959). In a survey of 19 studies on humans receiving dietary doses of caffeine (Table 8.1), 17 reported increases in epinephrine, 4 of which also showed increased norepinephrine. One study showed increased norepinephrine alone, while one found no effect on either catecholamine. The increase in epinephrine has been observed under resting conditions, in the laboratory and at home, and in the workplace during periods of normal work demand. It is seen following different methods of administration, dosages, and specimen sampling. It therefore appears that caffeine's predominant effect on catecholamine secretion is on epinephrine, while changes in norepinephrine appear to be less common and are perhaps related to specific circumstances.

It is noteworthy that the one study reporting no change in either norepinephrine or epinephrine (Onrot et al., 1985) was done on patients with autonomic failure. The lack of both neurosympathetic and sympathoadrenal responses in these autonomic failure patients suggests that caffeine was unable to cause a direct release of epinephrine at the adrenal medulla. It further suggests that when caffeine does act to increase catecholamine output, it depends on an intact sympathetic nervous system to do so. If the sympathetic nervous system is essential for caffeine's effect on catecholamine release, its relative specificity for epinephrine, while not absolute, indicates either a greater sensitivity of adenosine-mediated release of epinephrine than of norepinephrine, or a relative sensitivity due to the prevailing low concentrations of circulating epinephrine under basal conditions (20 to 50 µg/ml) in relation to norepinephrine's higher levels (450 µg/ml). In line with this possibility, a number of the studies reviewed showed some norepinephrine response to caffeine that did not meet usual criteria for statistical significance. The predominance of an epinephrine response also suggests that caffeine exerts a stresslike effect on catecholamine secretion, as it appears to do for cortisol secretion. Although caffeine exerts an acute effect on catecholamine secretion, its dietary effects are less well understood. Daily dosing of volunteers with 750 mg of caffeine was shown to produce at least partial tolerance to the acute effects of caffeine on catecholamine levels (Robertson et al., 1981). The degree of tolerance produced by daily caffeine intake and the doses necessary to produce either complete or partial caffeine tolerance remain unknown (James, 1991).

CAFFEINE'S CARDIOVASCULAR EFFECTS DURING REST

Caffeine's actions on the cardiovascular system are mediated by its competitive blockade of adenosine receptors. At the blood vessel wall, adenosine reduces the contraction of the smooth muscle cell (Suzuki et al., 2001), it enhances release by vascular endothelium of the regional vasodilator, nitric oxide (Buus et al., 2001), and it reduces release of the neurotransmitter norepinephrine (Fredholm and Dunwiddie, 1988). Adenosine therefore reduces vasomotor tone and vascular resistance to blood flow, and caffeine may be expected to oppose these effects.

Caffeine's Effects on Blood Pressure

Caffeine's ability to raise BP is perhaps its most extensively documented physiological effect, with publications dating back to the 1860s. While earlier studies caused uncertainty as to caffeine's actions due to a lack of placebo controls, established dosages, and the availability of reliable BP devices, the earliest reliable report of caffeine's pressor effect is by Wood (1912), who found an increase in BP and a slight fall in heart rate. Horst et al. (1934) confirmed caffeine's pressor effect and slight bradycardia, and these have since been reported by many other investigators (Robertson et al., 1978, 1981; Lane, 1983; Smits et al., 1983, 1985; Whitsett et al., 1980; Pincomb et al., 1985, 1987; James, 1990; Lane et al., 1990; Sung et al., 1990; Lovallo et al., 1991; Shepard et al., 2000), to cite only a few. In addition to studies in the laboratory, caffeine's pressor effect has also been measured outside the laboratory (Pincomb, 1987; Green and Suls, 1996; Shepard et al., 2000).

The BP response to a single dose of caffeine lasts approximately 3 h and is dependent on maintenance of elevated blood levels, declining as caffeine is cleared from the bloodstream (Robertson et al., 1978; Whitsett et al., 1984). In a representative study, 250 mg of caffeine was administered to healthy, caffeine-naïve subjects at rest, leading to BP increases (+14/10 mmHg) at 1 h, while heart rate fell slightly (Robertson et al., 1978). The nonconsuming status of these subjects indicates this may represent the upper limit of the BP response to caffeine among healthy persons. Comparison of BP responses to differing doses of caffeine spanning the range of usual daily intakes (2.2, 4.4, and 8.8 mg/kg, equivalent to about one to six cups of coffee) showed relatively little variation across doses and no tendency for a larger BP response to the highest dose (Whitsett et al., 1980). This suggests that caffeine's effect on BP has a relatively flat dose-response relationship. We have consistently failed to find a significant correlation between caffeine blood or saliva levels and BP responses to acute administration. These findings indicate that caffeine's occupancy of adenosine receptors in regard to BP regulation is complete at somewhere around 2 to 3 mg/kg and that higher doses, and consequently higher blood levels, do not exert a greater effect.

Despite the general understanding that caffeine elevates BP, there was longstanding uncertainty as to the underlying mechanism. It had been proposed that caffeine raised BP by increasing cardiac output (e.g., Rall, 1980). However, there was not good evidence for this view. A drawback to earlier investigations had been the difficulty of making measurements of cardiac output, particularly using reliable, noninvasive approaches that were not stressful for the subject or otherwise obtrusive on the test situation. In our first study (Pincomb et al., 1985), we administered oral doses of placebo vs. caffeine (3.3 mg/kg) to 15 medical students in a randomized, double-blind, crossover study and tracked their responses over 40 min. Along with BP, we measured cardiac output using impedance cardiography (Wilson et al., 1989; Sherwood et al., 1990), which allowed the calculation of total peripheral resistance, and therefore an estimation of the contributions to BP of changed blood flow and vascular resistance at each time point. Compared to placebo values, caffeine showed an increase in peripheral resistance that was statistically significant at 10 min after caffeine intake and increased progressively over the 40-min observation period, with a peak difference of +12% relative to placebo. This change preceded the rise in systolic/diastolic BP (+5%/+9%) and a decline in heart rate (–4.5%). Cardiac output did not change relative to the placebo condition (Pincomb et al., 1985). For this reason, it has been our position that caffeine's principal resting cardiovascular effect is an increase in vascular resistance, which is the cause of the rise in BP. This finding has been replicated using the same and other techniques to measure cardiac output (Pincomb et al., 1993) and by other investigators (Casiglia et al., 1990).

This interpretation is consistent with caffeine's adenosine antagonism. Competitive blockade of adenosine A_1 receptors in the peripheral vasculature could increase the responsiveness of vascular smooth muscle cells to the tonic norepinephrine release by prejunctional sympathetic terminals. Caffeine may also reduce local secretion of the smooth muscle cell relaxant nitric oxide, further contributing to an overall rise in vascular resistance (Freilich and Tepper, 1992). Consistent with this interpretation, adenosine infused into the forearm causes increased blood flow secondary to decreased vascular resistance (Smits et al., 1991a,b), and this effect is antagonized by caffeine (Smits et al., 1990). In our studies, the effect of caffeine on vascular resistance appears before any significant effect on BP or heart rate, suggesting a peripheral vascular effect. However, a 250-mg dose also lowers baroreflex sensitivity for up to 3 h (Mosqueda-Garcia et al., 1990), potentially contributing centrally to a rise in pressure.

CAFFEINE'S EFFECTS ON THE HEART

At the heart, adenosine decreases the firing rate of the sinoatrial node (Belloni et al., 1989), reducing the rate of atrial contraction and lowering heart rate. At the atrioventricular node, caffeine lengthens conduction time to the Hiss bundle (Conti et al., 1995), with no net effect on heart rate. At the ventricular wall, adenosine reduces contractility (Belardinelli et al., 1989), especially in the presence of beta-adrenergic stimulation (Isenberg and Belardinelli, 1984), therefore moderating the effects of myocardial ischemia (Freilich and Tepper, 1992; Sato et al., 1992; Song et al., 2002). Finally, adenosine improves coronary blood flow through its vascular actions. In contrast to the blood vessel, where peripheral factors are at least as important as central sympathetic outflow in determining vascular tonus, the heart at rest is governed more directly by a predominant parasympathetic drive (Berntson et al., 1994), augmented by sympathetic outflow under conditions of fight or flight, accompanied by a substantial responsiveness to circulating epinephrine (Sung et al., 1988). This interplay of dual central inputs combined with peripheral contributions in determining heart rate makes interpreting the present evidence difficult. For example, if adenosine increases the resting interval in the sinoatrial node, caffeine's opposing action would be expected to increase heart rate. For this reason, caffeine's bradycardic effect may be central in origin. At the brain stem, adenosine would be expected to reduce both parasympathetic and sympathetic outflow to the heart, with a greater effect on the parasympathetic branch, due to its greater gain factor at the sinoatrial node (Berntson et al., 1994), resulting in an increase in heart rate. Caffeine blockade of the above effects would therefore cause a net reduction of heart rate under resting conditions. There are no direct tests of this speculative explanation of caffeine's bradycardiac effects.

CAFFEINE AND HPA RESPONSES TO STRESS

Caffeine consumption often increases during times of stress (Conway et al., 1981; Ratliff-Crain and Kane, 1995). Previous experiments in mice have demonstrated pathogenic effects of caffeine when administered while mice were subjected to the chronic stress of living in a competitive social environment (Henry and Stephens, 1980). Caffeine added to drinking water, supplied over several months to rats living in crowded conditions, increased corticosterone, adrenal weight, plasma rennin activity, and BP, leading to significant increases in morbidity and mortality compared with control mice drinking water alone.

We and others have noted that caffeine elevates cortisol production during mental stress (Lane et al., 1990; al'Absi et al., 1995; Lovallo et al., 1996), suggesting important interactions between caffeine's CNS stimulation and endocrine components of the stress response. In a recent study, we evaluated the role of ACTH in the cortisol response to caffeine and stress (al'Absi et al., 1998). We measured ACTH, cortisol, and moods at several time points during placebo vs. caffeine on two resting control days and on two days with a behavioral stressor. During stress, the subjects worked on a reaction time (RT) task in alternation with mental arithmetic (MA). This simulated alternating work on mentally challenging tasks that people may engage in during daily life and that may be accompanied by consumption of caffeine. We found that caffeine and behavioral stress had a combined effect on both ACTH and cortisol (al'Absi et al., 1998).

In the presence of caffeine, behavioral challenges produce significant ACTH and cortisol responses, as documented in studies from our and others' laboratories measuring cortisol alone (Pincomb et al., 1988; Lane et al., 1990; al'Absi et al., 1995, 1998). In some studies neither caffeine nor the acute stress alone produced cortisol responses, although significant elevations occurred when confronting acute challenges in the presence of caffeine (Lane et al., 1990; al'Absi et al., 1995). Consistent with this, Lane et al. (1990) found that caffeine raised cortisol concentrations when combined with mental arithmetic, but not during rest. Other studies have shown that caffeine alone at a similar dose (250 mg) was capable of producing significant ACTH and cortisol rises in the total sample when measured over a longer period of time and when compared with a time-synchronized placebo during a resting control day (Lovallo et al., 1996).

In summary, caffeine enhances the adrenocortical response to behavioral stress in the laboratory and may potentiate cardiovascular responses during mental stress (Pincomb et al., 1988; al'Absi et al., 1995). Caffeine in dietary doses is therefore capable of producing a pattern of responses associated with mental stress and can enhance such responses when individuals confront acute challenge.

CAFFEINE'S EFFECTS ON CARDIOVASCULAR RESPONSES DURING STRESS

In an early animal study, Henry and Stephens (1980) exposed male mice to social crowding and provided caffeine (3.3 mg/kg/d) in their drinking water. These stressful living conditions increased BP, contributed to renal pathology, and increased death rates in the animals. These findings indicate the potential health relevance of studies of caffeine's influence on cardiovascular activity during states of stress in humans. In addressing this issue, three questions arise: Does caffeine act on stress responses in an additive or synergistic fashion? Does it interact differently based on the nature of the stressor? Does it act differentially in certain groups of persons? In the human literature, caffeine has been examined with regard to stressors that can broadly be classified as mental and cognitive, exercise and psychomotor, painful or aversive, and occupational.

MENTAL AND COGNITIVE STRESSORS

A common mental stressor in the psychological laboratory is mental arithmetic, calling for rapid accurate calculations, usually in the presence of an experimenter and often with corrections for errors. This involves significant cognitive effort, and it is a social stressor due to its evaluative nature (Lovallo, 1997). Other cognitive stressors involve demanding tasks such as the Stroop Color-Word Test, which calls for reading a list of color words printed in ink colors that differ from the word itself. The subject is instructed to read the ink color, ignoring the word, leading to a reliable stress response.

In most studies that use mental arithmetic or related stressors, caffeine consumed before stress raised systolic BP, diastolic BP, or both, during the prestress period. During the stressor, the most common finding has been that caffeine's pressure elevation persists, and it appears to be additive with the effects of the stressor. As a result, the BP increase to caffeine combined with the stressor is often nearly identical to their combined effects. Most of the studies in the literature have used placebo-controlled, crossover designs and caffeine doses of 130 to 500 mg, resulting in increased systolic and diastolic BP levels during stress along with decreases in heart rate (Lane, 1983; Goldstein and Shapiro, 1987; Ratliff-Crain et al., 1989; James, 1990; France and Ditto, 1992; Lane et al., 1990). Other studies report either increased diastolic BP alone (France and Ditto, 1988) or increased systolic BP, when diastolic BP was not measured (Greenberg and Shapiro, 1987; Myers et al., 1989).

Although these studies show a relatively uniform additive effect of caffeine on BP during stress, there are reports of nonadditive effects. One study found a smaller diastolic BP increase to mental arithmetic stress after caffeine relative to placebo (Lane and Williams, 1985). All other examples of nonadditive effects indicate a response enhancement by caffeine. MacDougall et al. (1988) found strongly enhanced effects of mental arithmetic and video game stress on heart rate and systolic, but not diastolic, BP following caffeine. France and Ditto (1992) also reported greater heart rate responses to MA after caffeine intake. In most studies of heart rate in relation to caffeine exposure, the effect of caffeine was to lower heart rate, and this effect was generally found to persist during stress. The studies reporting nonadditive effects of caffeine and stress have usually reported an enhanced heart rate response to stress during caffeine exposure, with occasional effects on systolic BP. One report that caffeine increased the response to stress during electric shock (Hasenfratz and Battig, 1992) suggests that aversive challenges may interact differentially with caffeine, particularly in relation to beta-adrenergic responses associated with anxiety or fear.

EXERCISE, PSYCHOMOTOR STRESS, AND COLD PRESSOR CHALLENGE

Exercise has different characteristics than mental stress. Dynamic exercise increases heart rate and cardiac output and it induces a degree of vasodilation in the exercising muscle (Lovallo, 1997). In a related fashion, psychomotor challenges, typically reaction time tasks, call for rapid and accurate motor responses and appear to engage a similar response pattern, probably through induction of exercise-related central command mechanisms (Hobbs, 1982). Although both forms of stress have some common central characteristics, in dynamic exercise, centrally induced changes and peripheral ones combine to produce the global hemodynamic response.

Studies of dynamic exercise have shown caffeine's pressor effect and modest heart rate decrease to be largely additive with the effects of exercise, such that BP is higher and heart rate the same or lower during exercise after caffeine is consumed. When caffeine (3.3 mg/kg) was administered to healthy male volunteers 1 h before a supine maximal bicycle ergometer test, it led to additive increases in systolic/diastolic BP and peripheral resistance at rest and during ergometry (Sung et al., 1990). This caused some subjects, especially those with a family history of hypertension, to have exaggerated BP responses to the exercise (see also Pincomb et al., 1991). A higher dose of caffeine (6 mg/kg) increased systolic BP to an equal degree at rest and during bicycle ergometer exercise (65% VO_2 max) (Daniels et al., 1998). Caffeine did not affect heart rate, although it attenuated the increase in forearm blood flow otherwise seen during placebo. Similar BP and heart rate effects were seen when caffeine (5 mg/kg) was consumed before very mild (30 W) bicycle exercise (Perkins et al., 1994). Studies of isometric or static exercise have also shown predominantly additive effects with caffeine (France and Ditto, 1992). Performance of a simple manual reaction time task is also primarily additive with the effects of caffeine on systolic/diastolic BP (Pincomb et al., 1988), although during the reaction time task caffeine increased cardiac output, an effect not seen at rest.

Some authors have tested the effects of habitual or background caffeine consumption on BP during exercise and related challenges. Instructions to drink six cups of coffee per day vs. none for 8 weeks resulted in volunteers having higher systolic/diastolic BP during orthostatic challenge than under no coffee, with no difference in the magnitude of the change (van Dusseldorp et al., 1992). Women who had consumed coffee on the day of dynamic and static exercise testing had a modest increment in systolic BP relative to those who had not (Hofer and Battig, 1993). In a study of the effect of caffeine intake on the results of diagnostic adenosine perfusion scans, caffeine blood levels ranging from 0.1 to 8.8 mg/l (equal to recent intake of up to five cups of coffee) were considered to be indicators of the recency and quantity of dietary caffeine intake (Majd-Ardekani et al., 2000). Higher blood levels of caffeine (> 2.9 mg/l) were correlated with higher systolic/diastolic BP before and during adenosine perfusion.

The majority of the studies reviewed above indicate that caffeine's effects on BP and heart rate are additive with the effects of a variety of challenges. A small number of studies indicate that in some persons and under certain conditions, nonadditive or synergistic effects may also occur. Several of these studies have used psychomotor challenge (a simple reaction time task), some have used dynamic exercise, one used aversive stimulation, and some of these have compared persons at differing risk for hypertension. Caffeine therefore appears to have additive effects with acute mental stress, and both may have synergistic effects as well.

WORKPLACE AND RELATED SETTINGS OUTSIDE THE LABORATORY

Workplace studies have also found caffeine to elevate BP. Systolic BP increased after caffeine during telemarketing work (France and Ditto, 1989), and ad libitum coffee intake increased ambulatory diastolic BP in university workers relative to ad libitum decaffeinated coffee intake (Jeong and Dimsdale, 1990). Similarly, Lane et al. (1998, 2002) found caffeine increased ambulatory BP and heart rate at work and at home relative to placebo. Caffeine was examined in relation to the effects of variations in occupational stress by comparing BPs on lecture vs. exam days in medical students exposed to placebo and caffeine at both times. Caffeine (3.3 mg/kg) increased systolic/diastolic BP on mornings of lecture days, and it further elevated BP on mornings of exams, showing an additive effect (Pincomb et al., 1987). This additive relationship was confirmed in a more extensive study using ambulatory monitoring during extended portions of lecture and exam days (Shepard et al., 2000).

Two studies of Italian employees monitored in the workplace reported an inverse relationship between BP and self-reported coffee intake (Periti et al., 1987; Salvaggio et al., 1990). However, these studies explicitly requested all employees to refrain from coffee intake the mornings of BP screenings, raising the possibility of a rebound BP effect in the regular consumers. Several other studies have found a positive association between casual BPs and reported caffeine intake among students attending a college campus BP screening (McCubbin et al., 1991). In 2436 Canadians being surveyed, reported coffee intake had a significant positive association with diastolic BP, although the effect was small (Birkett and Logan, 1988). In Denmark, coffee intake was significantly related to BP in 3608 adults (Kirchoff et al., 1994). In France, systolic/diastolic BPs were higher in 5430 coffee drinkers than in 891 abstainers (Lang et al., 1983b). In Algeria, diastolic BP was positively associated with coffee consumption in 1491 women and men (Lang et al., 1983a).

The nonlaboratory studies are therefore generally in agreement with those in the laboratory; BP is elevated by caffeine intake, either explicitly administered or by self-reported intake in the diet, and these effects are additive with the effects of stress when stress was manipulated. In general, the effects on BP are larger after specific dosing and smaller in relation to self-reported intake, perhaps due to lower reliability of self-reports and varying intervals between consumption and BP measurements. Although no mechanistic studies have been done, the results are generally in agreement with laboratory studies suggesting caffeine's effect on the constrictor status of the blood vessel wall.

INDIVIDUAL DIFFERENCES IN THE EFFECTS OF CAFFEINE

Caffeine's effects on the stress axis are variable across individuals. One of the individual differences that has been found to influence effects of caffeine on adrenocortical and sympathetic responses is risk for hypertension. Previous research has documented that risk for hypertension is accompanied by increased activation of the autonomic nervous system (Anderson et al., 1989) and the cardiovascular control centers of the hypothalamus and medulla (Folkow, 1982; Julius et al., 1988). Persons at genetic risk for hypertension may show greater cardiovascular reactivity to psychological stress (Fredrikson and Matthews, 1990; Lovallo and Wilson, 1992; al'Absi et al., 1996). The genetic disposition combined with the relatively elevated BP and exaggerated reactivity in these individuals may put them at a higher risk of becoming hypertensive (Borghi et al., 1986; Fredrikson and Matthews, 1990). The HPA axis is activated by exposure to novel and stressful situations, resulting in the secretion of cortisol from the adrenal cortex (Mason, 1975; al'Absi and Lovallo, 1993; al'Absi and Arnett, 2001). This activation may be exaggerated in persons at risk for hypertension (al'Absi et al., 1994; al'Absi and Wittmers, 2003).

Persons at risk for hypertension may be especially sensitive to caffeine's effects on the HPA axis. For example, cortisol responses to caffeine at rest and following behavioral stress are greater and more persistent in persons at high risk for hypertension than in low-risk normotensives (Lovallo et al., 1989; al'Absi et al., 1995). Our research with individuals at high risk for hypertension also shows that ACTH and cortisol values at rest after caffeine were as high as those shown by the low-risk group in response to caffeine plus the tasks (al'Absi et al., 1995; Lovallo et al., 1996).

This indicates that caffeine may produce stresslike responses in high-risk persons even in the absence of behavioral demands, consistent with the hypothesis that risk for hypertension may be associated with increased activation of the HPA axis (al'Absi and Lovallo, 1993). Persons at risk for hypertension have increased activation of the autonomic nervous system (Julius et al., 1988) and the cardiovascular control centers of the hypothalamus and medulla (Folkow, 1982; Wyss et al., 1990). This may be paralleled by enhanced responses of the HPA responses to stimulant agents, such as caffeine.

In light of caffeine's pressor effect and cortisol's effects in enhancing cardiovascular activation, this line of research argues for further attention to the effect of caffeine in individuals at high risk for hypertension. For example, when at rest in a novel experimental environment, borderline hypertensives show enhanced adrenocortical activation relative to low-risk controls (Fredrikson et al., 1991; al'Absi and Wittmers, 2003) and they have larger responses during work on mental arithmetic and psychomotor stress (al'Absi et al., 1994). These tendencies are exaggerated in the presence of caffeine. Compared to low-risk controls, caffeine can differentially increase cortisol secretion in normotensive men at high risk for hypertension during work on a demanding psychomotor task (Lovallo et al., 1989) and in response to a combination of mental arithmetic and reaction time tasks (al'Absi et al., 1995). This suggests that persons at risk of hypertension may be especially sensitive to caffeine's pituitary–adrenocortical effects when under stressful conditions.

In relation to caffeine's predominant pattern of increased BP and bradycardia, the effects of caffeine may show important individual differences. Our initial finding that caffeine elevated vascular resistance led us to target persons for study who were at elevated risk for hypertension. Hypertension development may be accompanied by early vascular remodeling, with progressive thickening of the blood vessel wall (Folkow, 1990), causing an increase in vascular resistance while BP is still within the normal range (Lovallo et al., 1991; Marrero et al., 1997; Lovallo and al'Absi, 1998), and consequently a greater sensitivity to agents that might elevate vascular resistance. We have therefore observed caffeine's hemodynamic effects in young adults at differing levels of risk for hypertension, ranging from low risk (negative parental history and resting systolic BP < 125) to high risk (positive parental history and systolic BP > 125) to borderline hypertensive (BP > 135/85 and periodic elevations > 140/90) and medicated hypertensive patients. Persons at successively higher levels of hypertension risk have greater BP responses to caffeine (Hartley et al., 2000). This is especially apparent in borderline hypertensives and medicated hypertensives who have both a greater and more prolonged BP response to caffeine along with larger vascular resistance increases compared to controls (Sung et al., 1994, 1995; Pincomb et al., 1996). This same dose of caffeine in borderline hypertensives led to a slightly greater increase in BP to the combination of task plus caffeine than in normotensive controls (Lovallo et al., 1996). This greater pressor effect in the borderline hypertensives was persistent over a 1-h task period involving both unsignaled reaction time and alternations with mental arithmetic in the latter half of the stress period (Lovallo and Thomas, 2000). Lovallo et al. (1991) found that caffeine (3.3 mg/kg) caused consistent increases in systolic/diastolic BP at rest due to increased vascular resistance. During an unsignaled simple reaction time task, caffeine caused a rise in cardiac output larger than its response to the task alone, contributing to a superadditive increase in the systolic BP response to the task.

Sung et al. (1995) compared normotensive and unmedicated hypertensive men during caffeine exposure and bicycle ergometer exercise. Caffeine (3.3 mg/kg) had an additive effect on systolic/diastolic BP during exercise in both groups and caused an elevation in vascular resistance and a small but significant decline in heart rate. During caffeine exposure, the diastolic BP rise to exercise diminished after 15 min in controls but persisted until 30 min in hypertensives, indicating a greater sensitivity to caffeine's pressor effects in the hypertensives. This led a greater number of hypertensive mens having an exaggerated BP response at some point during exercise (> 230/120 mmHg). The studies showing nonadditive effects of caffeine and stress fall into two groups. One has found predominantly enhanced cardiac function as a result of psychomotor tasks, exercise, or aversive challenge. The second group includes studies of hypertensives or persons at high risk for future hypertension in whom increases in vascular responsiveness combined with increased cardiac output appear to have induced relatively greater BP rises than those found in normotensive controls.

In summary, persons at high risk for hypertension had greater pituitary–adrenocortical responsivity than low-risk men to the task challenges in the absence of caffeine compared with time-synchronized resting placebo control values. Caffeine alone elevated ACTH and cortisol concentrations relative to placebo in both groups. Both groups showed significant ACTH and cortisol rises to the combined tasks in the presence of caffeine, and the high-risk group showed earlier and more persistent rises throughout the tasks and the highest ACTH and cortisol levels seen in the experiment.

OTHER INDIVIDUAL DIFFERENCES

Caffeine's BP effects are larger in older persons (Izzo et al., 1983) and men (James, 1990) and are more prolonged in blacks (Myers et al., 1989). Because there are on the order of 100 studies of caffeine's pressor effect, a detailed review would be excessively long for the present purposes. A good, brief review is provided by James (1991).

CONCLUDING REMARKS AND FUTURE DIRECTIONS

Although research has demonstrated significant adrenocortical and sympathetic nervous system changes in response to caffeine, most studies produced findings that are limited to the examination of the effects of a single dose in regular users of caffeine who had abstained overnight. The acute and chronic tolerance effects of caffeine on adrenocortical responses to caffeine use on and behavioral challenges remain to be tested (see Robertson et al., 1981; James, 1993). Research so far is also limited by the exclusive use of men in most studies. Future studies could profitably address sex differences and should examine changes in response to caffeine and stress as a function of age and previous caffeine use.

The research conducted by our group showing an enhanced influence of caffeine in samples of normotensive men at elevated risk for hypertension suggests that similar studies should be conducted on other groups at elevated risk for cardiovascular disease, including smokers, blacks, and postmenopausal women. In addition, the responsiveness of the HPA axis to caffeine indicates that caffeine may also elevate levels of beta-endorphin, a potent ligand for opioid receptors located in the amygdala and elsewhere. Beta-endorphin shares an identical mode of release with ACTH at the pituitary (Guillemin et al., 1977). Future research into the role of neuroendocrine factors mediating the possible reinforcing effects of caffeine use could benefit from the examination of the HPA axis and endogenous function as potential mechanisms. Related systems that have been implicated in the addictive liability in caffeine, such as the GABAergic and dopaminergic systems, may also need to be further probed in terms of how their effects interact with HPA responses to caffeine and stress. Caffeine's habitual intake may be encouraged by a mild experience of reward due to activation of central GABAergic neurons and release of dopamine from parts of the mesopontine reticular system (Daly and Fredholm, 1998; Nehlig, 1999). These effects may underlie a reported inverse relationship between caffeine use and suicide in women (Kawachi et al., 1996) and the inverse risk of Parkinson's disease in caffeine consumers (Ross and Petrovich, 2001). Work to examine the minimum effective dose of caffeine on these systems is still needed. Furthermore, the clinical significance of these changes needs to be clarified within the several lines of research that have been reviewed in this chapter.

In conclusion, current research indicates that caffeine, when combined with behavioral stress, enhances the effects of stress on ACTH and cortisol concentrations in humans. Individual differences in the effects of caffeine have also been shown. Specifically, research on persons at risk for hypertension indicates that these individuals may have greater pituitary–adrenocortical and sympathetic responses to caffeine, alone and combined with behavioral stress. In light of previously reported greater cardiovascular responses to stress in hypertension-prone individuals, this set of findings indicates a potential for greater negative impacts of caffeine in this group.

REFERENCES

al'Absi, M. and Arnett, D.K. (2001) Adrenocortical responses to psychological stress and risk for hypertension. *Biomedical Pharmacotherapy*, 54, 234–244.

al'Absi, M., Bongard, S., Buchanan, T., Pincomb, G.A., Licinio, J. and Lovallo, W.R. (1997) Cardiovascular and neuroendocrine adjustment to public speaking and mental arithmetic stressors. *Psychophysiology*, 34, 266–275.

al'Absi, M., Buchanan, T.W. and Lovallo, W.R. (1996) Cardiovascular and pain responses to cold pressor in persons with positive parental history of hypertension. *Psychophysiology*, 33, 655–661.

al'Absi, M. and Lovallo, W.R. (1993) Cortisol concentrations in serum of borderline hypertensive men exposed to a novel experimental setting. *Psychoneuroendocrinology*, 18, 355–363.

al'Absi, M., Lovallo, W.R., McKey, B.S. and Pincomb, G.A. (1994) Borderline hypertensives produce exaggerated adrenocortical responses to sustained mental stress. *Psychosomatic Medicine*, 56, 245–250.

al'Absi, M., Lovallo, W.R., McKey, B., Sung, B.H., Whitsett, T.L. and Wilson, M.F. (1998) Hypothalamic-pituitary-adrenocortical responses to psychological stress and caffeine in men at high and low risk for hypertension. *Psychosomatic Medicine*, 60, 521–527.

al'Absi, M., Lovallo, W.R., Pincomb, G.A., Sung, B.H. and Wilson, M.F. (1995) Adrenocortical effects of caffeine at rest and during mental stress in borderline hypertensive men. *International Journal of Behavioral Medicine*, 2, 263–275.

al'Absi, M., Lovallo, W.R., Sung, B.H. and Wilson, M.F. (1993) Persistent adrenocortical sensitivity to caffeine in borderline hypertensive men. *FASEB Journal*, 7, A552.

al'Absi, M. and Wittmers, L.E. (2003) Enhanced adrenocortical responses to stress in hypertension-prone men and women. *Annals of Behavioral Medicine*, 25, 25–33.

Anderson, E.A., Sinkey, C.A., Lawton, W.J. and Mark, A.L. (1989) Elevated sympathetic nerve activity in borderline hypertensive humans. *Hypertension*, 14, 177–183.

Arnold, M.A., Carr, D.B., Togasaki, D.M., Pian, M.C. and Martin, J.B. (1982) Caffeine stimulates beta-endorphin release in blood but not in cerebrospinal fluid. *Life Sciences*, 32, 1017–1024.

Avogaro, P., Capri, C. and Pais, M. (1973) Plasma and urine cortisol behavior and fat mobilization in man after coffee ingestion. *Israel Journal of Medical Sciences*, 9, 114.

Belardinelli, L., Linden, J. and Berne, R.M. (1989) The cardiac effects of adenosine. *Progress in Cardiovascular Diseases*, 32, 73–97.

Belardinelli, L., Shryock, J.C., Song, Y., Wang, D. and Srinivas, M. (1995) Ionic basis of the electrophysiological actions of adenosine on cardiomyocytes. *FASEB Journal*, 9, 359–365.

Bellet, S., Kostis, J., Roman, L. and DeCastro, O. (1969) Effect of coffee ingestion on catecholamine release. *Metabolism*, 18, 288–291, related articles, links.

Belloni, F.L., Belardinelli, L., Halperin, C. and Hintze, T.H. (1989) An unusual receptor mediates adenosine-induced SA nodal bradycardia in dogs. *American Journal of Physiology*, 256, H1553–H1564.

Berntson, G.G., Cacioppo, J.T. and Quigley, K.S. (1994) Autonomic cardiac control. I. Estimation and validation from pharmacological blockades. *Psychophysiology*, 31, 572–585.

Birkett, N.J. and Logan, A.G. (1988) Caffeine-containing beverages and the prevalence of hypertension. *Journal of Hypertension*, Suppl. 6, S620–S622.

Bonham, G.S. and Leaverton, P.E. (1979). Use habits among adults of cigarettes, coffee, aspirin, and sleeping pills. *Vital Health Statistics*, 10, 1–48.

Borghi, C., Costa, F.V., Boschi, S., Mussi, A. and Ambrosioni, E. (1986) Predictors of stable hypertension in young borderline subjects: a five-year follow-up study. *Journal of Cardiovascular Pharmacology*, 8 (Suppl. 5), S138–S141.

Buus, N.H., Bottcher, M., Hermansen, F., Sander, M., Nielsen, T.T. and Mulvany, M.J. (2001) Influence of nitric oxide synthase and adrenergic inhibition on adenosine-induced myocardial hyperemia. *Circulation*, 104, 2305–2310.

Casiglia, E., Paleari, C.D., Daskalakis, C., Petucco, S., Bongiovi, S. and Pessina, A.C. (1990) Hemodynamic effects of "expresso" Italian coffee and pure caffeine on healthy volunteers. *Cardiologia*, 35, 575–580.

Conti, J.B., Belardinelli, L., Utterback, D.B. and Curtis, A.B. (1995) Endogenous adenosine in an antiarrhythmic agent. *Circulation*, 91, 1761–1767.

Conway, T.L., Vickers, R.R., Ward, H.W. and Rahe, R. (1979) Occupational stress and variation in cigarette, coffee, and alcohol consumption (Report No. 79-32), Bureau of Medicine and Surgery, Department of the Navy, Research Work Unit, ZF51.524.002–5020.

Conway, T.L., Vickers, R.R., Ward, H.W. and Rahe, R.H. (1981) Occupational stress and variation in cigarette, coffee, and alcohol consumption. *Journal of Health and Social Behavior*, 22, 155–165.

Daly, J.W. and Fredholm, B.B. (1998) Caffeine: an atypical drug of dependence. *Drug and Alcohol Dependence*, 51, 199–206.

Daniels, J.W., Mole, P.A., Shaffrath, J.D. and Stebbins, C.L. (1998) Effects of caffeine on blood pressure, heart rate, and forearm blood flow during dynamic leg exercise. *Journal of Applied Physiology*, 85, 154–159.

Daubresse, J.C., Luyckx, A., Demey-Ponsart, E. and Franchimont, P. (1973) Effects of coffee and caffeine on carbohydrate metabolism and cortisol plasma levels in man. *Acta Diabetica Latina*, 10, 1069.

Debrah, K., Haigh, R., Sherwin, R., Murphy, J. and Kerr, D. (1995) Effect of acute and chronic caffeine use on the cerebrovascular, cardiovascular and hormonal responses to orthostasis in healthy volunteers. *Clinical Science*, 89, 475–480.

Deschaepdryver, A.F. (1959) Physio-pharmacological effects on suprarenal secretion of adrenaline and noradrenaline in dogs. *Archives Internationales de Pharmacodynamie et de Thérapie*, 119, 517–518.

Diorio, D., Viau, V. and Meaney, M. (1993) The role of the medial prefrontal cortex (cingulate gyrus) in the regulation of hypothalamic-pituitary-adrenal responses to stress. *The Journal of Neuroscience*, 13, 3839–3847.

Fink, G., Robinson, I.C. and Tannahill, L.A. (1988) Effects of adrenalectomy and glucocorticoids on the peptides CRF-41, AVP and oxytocin in rat hypophysial portal blood. *Journal of Physiology*, 401, 329–345; erratum in *Journal of Physiology*, 1988, 405, 785.

Folkow, B. (1982) Physiological aspects of primary hypertension. *Physiological Reviews*, 62, 347–504.

Folkow, B. (1990) "Structural factor" in primary and secondary hypertension. *Hypertension*, 16, 89–101.

France, C. and Ditto, B. (1988) Caffeine effects on several indices of cardiovascular activity at rest and during stress. *Journal of Behavioral Medicine*, 11, 473–482.

France, C. and Ditto, B. (1989) Cardiovascular responses to occupational stress and caffeine in telemarketing employees. *Psychosomatic Medicine*, 51, 145–151.

France, C. and Ditto, B. (1992) Cardiovascular responses to the combination of caffeine and mental arithmetic, cold pressor, and static exercise stressors. *Psychophysiology*, 29, 272–282.

Fredholm, B.B. (1980) Are methylxanthine effects due to antagonism of endogenous adenosine? *Trends in Pharmacological Sciences*, 1, 129–132.

Fredholm, B.B. (1995) Astra Award Lecture: Adenosine, adenosine receptors and the actions of caffeine. *Pharmacology and Toxicology*, 76, 93–101.

Fredholm, B.B. and Dunwiddie, T.V. (1988) How does adenosine inhibit transmitter release? *Trends in Pharmacological Sciences*, 9, 130–134.

Fredholm, B.B., Battig, K., Holmen, J., Nehlig, A. and Zvartau, E.E. (1999) Actions of caffeine in the brain with special reference to factors that contribute to its widespread use. *Pharmacological Reviews*, 51, 83–127.

Fredrikson, M. and Matthews, K.A. (1990) Cardiovascular responses to behavioral stress and hypertension: a meta-analytic review. *Annals of Behavioral Medicine*, 12, 30–39.

Fredrikson, M., Tuomisto, M. and Bergman-Losman, B. (1991) Neuroendocrine and cardiovascular stress reactivity in middle-aged normotensive adults with parental history of cardiovascular disease. *Psychophysiology*, 28, 656–664.

Freilich, A. and Tepper, D. (1992) Adenosine and its cardiovascular effects. *American Heart Journal*, 123, 1324–1328.

Goldstein, I.B. and Shapiro, D. (1987) The effects of stress and caffeine on hypertensives. *Psychosomatic Medicine*, 49, 226–235.

Green, P.J. and Suls, J. (1996) The effects of caffeine on ambulatory blood pressure, heart rate, and mood in coffee drinkers. *Journal of Behavioral Medicine*, 19, 111–128.

Greenberg, W. and Shapiro, D. (1987) The effects of caffeine and stress on blood pressure in individuals with and without a family history of hypertension. *Psychophysiology*, 24, 151–156.

Guillemin, R., Vargo, T., Rossier, J., Minick, S., Ling, N., Rivier, C. et al. (1977) Beta-endorphin and adrenocorticotropin are secreted concomitantly by the pituitary gland. *Science*, 197, 1367–1369.

Hartley, T.R., Sung, B.H., Pincomb, G.A., Whitsett, T.L., Wilson, M.F. and Lovallo, W.R. (2000) Hypertension risk status and caffeine's effect on blood pressure. *Hypertension*, 36, 137–141.

Hasenfratz, M. and Battig, K. (1992) No psychophysiological interactions between caffeine and stress? *Psychopharmacology*, 109, 283–290.

Henry, J.P. and Stephens, P.M. (1980) Caffeine as an intensifier of stress-induced hormonal and pathophysiologic changes in mice. *Pharmacology, Biochemistry and Behavior*, 13, 719–727.

Hobbs, S. (1982) Central command during exercise: parallel activation of the cardiovascular and motor systems by descending command signals, in *Circulation, Neurobiology and Behavior*, Smith, O.A., Galosy, R.A. and Weiss, S.M., Eds., Elsevier, Amsterdam, pp. 217–231.

Hofer, I. and Battig, K. (1993) Coffee consumption, blood pressure tonus and reactivity to physical challenge in 338 women. *Pharmacology, Biochemistry and Behavior*, 44, 573–576.

Horst, K., Robinson, W.D., Jenkins, W.L. and Bao, D.L. (1934) The effect of caffeine, coffee and decaffeinated coffee upon blood pressure, pulse rate and certain motor reactions of normal young men. *Journal of Pharmacology and Experimental Therapeutics*, 52, 307–321.

Hoyle, C.H.V. (1992) Transmission: purines, in *Autonomic Neuroeffector Mechanisms*, Burnstock, G. and Hoyle, C.H.V., Eds., Harwood Academic Publishers, New York, pp. 367–408.

Isenberg, G. and Belardinelli, L. (1984) Ionic basis for the antagonism between adenosine and isoproterenol on isolated mammalian ventricular myocytes. *Circulation Research*, 55, 309–325.

Izzo, J.L., Jr., Ghosal, A., Kwong, T., Freeman, R.B. and Jaenike J.R. (1983). Age and prior caffeine use alter the cardiovascular and adrenomedullary responses to oral caffeine. *American Journal of Cardiology*, 52, 769–773

Jackson, S. (1849) On the influence upon health of the introduction of tea and coffee in large proportion into the dietary of children and the laboring classes. *American Journal of the Medical Sciences*, 18, 79–86.

James, J.E. (1990) The influence of user status and anxious disposition on the hypertensive effects of caffeine. *International Journal of Psychophysiology*, 10, 171–179.

James, J.E. (1991) *Caffeine and Health*, Academic Press, London, pp. 111–112.

James, J.E. (1993) Caffeine and ambulatory blood pressure. *American Journal of Hypertension*, 6, 91–96.

Jeong, D.U. and Dimsdale, J.E. (1990) The effects of caffeine on blood pressure in the work environment. *American Journal of Hypertension*, 3, 749–753.

Johnson, C.D., Coney, A.M. and Marshall, J.M. (2001) Roles of norepinephrine and ATP in sympathetically evoked vasoconstriction in rat tail and hindlimb in vivo. *American Journal of Physiology*, 281, H2432–H2440.

Julius, S., Schork, N. and Schork A. (1988) Sympathetic hyperactivity in early stages of hypertension: the Ann Arbor data set. *Journal of Cardiovascular Pharmacology*, 12 (Suppl.), S121–S129.

Jung, R.T., Shetty, P.S., James, W.P., Barrand, M.A. and Callingham, B.A. (1981) Caffeine: its effect on catecholamines and metabolism in lean and obese humans. *Clinical Science*, 60, 527–535.

Kawachi, I., Willett, W.C., Colditz, G.A., Stampfer, M.J. and Speizer, F.E. (1996) A prospective study of coffee drinking and suicide in women. *Archives of Internal Medicine*, 156, 521–525.

Kirchoff, M., Torp-Pedersen, C., Hougaard, K., Jacobsen, T.J., Sjol, A., Munch, M. et al. (1994) Casual blood pressure in a general Danish population: relation to age, sex, weight, height, diabetes, serum lipids and consumption of coffee, tobacco and alcohol. *Journal of Clinical Epidemiology*, 47, 469–474.

Lane, J.D. (1983) Caffeine and cardiovascular responses to stress. *Psychosomatic Medicine*, 45, 447–451.

Lane, J.D. (1994) Neuroendocrine responses to caffeine in the work environment. *Psychosomatic Medicine*, 56, 267–270.

Lane, J.D. and Williams, R.B., Jr. (1985) Caffeine affects cardiovascular responses to stress. *Psychophysiology*, 22, 648–655.

Lane, J.D., Adcock, R.A., Williams, R.B. and Kuhn, C.M. (1990) Caffeine effects in cardiovascular and neuroendocrine responses to acute psychosocial stress and their relationship to level of habitual coffee consumption. *Psychosomatic Medicine*, 52, 320–336.

Lane, J.D., Phillips-Bute, B.G. and Pieper, C.F. (1998) Caffeine raises blood pressure at work. *Psychosomatic Medicine*, 60, 327–330.

Lane, J.D., Pieper, C.F., Phillips-Bute, B.G., Bryant, J.E. and Kuhn, C.M. (2002) Caffeine affects cardiovascular and neuroendocrine activation at work and home. *Psychosomatic Medicine*, 64, 595–603.

Lang, T., Bureau, J.F., Degoulet, P., Salah, H. and Benattar, C. (1983a) Blood pressure, coffee, tea and tobacco consumption: an epidemiogical study in Algiers. *European Heart Journal*, 4, 602–607.

Lang, T., Degoulet, P., Aime, F., Fouriaud, C., Jacquinet-Salord, M.C., Laprugne, J. et al. (1983b) Relation between coffee drinking and blood pressure: analysis of 6,321 subjects in the Paris region. *American Journal of Cardiology*, 52, 1238–1242.

Laurent, D., Schneider, K.E., Prusaczyk, W.K., Franklin, C., Vogel, S.M., Krssak, M. et al. (2000) Effects of caffeine on muscle glycogen utilization and the neuroendocrine axis during exercise. *Journal of Clinical Endocrinology and Metabolism*, 85, 2170–2175.

Levi, L. (1967) The effect of coffee on the function of the sympatho-adrenomedullary system in man. *Acta Medicina Scandinavica*, 181, 431–438.

Lovallo, W.R. (1997) *Stress and Health: Biological and Psychological Interactions*. Sage, Thousand Oaks, CA.

Lovallo, W.R. and al'Absi, M. (1998) Hemodynamics during rest and behavioral stress in normotensive men at high risk for hypertension. *Psychophysiology*, 35, 47–53.

Lovallo, W.R., al'Absi, M., Blick, K., Whitsett, T.L. and Wilson, M.F. (1996) Stress-like adrenocorticotropin responses to caffeine in young healthy men. *Pharmacology, Biochemistry and Behavior*, 55, 365–369.

Lovallo, W.R., Pincomb, G.A., Sung, B.H., Everson, S.A., Passey, R.B. and Wilson, M.F. (1991) Hypertension risk and caffeine's effect on cardiovascular activity during mental stress in young men. *Health Psychology*, 10, 236–243.

Lovallo, W.R., Pincomb, G.A., Sung, B.H., Passey, R.B., Sausen, K.P. and Wilson, M.F. (1989) Caffeine may potentiate adrenocortical stress responses in hypertension-prone men. *Hypertension*, 14, 170–176.

Lovallo, W.R. and Thomas, T. (2000) Stress hormones in psychophysiological research, in *Handbook of Psychophysiology*, Cacioppo, J.T., Tassinary, L.G. and Berntson, G., Eds., Cambridge University Press, Cambridge.

Lovallo, W.R. and Wilson, M.F. (1992) The role of cardiovascular reactivity in hypertension risk, in *Individual Differences in Cardiovascular Responses to Stress*, Turner, J.R., Sherwood, A. and Light, K.C., Eds., Plenum Press, New York, pp. 165–186.

MacDougall, J.M., Musante, L., Castillo, S. and Acevedo, M.C. (1988) Smoking, caffeine, and stress: effects on blood pressure and heart rate in male and female college students. *Health Psychology*, 7, 461–478.

Majd-Ardekani, J., Clowes, P., Menash-Bonsu, V. and Nunan, T.O. (2000) Time for abstention from caffeine before an adenosine myocardial perfusion scan. *Nuclear Medicine Communications*, 21, 361–364.

Marrero, A.F., al'Absi, M., Pincomb, G.A. and Lovallo, W.R. (1997) Men at risk for hypertension show elevated vascular resistance at rest and during mental stress. *International Journal of Psychophysiology*, 25, 185–192.

Marzouk, H.F.A.I., Zuyderwijk, J., Uitterlinden, P., van Koetsveld, P., Blijd, J.J., Abou-Hashim, E.M. et al. (1991) Caffeine enhances the speed of recovery of the hypothalamo-pituitary-adrenocortical axis after chronic prednisolone administration in the rat. *Neuroendocrinology*, 54, 439–446.

Mason, J.W. (1975) A historical view of the stress field. *Journal of Human Stress*, 1, 22–36.

McCubbin, J.A., Wilson, J.F., Bruehl, S., Brady, M., Clark, K. and Kort, E. (1991) Gender effects on blood pressures obtained during an on-campus screening. *Psychosomatic Medicine*, 53, 909–100.

Mosqueda-Garcia, R., Tseng, C.J., Biaggioni, I., Robertson, R.M. and Robertson, D. (1990) Effects of caffeine on baroreflex activity in humans. *Clinical Pharmacology and Experimental Therapeutics*, 48, 568–574.

Myers, H.F., Shapiro, D., McClure, F. and Daims, R. (1989) Impact of caffeine and psychological stress on blood pressure in black and white men. *Health Psychology*, 8, 597–612.

Nehlig, A. (1999) Are we dependent upon coffee and caffeine?: a review on human and animal data. *Neuroscience and Biobehavioral Reviews*, 23, 563–576.

Nehlig, A., Daval, J.L. and Debry, G. (1992) Caffeine and central nervous system: mechanisms of action, biochemical, metabolic, and psychostimulant effects. *Brain Research Reviews*, 17, 139–170.

Neville, A.M. and O'Hare, M.J., Eds. (1984) *The Human Adrenal Cortex*, Springer Verlag, New York, pp. 211–231.

Oberman, Z., Herzberg, M., Jaskolka, H., Harell, A., Hoerer, E. and Laurian L. (1975). Changes in plasma cortisol, glucose and free fatty acids after caffeine ingestion in obese women. *Israel Journal of Medical Science*, 11, 33–36.

Onrot, J., Goldberg, M.R., Biaggioni, I., Hollister, A.S., Kingaid, D. and Robertson, D. (1985) Hemodynamic and humoral effects of caffeine in autonomic failure: therapeutic implications for postprandial hypotension. *New England Journal of Medicine*, 313, 549–554.

Patwardhan, R.V., Desmond, P.V., Johnson, R.F., Dunn, G.D., Robertson, D.H., Hoyumpa, A.M., Jr. and Schenker, S. (1980) Effects of caffeine on plasma free fatty acids, urinary catecholamines and drug binding. *Clinical Pharmacology and Experimental Therapeutics*, 28, 398–403.

Periti, M., Salvaggio, A., Quaglia, G. and Di Marzio, L. (1987) Coffee consumption and blood pressure: an Italian study. *Clinical Science*, 72, 443–447.

Perkins, K.A., Sexton, J.E., Stiller, R.L., Fonte, C., DiMarco, A., Goettler, J. and Scierka, A. (1994). Subjective and cardiovascular responses to nicotine combined with caffeine during rest and casual activity. *Psychopharmacology*, 113, 438–444.

Petrusz, P. and Merchenthaler, I. (1992) The corticotropin releasing factor system, in *Neuroendocrinology*, Nemeroff, C.B., Ed., CRC Press, Boca Raton, FL, pp. 129–183.

Pincomb, G.A., Lovallo, W.R., McKey, B.S., Sung, B.H., Passey, R.B., Everson, S.A. and Wilson, M.F. (1996) Acute blood pressure effects of caffeine in men with borderline to mild hypertension. *American Journal of Cardiology*, 77, 270–274.

Pincomb, G.A., Lovallo, W.R., Passey, R.B., Brackett, D.J. and Wilson, M.F. (1987) Caffeine enhances the physiological response to occupational stress in medical students. *Health Psychology*, 6, 101–112.

Pincomb, G.A., Lovallo, W.R., Passey, R.B., Whitsett, T.L., Silverstein, S.M. and Wilson, M.F. (1985) Effects of caffeine on vascular resistance, cardiac output and myocardial contractility in young men. *American Journal of Cardiology*, 56, 119–122.

Pincomb, G.A., Lovallo, W.R., Passey, R.B. and Wilson, M.F. (1988) Effect of behavior state on caffeine's ability to alter blood preasure. *American Journal of Cardiology*, 61, 798–802.

Pincomb, G.A., Sung, B.H., Sausen, K.P., Lovallo, W.R. and Wilson, M.F. (1993) Consistency of cardiovascular response pattern to caffeine across multiple studies using impedance and nuclear cardiography. *Biological Psychology*, 36, 131–138.

Pincomb, G.A., Wilson, M.F., Sung, B.H., Passey, R.B. and Lovallo, W.R. (1991) Effects of caffeine and exercise on pressure regulation in men at risk for hypertension. *American Heart Journal*, 122, 1107–1115.

Porkka-Heiskanen, T., Strecker, R.E., Thakkar, M., Bjorkum, A.A., Greene, R.W. and McCarley, R.W. (1997) Adenosine: a mediator of the sleep-inducing effects of prolonged wakefulness. *Science*, 276, 1265–1268.

Rall, T.W. (1980) Central nervous system stimulants: the xanthines, in *Pharmacological Basis of Therapeutics*, 6th ed., Goodman, L.S. and Gilman, A.G., Eds., MacMillan, New York, pp. 589–603.

Ratliff-Crain, J. and Kane, J. (1995) Predictors for altering caffeine consumption during stress. *Addictive Behavior*, 20, 509–516.

Robertson, D., Frolich, J.C., Carr, R.K., Watson, J.T., Hollifield, J.W., Shand, D.G. and Oates, J.A. (1978) Effects of caffeine on plasma renin activity, catecholamines and blood pressure. *New England Journal of Medicine*, 298, 181–186.

Robertson, D., Johnson, G.A., Robertson, R.M., Nies, A.S., Shand, D.G. and Oates J.A. (1979) Comparative assessment of stimuli that release neuronal and adrenomedullary catecholamines in man. *Circulation*, 59, 637–643.

Robertson, D., Wade, D., Workman, R., Woosley, R.L. and Oates, J.A. (1981) Tolerance to the humoral and hemodynamic effects of caffeine in man. *Journal of Clinical Investigation*, 67, 1111–1117.

Ross, G.W. and Petrovitch, H. (2001) Current evidence for neuroprotective effects of nicotine and caffeine against Parkinson's disease. *Drugs & Aging*, 18, 797–806.

Ross G.W., Abbott, R.D., Petrovitch H., White, L.R. and Tanner, C.M. (2000) Association of coffee and caffeine intake with the risk of Parkinson's disease. *Journal of the American Medical Association*, 283, 2674–2679.

Salvaggio, A., Periti, M., Miano, L. and Zambelli, C. (1990) Association between habitual coffee consumption and blood pressure levels. *Journal of Hypertension*, 8, 585–590.

Sato, H., Hori, M., Kitakaze, M., Takashima, S., Inoue, M., Kitabatake, A. and Kamada, T. (1992) Endogenous adenosine blunts beta-adrenoceptor-mediated inotropic response in hypoperfused canine myocardium. *Circulation*, 85, 1594–1603.

Shepard, J.D., al'Absi, M., Whitsett, T.L. and Lovallo, W.R. (2000) Additive pressor effects of caffeine and stress in male medical students at risk for hypertension. *American Journal of Hypertension*, 13, 475–481.

Sherwood, A., Allen, M.T., Fahrenberg, J., Kelsey, R.M., Lovallo, W.R. and van Doornenm, L.J. (1990) Methodological guidelines for impedance cardiography. *Psychophysiology*, 27, 1–23.

Shinozuka, K., Mizuno, H., Nakamura, K. and Kunitomo, M. (2002) Purinergic modulation of vascular sympathetic neurotransmission. *Japanese Journal of Pharmacology*, 88, 19–25.

Smits, P., Boekema, P., De Abreu, R., Thien, T. and van't Laar, A. (1987) Evidence for an antagonism between caffeine and adenosine in the human cardiovascular system. *Journal of Cardiovascular Pharmacology*, 10, 136–143.

Smits, P., Hoffmann, H., Thien, T., Houben, H. and van't Laar, A. (1983) Hemodynamic and humoral effects of coffee after beta 1-selective and nonselective beta-blockade. *Clinical Pharmacology and Therapeutics*, 34, 153–158.

Smits, P., Schouten, J. and Thien, T. (1989) Cardiovascular effects of two xanthines and the relation to adenosine antagonism. *Clinical Pharmacology and Therapeutics,* 45, 593–599.

Smits, P., Lenders, J.W. and Thien, T. (1990) Caffeine and theophylline attenuate adenosine-induced vasodilation in humans. *Clinical Pharmacology and Therapeutics,* 48, 410–418.

Smits, P., Lenders, J.W., Willemsen, J.J., den Arend, J.A. and Thien, T. (1991a) Adenosine attenuates the vasoconstrictor response to the cold pressor test in humans. *Journal of Cardiovascular Pharmacology,* 17, 1019–1022.

Smits, P., Lenders, J.W., Willemsen, J.J. and Thien, T. (1991b) Adenosine attenuates the response to sympathetic stimuli in humans. *Hypertension,* 18, 216–223.

Smits, P., Thien, T. and van't Laar, A. (1985) The cardiovascular effects of regular and decaffeinated coffee. *British Journal of Clinical Pharmacology,* 19, 852–854.

Smits, P., Pieters, G. and Thien, T. (1986) The role of epinephrine in the circulatory effects of coffee. *Clinical Pharmacology and Therapeutics,* 40, 431–437.

Snyder, S.H., Katims, J.J., Annau, Z., Bruns, R.F. and Daly, J.W. (1981). Adenosine receptors and behavioral actions of methylxanthines. *Proceedings of the National Academy of Sciences of the USA,* 78, 3260–3264.

Song, Y., Wu, L., Shryock, J.C. and Belardinelli, L. (2002) Selective attenuation of isoproterenol-stimulated arrhythmic activity by a partial agonist of adenosine A1 receptor. *Circulation,* 105, 118–123.

Spindel, E.R., Wurtman, R.J., McCall, A., Carr, D.B., Conlay, L., Griffith, L. and Arnold, M.A. (1984) Neuroendocrine effects of caffeine in normal subjects. *Clinical Pharmacology and Therapeutics,* 36, 402–407.

Sung, B.H., Lovallo, W.R., Pincomb, G.A. and Wilson, M.F. (1990) Effects of caffeine on blood pressure response during exercise in normotensive healthy young men. *American Journal of Cardiology,* 65, 909–913.

Sung, B.H., Lovallo, W.R., Whitsett, T. and Wilson, M.F. (1995) Caffeine elevates blood pressure response to exercise in mild hypertensive men. *American Journal of Hypertension,* 8, 1184–1188.

Sung, B.H., Whitsett, T.L., Lovallo, W.R., al'Absi, M., Pincomb, G.A. and Wilson, M.F. (1994) Prolonged increase in blood pressure by a single oral dose of caffeine in mild hypertensive men. *American Journal of Hypertension,* 7, 755–758.

Sung, B.H., Wilson, M.F., Robinson, C., Thadani, U. and Lovallo, W.R. (1988) Mechanisms of myocardial ischemia induced by epinephrine: comparison with exercise-induced ischemia. *Psychosomatic Medicine,* 50, 381–393.

Suzuki, M., Li, R.A., Miki, T., Uemura, H., Sakamoto, N., Ohmoto-Sekine, Y. et al. (2001) Functional roles of cardiac and vascular ATP-sensitive potassium channels clarified by Kir6.2-knockout mice. *Circulation Research,* 88, 570–577.

Swanson, L.W. and Simmons, D.M. (1989) Differential steroid hormone and neural influences on peptide mRNA levels in CRH cells of the paraventricular nucleus: a hybridization histochemical study in the rat. *Journal of Comparative Neurology,* 285, 413–435.

Takiyyuddin, M.A., Brown, M.R., Dinh, T.Q., Cervenka, J.H., Braun, S.D., Parmer, R.J. et al. (1994) Sympathoadrenal secretion in humans: factors governing catecholamine and storage vesicle peptide co-release. *Journal of Autonomic Pharmacology,* 1, 187–200.

Uhde, T.W., Bierer, L.M. and Post, R.M. (1985) Caffeine-induced escape from dexamethasone suppression. *Archives of General Psychiatry,* 42, 737–738.

van Dusseldorp, M., Smits, P., Lenders, J.W., Temme, L., Thien, T. and Katan, M.B. (1992) Effects of coffee on cardiovascular responses to stress: a 14-week controlled trial. *Psychosomatic Medicine,* 54, 344–353.

Vander, A.J., Sherman, J.H. and Luciano, D.S. (1994) *Human Physiology,* 6th ed., McGraw-Hill, New York.

Vernikos-Danellis, J. and Harris, C.G. (1968) The effect of *in vitro* and *in vivo* caffeine, theophylline, and hydrocortisone on the phosphodiesterase activity of the pituitary, median eminence, heart and cerebral cortex of the rat. *Proceedings of the Society for Experimental Biology and Medicine,* 128, 1016–1021.

Whitsett, T.L., Christensen, H.D. and Hirsh, K.R. (1980) Cardiovascular effects of caffeine in humans, in *Central Control Mechanisms and Related Topics,* Wang, H.H., Blumenthal, M.R. and Ngai, S.H., Eds., Futura Publishing, Mount Kisco, NY.

Shitsett, T.L., Manion, C.V. and Christensen, H.D. (1984) Cardiovascular effects of coffee and caffeine. *American Journal of Cardiology,* 53, 918–922.

Wilson, J.D. and Foster, D.W. (1992) *Williams Textbook of Endocrinology,* 8th ed., Saunders, Philadelphia.

Wilson, M.F., Sung, B.H., Pincomb, G.A. and Lovallo, W.R. (1989) Simultaneous stroke volume measurement by impedance and nuclear ventriculography: comparison at rest and during interventions. *Annals of Biomedical Engineering*, 17, 475–482.

Wood, H.C. (1912) The effects of caffeine on the circulatory and muscular systems. *Therapeutic Gazette*, 36, 6–12.

Wyss, J.M., Oparil, S. and Chen, Y. (1990) The role of the central nervous system in hypertension, in *Hypertension: Pathophysiology, Diagnosis, and Management*, Laragh, J.H. and Brenner, B.M., Eds., Raven Press, New York, pp. 679–701.

9 Dependence upon Coffee and Caffeine: An Update

Astrid Nehlig

CONTENTS

Introduction 133
Coffee and Caffeine Consumption 134
    Coffee Consumption 134
    Caffeine Consumption 134
Addiction and Drug Dependence 134
Caffeine Withdrawal 135
    Caffeine Withdrawal in Animals 135
    Characterization of Withdrawal Symptoms in Humans 135
Tolerance to the Effects of Caffeine 136
    Tolerance to Caffeine in Animals 136
    Tolerance to Caffeine in Humans 137
Reinforcement Properties of Caffeine 137
    Reinforcing Effects of Caffeine in Animals 137
    Reinforcing Effects of Caffeine in Humans 138
Is Caffeine Activating the Brain Circuits Underlying Dependence to Drugs? 139
Conclusion 140
References 140

INTRODUCTION

Caffeine is the most widely used psychoactive substance in the world. Most of the caffeine consumed comes from dietary sources such as coffee, tea, cola drinks, and chocolate. The most notable behavioral effects of caffeine occur after low to moderate doses (50 to 300 mg) and include increased alertness, energy, and ability to concentrate. Moderate caffeine consumption leads very rarely to health risks (Curatolo and Robertson, 1983; Benowitz, 1990; Fredholm et al., 1999; Nawrot et al., 2003). Higher doses of caffeine, however, induce negative effects such as anxiety, restlessness, insomnia, and tachycardia, primarily in a small portion of caffeine-sensitive individuals. Four attitudes are positively linked to the quantity of coffee consumed (SECED, 1980). They are, in decreasing order of importance: (1) the need for a stimulant, (2) the preference for strong coffee, (3) the knowledge of coffee, and (4) the preference for the coffee roasting shop. On the other hand, caffeine has also been considered as a potential drug of abuse (Gilliland and Bullock, 1984; Holtzman, 1990), and the possibility that caffeine withdrawal — but not abuse and dependence — should be added to diagnostic manuals has been considered in the U.S. (Hughes et al., 1992). Two recent reviews have detailed the various aspects linked to caffeine dependence (Fredholm et al., 1999; Nehlig, 1999). In this chapter, we will consider whether or not caffeine can be considered a drug of dependence.

COFFEE AND CAFFEINE CONSUMPTION

COFFEE CONSUMPTION

Consumption of coffee varies largely among the different countries. The highest consumption (more than 10 kg/person/year) occurs in Scandinavia, Austria, and the Netherlands. In most Western European countries, coffee consumption ranges from 6 to 9 kg/person/year. The lowest consumption (less than 5 kg/person/year) is found in the U.S. and Italy (D'Amicis and Viani, 1993; Debry, 1994).

The content of caffeine per cup of coffee varies with the size of the serving, the mode of preparation of coffee (boiled, filtered, percolated, espresso, or instant), and the type of coffee, Arabica or Robusta (D'Amicis and Viani, 1993; Debry, 1994). The size of the cup ranges from 50 to 190 ml and the content of caffeine in a 150-ml cup of coffee is as low as 19 mg/cup in instant coffee and as high as 177 mg/cup in boiled coffee.

CAFFEINE CONSUMPTION

Caffeine is present in a number of dietary sources consumed worldwide: tea, coffee, cocoa beverages, candy bars, and soft drinks. The content of caffeine in these food items ranges from 71 to 220 mg/150 ml for coffee to 32 to 42 mg/150 ml for tea, 32 to 70 mg/330 ml for cola, and 4 mg/150 ml for cocoa (Debry, 1994). Caffeine consumption from all sources can be estimated as 76 mg/person/d but reaches 210 to 238 mg/d in the U.S. and Canada and more than 400 mg/person/d in Sweden and Finland, where 80 to 100% of the caffeine intake comes only from coffee (Viani, 1993, 1996; Debry, 1994; Barone and Roberts, 1996). In the U.K., 72% of the caffeine consumed comes from tea (Debry, 1994). In adult consumers, the daily caffeine consumption from all sources in the U.S. reaches a value of 2.4 to 4.0 mg/kg (170 to 300 mg) in a 60 to 70-kg individual (Barone and Roberts,1996). In children, soft drinks represent 55%, chocolate foods and beverages 35 to 40%, and tea 6 to 10% of the total caffeine intake (Ellison et al., 1995).

ADDICTION AND DRUG DEPENDENCE

Drug dependence has been defined as "a pattern of behavior focused on the repetitive and compulsive seeking and taking a psychoactive drug" (Heishman and Henningfield, 1992). However, it is necessary to demonstrate psychoactive effects to differentiate drug dependence from other habitual or controlled behaviors, such as the daily ingestion of medications such as aspirin or vitamins. Moreover, it is necessary to demonstrate that the drug is positively reinforcing its own ingestion.

The recent diagnostic manuals from the World Health Organization (WHO, 1994) and the American Psychiatric Association (APA, 1987, 1992) proposed a new set of criteria for dependence based on the fulfillment of three (unspecified) of the six WHO or seven APA criteria. The seven criteria of dependence as proposed by the APA (1992) in DSM-IV (*Diagnostic and Statistical Manual of Mental Disorders*, 4th ed.) are: (1) tolerance (not specified for severity), (2) substance-specific withdrawal syndrome (psychic or physiological, not specified for severity), (3) substance is often taken in larger amounts or over a longer period than intended, (4) persistent desire or unsuccessful efforts to cut down or control use, (5) a great deal of time spent in activities necessary to obtain, use, or recover from the effects of the substance, (6) important social, occupational, or recreational activities given up or reduced because of substance use, (7) use continued despite knowledge of a persistent or recurrent physical or psychological problem that is likely to have been caused or exacerbated by the substance. The six criteria proposed by the WHO (1994) differ only slightly from those of the APA, mainly by a different sequence, slightly different formulations, and the combination of criteria (5) and (6) into a single one. The only possibility to differentiate between substances that can lead to dependence is to classify them according to the number of criteria met and specify severity of symptoms and frequency of occurrence.

The possible dependence on caffeine has been considered by several groups for almost two decades (Gilliland and Bullock, 1984; Griffiths et al., 1986, 1990; Griffiths and Woodson, 1988a; Heishman and Henningfield, 1992, 1994; Griffiths and Mumford, 1995; Strain and Griffiths, 1995), and caffeine has even been postulated as a "potential model of drug of abuse" (Holtzman, 1990). In the last decade, three studies used the criteria cited above and reported a dependence on caffeine in a subset of the general population. As a result of a random telephone survey in Vermont, Hughes et al. (1993) reported that 14 (3%) of the 166 caffeine users interviewed met the criteria for moderate and severe caffeine dependence, respectively. After telephone screening performed on 99 subjects in the U.S., Strain et al. (1994) found 16 individuals that fulfilled four criteria [(1), (2), (4), and (7) of the DSM-IV] out of the seven criteria cited above and were thus considered dependent on caffeine. The daily caffeine intake of the caffeine-dependent subjects ranged from 129 to 2548 mg/d, with a median daily caffeine intake of 360 mg (Strain et al., 1994). However, in spite of the absence of current psychiatric disorders at the time of the study in 14 out of 16 individuals, 11 of the 16 persons diagnosed with "caffeine dependence" had a history of psychiatric disorders, mainly substance abuse disorders (ten subjects) and mood disorders (seven subjects), which represents a higher prevalence of these disorders than in the general population (i.e., 50%) (Kessler et al., 1994). More recently, dependence was reported to occur in a subgroup of 36 teenagers recruited through newspaper advertisements and posters. There was no difference in the amount of caffeine consumed daily by caffeine-dependent vs. caffeine-nondependent teenagers. However, due to the method of recruitment, this study does not indicate how often these problems may occur in the teenage population (Bernstein et al., 2002; Oberstar et al., 2002).

In the present review, the consequences of coffee and caffeine consumption on various criteria possibly leading to the diagnosis of dependence will be considered. Among the seven criteria for drug dependence that have been cited above, the four main factors considered here will be withdrawal, tolerance, reinforcement, and dependence.

CAFFEINE WITHDRAWAL

Caffeine Withdrawal in Animals

Signs of caffeine withdrawal in rats, cats, and monkeys include decreases in locomotor activity (Holtzman, 1983; Finn and Holtzman, 1986), operant behavior (Carney, 1982; Mumford et al., 1988; Carroll et al., 1989), reinforcement threshold for electrical brain stimulation (Mumford et al., 1988), and time spent in slow-wave sleep (Sinton and Petitjean, 1989). The severity and duration of caffeine withdrawal symptoms depend on the dose and duration of the treatment. Decreases in locomotor activity appear only when very high caffeine doses, at least 67 mg/kg/d, are substituted by water (Holtzman, 1983; Finn and Holtzman, 1986). Caffeine withdrawal symptoms peak around 24 to 48 h and do not usually last longer than a few days (Vitiello and Woods, 1977; Carney, 1982; Holtzman, 1983; Finn and Holtzman, 1986; Mumford et al., 1988; Carroll et al., 1989; Sinton and Petitjean, 1989; Griffiths and Mumford, 1996).

Characterization of Withdrawal Symptoms in Humans

In humans, the caffeine withdrawal symptoms most often reported are headaches; feelings of weariness, weakness, and drowsiness; impaired concentration; fatigue and work difficulty; depression; anxiety; irritability; increased muscle tension; and occasionally tremor, nausea, or vomiting; as well as withdrawal feelings (Griffiths et al., 1990; Silverman et al., 1992; Hughes et al., 1993; Nehlig and Debry, 1994; Strain et al., 1994, 1995). Interestingly, some of the same symptoms have been reported following excessive intake of caffeine (Dews et al., 2002). Withdrawal symptoms generally begin about 12 to 24 h after sudden cessation of caffeine consumption and reach a peak after 20 to 48 h. However, in some individuals, these symptoms can appear within only 3 to 6 h and last for up to 1 week (James, 1991; Nehlig and Debry, 1994; Lane, 1997). When symptoms of caffeine withdrawal continue for days, weeks, or even months, it appears possible that these individuals may miss the enhancement of performance they customarily receive at appropriate times from consumption of caffeine. These subjects would then report their lower (noncaffeine) level of performance at times as continuance of withdrawal symptoms (Smith et al., 1993). Withdrawal symptoms do not seem to relate to the quantity of caffeine ingested daily (Griffiths and Woodson, 1988a,b,c; Hughes et al., 1992; Strain et al., 1994) and do not occur when caffeine consumption is progressively decreased (James et al., 1988).

Withdrawal symptoms vary with the subjects, and their frequency reported in different studies varies considerably, most often from 11 to 22%, but it can reach 0% in studies not focused on withdrawal symptoms and have reached 100% in a study in which the subjects knew when they were withdrawn (for review, see Dews et al., 2002). In a population of 144 students that discriminated caffeine only at chance level, the frequency of headaches was similar in the group in which caffeine was withdrawn and the one receiving caffeinated coffee over the 3-d protocol. This effect was interpreted as an expectancy effect that might not have occurred in other studies where subjects were able to discriminate caffeine (Smith, 1996). Caffeine withdrawal was also reported recently in schoolchildren, either only as a deterioration of performance (Bernstein et al., 1998) or as physical symptoms (Bernstein et al., 2002; Oberstar et al., 2002). Caffeine withdrawal symptoms can also occur in newborns whose mothers were heavy coffee drinkers during pregnancy. The infants display irritability, high emotivity, and, eventually, vomiting. Symptoms begin at birth but spontaneously disappear after a few days (McGowan et al., 1988).

There is a relationship between caffeine withdrawal, the development of headaches, and changes in cerebral blood flow. Cerebral blood flow velocities increase during withdrawal headaches, significantly decrease within 30 min after caffeine intake in all subjects, and return to baseline values after 2 h (Couturier et al., 1997). Thus, as suggested in previous studies, increased blood volume may be involved in caffeine withdrawal headache (Dreisbach and Pfeiffer, 1943; Von Borstel et al., 1983; Hirsch, 1984; Matthew and Wilson, 1985). Caffeine withdrawal symptoms disappear soon after absorption of caffeine. This effect is strongly linked to the psychological satisfaction related to the ingestion of caffeine, especially the first cup of the day. The potential reversal of caffeine withdrawal-induced headache and other symptoms by the absorption of caffeine alone has been known for over 50 years and shown repeatedly (Dreisbach and Pfeiffer, 1943; Goldstein et al., 1969a,b; Griffiths and Woodson, 1988b; Hughes et al., 1991; Watson et al., 2000).

TOLERANCE TO THE EFFECTS OF CAFFEINE

Tolerance to a drug refers to an acquired change in responsiveness of a subject repeatedly exposed to the substance and can be considered in two ways. First, tolerance might indicate that the dose necessary to achieve the desired euphoric or reinforcing effects increases with time, thus inciting people to gradually consume more drug. Second, tolerance to the aversive effects of high doses of the drug may occur, hence leading people to consume higher doses over time.

TOLERANCE TO CAFFEINE IN ANIMALS

In mice, cats, and squirrel monkeys chronically treated with the methylxanthine, tolerance to caffeine-induced locomotor stimulation, cerebral electrical activity, reinforcement thresholds for electrical brain stimulation, schedule-controlled responding maintained by presentation of food and electric shock, and thresholds for caffeine- or NMDA-induced seizures have been described (for review, see Holtzman and Finn, 1988; Griffiths and Mumford, 1996). The development of tolerance to caffeine in animals is rapid, usually insurmountable, and shows cross-tolerance with the other methylxanthines but not with psychomotor stimulants such as amphetamine and methylphenidate (Finn and Holtzman, 1987, 1988). However, most of these studies used high doses of caffeine, not relevant to human consumption, mainly because the development of tolerance was suggested to become more complete at higher doses of caffeine (Holtzman and Finn, 1988; Lau and Falk, 1995). At those doses, caffeine stimulates nonspecific functional activity in all brain regions (Nehlig et al., 1984, 1986). Tolerance to behavioral effects of caffeine in animals does not seem to involve adaptive changes in adenosine receptors (Holtzman et al., 1991; Georgiev et al., 1993) and may result from compensatory changes in the dopaminergic system as a result of chronic adenosine receptor blockade (Garrett and Holtzman, 1994) linked to a reduced effect of caffeine on striatopallidal neurons that are enriched in A_{2A} receptors (Svenningsson et al., 1999).

TOLERANCE TO CAFFEINE IN HUMANS

In humans, the tolerance to some peripheral actions of caffeine has been shown to occur. This is the case for the effect of caffeine on blood pressure and heart rate (Robertson et al., 1981; Ammon et al., 1983; Denaro et al., 1991; Shi et al., 1993; Satoh and Tanaka, 1997), diuresis (Eddy and Downs, 1928), plasma adrenaline and noradrenaline levels, and renin activity (Robertson et al., 1981). Tolerance to some subjective effects of caffeine, such as increases in tension/anxiety, jitteriness/nervousness, activity/stimulation/energy, and the strength of the drug's effect was reported to occur (Evans and Griffiths, 1992). Although tolerance to the enhancement of arithmetic skills by caffeine was shown (Satoh and Tanaka, 1997), there is only limited evidence for tolerance to caffeine-induced alertness and wakefulness (Goldstein et al., 1965; Colton et al., 1968; Höfer and Bättig, 1994a,b). These effects are paralleled by the lack of tolerance of cerebral energy metabolism to caffeine; an acute administration of 10 mg/kg caffeine induces the same metabolic increases whether rats have been exposed to a previous daily chronic treatment of caffeine or saline for 15 d (Nehlig et al., 1986). These data show that every single exposure to caffeine is able to produce cerebral stimulant effects, and this is mainly true in the areas that control locomotor activity (caudate nucleus) and the structures involved in the sleep–wake cycle (locus coeruleus, raphe nuclei, and reticular formation) that are especially sensitive to the effects of caffeine (Nehlig et al., 1986).

In humans, sleep seems to be the physiological function most sensitive to the effects of caffeine, as reviewed by Snel (1993). Generally more than 200 mg of caffeine are needed to affect sleep significantly. It is not clearly established yet whether or not the difference in the sensitivity to the effects of coffee on sleep could be attributable to tolerance. According to some studies, this difference could reflect interindividual sensitivity to caffeine, possibly related to lower rates of caffeine metabolism in poor sleepers (Levy and Zylber-Katz, 1983; Tiffin et al., 1995). The development of tolerance to sleep disturbances seems to relate to caffeine intake; heavy coffee drinkers appear less sensitive to caffeine-induced sleep disturbances than light coffee drinkers (Colton et al., 1968). However, the tolerance is not complete, and their sleep efficiency remains below 90% of the baseline value (Bonnet and Arand, 1992).

Thus, in humans, tolerance to behavioral effects of caffeine appears to be of low magnitude and incomplete (Fredholm et al., 1999; Dews et al., 2002; Watson et al., 2002). This may underlie individual differences in the susceptibility and tolerance to caffeine-induced effects. Moreover, mechanisms of tolerance may be overwhelmed by the nonlinear accumulation of caffeine and its main metabolites in the human body when caffeine metabolism becomes saturable under multiple dosing conditions (Denaro et al., 1990, 1991).

REINFORCEMENT PROPERTIES OF CAFFEINE

REINFORCING EFFECTS OF CAFFEINE IN ANIMALS

Reinforcing efficacy of a drug refers to the relative efficacy in establishing or maintaining a behavior on which the delivery of the drug is dependent. In animals, intravenous self-administration and behavioral reinforcement of caffeine have been studied in a paradigm allowing the animals to self-administer the drug by pressing a lever (Griffiths and Mumford, 1995). According to the studies, caffeine was shown to be self-injected in all (Griffiths et al., 1979; Dworkin et al., 1993; Griffiths and Mumford, 1995) or only in a limited subset of animals (Schuster et al., 1969; Atkinson and Enslen, 1976). A sporadic pattern of caffeine self-administration, characterized by periods with high rates of self-injection alternating with periods of rather low intake, was found in nonhuman primates (Deneau et al., 1969; Griffiths et al., 1979; Griffiths and Mumford, 1995). Thus, although caffeine seems to be able to act as a reinforcer in some conditions, there is a marked difference between caffeine and classic drugs of abuse such as amphetamine and cocaine, whose self-administration is maintained across species and conditions (Griffiths et al., 1979; Pontieri et al., 1995). Caffeine was shown to be able to reinstate estinguished cocaine-taking behavior in rats. This effect was more marked when caffeine was given 1 d compared to 4 d following the last cocaine self-administration session. Thus, extended withdrawal is able to increase the priming effects of caffeine (Worley et al., 1994; Schenk et al., 1996). However, it must be remembered that these animal studies used intravenous self-administration, whereas in humans, caffeine is always consumed orally. It is known that the former mode of administration is by far more addictive than the latter one (Heishman and Henningfield, 1994). Indeed, recent studies reported that caffeine injection increases the amount of self-administered cocaine while caffeine drinking reduces it (Kuzmin et al., 1999, 2000). Thus, caffeine does not appear to be a very robust reinforcer in animals.

REINFORCING EFFECTS OF CAFFEINE IN HUMANS

In humans, the widely recognized behavioral stimulant and mildly reinforcing properties of caffeine are probably responsible for the maintenance of caffeine self-administration, primarily in the form of caffeinated beverages such as coffee, tea, and cola (Nehlig and Debry, 1994; Griffiths and Mumford, 1995). In some studies, the choice of caffeine has been shown to be more potently controlled by avoiding withdrawal than by its positive effects (Rogers and Richardson, 1993; Shi et al., 1993; Yeomans et al., 2002), while other data support the hypothesis that the true performance-enhancing effects of caffeine are responsible for its self-administration (Rogers et al., 1995). Most data showed that caffeine reinforcement occurs in 100% of heavy caffeine consumers (1020 to 1530 mg/d) that also had histories of alcohol or drug abuse (Griffiths et al., 1986, 1989). For moderate caffeine users (128 to 595 mg/d), caffeine reinforcement occurs in about 45% (Griffiths and Woodson, 1988a; Hughes et al., 1991, 1992; Oliveto et al., 1991, 1992) to possibly 80 to 100% of the subjects (Evans et al., 1994).

Caffeine reinforcement varies with the dose. Doses of caffeine encountered in tea and coffee are high enough to act as reinforcers, since people look for them in case of withdrawal symptoms (Hughes et al., 1991). A dose of 25 to 50 mg of caffeine per cup of coffee acts as a reinforcer, while increasing doses beyond 50 or 100 mg tends to decrease the choice of caffeine or the frequency of caffeine self-administration (Griffiths and Mumford, 1995). High doses of caffeine (400 to 600 mg in a single dose) are avoided (Griffiths and Woodson, 1988b). The choice of caffeine seems also to be oriented towards avoiding withdrawal rather than its positive effects (Shuh and Griffiths, 1997). Subjects that consistently suffered from caffeine withdrawal headache increased their chance to select caffeinated coffee (containing 100 mg caffeine) by 2.6-fold (Hughes et al., 1993).

The conditions under which caffeine acts as a reinforcer are not clearly understood. However, the possible reinforcing effects of coffee unrelated to caffeine, but related to its smell, taste, and the social environment usually accompanying coffee consumption should be taken into account in the everyday motivations for caffeine-containing or caffeine-free coffee consumption. In subjects with a mean habitual coffee consumption of six cups/d who switched to the consumption of 600 mg of caffeine either in tablets or to decaffeinated instant coffee for 3 d, the desire for coffee in the next 3 d largely increased in the group given caffeine tablets but remained unchanged in the group given decaffeinated instant coffee, although the latter group experienced marked symptoms of caffeine withdrawal (Höfer and Bättig, 1994b).

Taken together, all these data confirm the fact that caffeine is less reinforcing than amphetamines and related psychomotor stimulants (Johanson et al., 1983; Chait et al., 1987; Stern et al., 1989).

IS CAFFEINE ACTIVATING THE BRAIN CIRCUITS UNDERLYING DEPENDENCE TO DRUGS?

The critical role of the mesolimbic dopamine system has been emphasized as underlying reinforcement and drug dependence (Koob, 1992; Self and Nestler, 1995). This system consists of the dopaminergic neurons originating in the ventral tegmental area and ending in the nucleus accumbens and prefrontal cortex. The latter nucleus that plays a central role in the mechanism of drug dependence is functionally and morphologically divided into a core and a shell. The medioventral shell is related to the limbic "extended amygdala" assumed to play a role in emotional, motivational, and reward functions, whereas the laterodorsal core regulates somatomotor functions (Heimer et al., 1991).

The specificity of cocaine, amphetamines, morphine, alcohol, cannabinoids, and nicotine is to activate selectively and at low doses the dopaminergic neurotransmission in the shell of the nucleus accumbens (Pontieri et al., 1995, 1996; Tanda et al., 1997), a property that has been related to the strong addictive properties of these drugs (Koob, 1992; Self and Nestler, 1995). Simultaneously, amphetamines, cocaine, and nicotine also induce specific increases in the rates of cerebral glucose utilization and blood flow in the shell of the nucleus accumbens but no changes in the core of the same nucleus (Porrino et al., 1984, 1988; Stein and Fuller, 1992; Orzi et al., 1993; Pontieri et al., 1996). The doses of addictive drugs that elicit the increase in functional activity in the shell of the nucleus accumbens are rather low and activate only a very limited number of brain regions together with the latter structure, such as the caudate nucleus in the case of amphetamines (Porrino et al., 1984).

Unlike the drugs of abuse, caffeine increases dopamine release in the caudate nucleus (Okada et al., 1996, 1997), which reflects the stimulatory properties of caffeine on locomotor activity (for review, see Nehlig et al., 1992; Nehlig and Debry, 1994) but does not induce any release of dopamine in the shell of the nucleus accumbens when injected at doses ranging from 0.5 to 5.0 mg/kg (Acquas et al., 2002; Solinas et al., 2002), which represent the normal human daily consumption (Debry, 1994; Barone and Roberts, 1996). It is necessary to reach quite high doses of caffeine (10 or 30 mg/kg) in rats, no longer representative of human consumption — 5 to 15 times the daily human consumption in one dosage (Debry, 1994; Barone and Roberts, 1996) — to elicit a release of dopamine from the shell of the nucleus accumbens (Solinas et al., 2002). Likewise, the acute administration of caffeine does not lead to a metabolic increase in the shell of the nucleus accumbens at doses ranging from 1 to 5 mg/kg in the rat. Only the dose of 10 mg/kg caffeine elicits metabolic increases in the vast majority of brain regions including the shell and the core of the nucleus accumbens, the extrapyramidal motor system, most limbic and thalamic regions, and the cerebral cortex (Nehlig et al., 1984, 1986; Nehlig and Boyet, 2000).

Taken together, these data show that the psychostimulant properties of caffeine are unrelated to the stimulation of endogenous dopamine transmission in the shell of the nucleus accumbens, which is the case for the drugs of dependence. Caffeine acts primarily on the extrapyramidal motor system, leading to a release of dopamine (Okada et al., 1995, 1996), and increases functional activity in the caudate nucleus and cerebral structures related to the sleep–wake cycle, such as the reticular formation, raphe nuclei, and locus coeruleus (Nehlig et al., 1984, 1986, 2000). These data are in good accordance with the facilitated motor output (James, 1991; Lorist et al., 1994), the increase in wakefulness reported in humans after caffeine ingestion (James, 1991), and the caffeine-induced dopamine release in the caudate nucleus and the prefrontal cortex (Okada et al., 1996, 1997; Acquas et al., 2002). The psychostimulant properties of caffeine are likely to be the result of blockade of adenosine A_{2A} receptors in the striatum, including the nucleus accumbens (Palmer and Stiles, 1995; Dixon et al., 1996). The inability to activate dopamine transmission in the nucleus accumbens reflects the weak reinforcing properties of caffeine. Thus, caffeine, in spite of its ability to induce tolerance and withdrawal symptoms, does not consistently induce the behavioral abnormalities typical of addictive drugs, as reported in DSM-IV (APA, 1994). Conversely, the property of amphetamines, cocaine, and nicotine of increasing dopamine release in the shell of the nucleus accumbens is quite specific and occurs at doses that do not lead to the activation of other brain regions (Porrino et al., 1988; Stein and Fuller, 1992; Pontieri et al., 1996).

CONCLUSION

Among the factors considered in the present review to try to assess the possible dependence potential of caffeine, it appears that tolerance does not develop completely. Withdrawal symptoms are only present in a subset of the general population after abrupt cessation of caffeine, while progressive dosage decrease avoids the physical symptoms of withdrawal (James et al., 1988). Finally, caffeine does not share with the common drugs of abuse the ability to stimulate functional activity and dopamine release in the shell of the nucleus accumbens, the anatomical substrate of addiction.

Moreover, at present, withdrawal syndroms are no longer given the same importance as previously when considering maintenance of drugs of abuse. Formerly, it was considered that the desire of preventing withdrawal symptoms was the main reason for maintaining sustained drug abuse. This was the case for heroin. However, cocaine or alcohol abusers experience only minimal withdrawal symptoms, and it appears that abuse behavior can be strongly maintained when abstinence symptoms are inconsequential, as with cocaine, or rare as with alcohol. Conversely, many therapeutic agents, such as antihypertensive medication, produce tolerance and, if stopped abruptly, lead to withdrawal symptoms but are not abused. Thus, dependence and abuse should not be associated, and, as pointed out recently, discussing caffeine in terms of drugs of abuse might in fact trivialize the dangers of drugs such as cocaine (Dews et al., 2002).

Altogether, it appears that although caffeine partly fulfills some of the criteria defined for drug dependence, the relative risk of addiction of caffeine is quite low, and as reported caffeine is the lowest among seven drugs or drug classes considered (Goldstein and Kalant, 1990). Moreover, with the exception of poisoning by very high doses of caffeine, the consumption of moderate doses of methylxanthine has not been reported as presenting potential harmful effects (Nawrot et al., 2003).

REFERENCES

Acquas, E., Tanda, G. and Di Chiara, G. (2002) Differential effects of caffeine on dopamine and acetylcholine transmission in brain areas of drug-naive and caffeine-pretreated rats. *Neuropsychopharmacology*, 27, 182–193.

American Psychiatric Association (1987) *Diagnostic and Statistical Manual of Mental Disorders, Revised Third Edition*. American Psychiatric Association, Washington, DC.

American Psychiatric Association (1992) *Diagnostic and Statistical Manual of Mental Disorders, Fourth Edition*. American Psychiatric Association Washington, DC.

Ammon, H.P.T., Bieck, P.R., Mandalaz, D. and Verspohl, E.J. (1983) Adaptation of blood pressure to continuous heavy coffee drinking in young volunteers: a double-blind crossover study. *British Journal of Clinical Pharmacology*, 15, 701–706.

Atkinson, J. and Ensslen, M. (1976) Self-administration of caffeine by the rat. *Arzneimittelforschung*, 26, 2059–2061.

Barone, J.J. and Roberts, H.R. (1996) Caffeine consumption. *Food and Chemical Toxicology*, 34, 119–126.

Benowitz, N.L. (1990) Clinical pharmacology of caffeine. *Annual Review of Medicine*, 41, 277–288.

Bernstein, G.A., Carroll, M.E., Dean, N.W., Crosby, R.D., Perwien, A.R. and Benowitz, N.L. (1998) Caffeine withdrawal in normal school-age children. *Journal of the American Academy of Child and Adolescent Psychiatry*, 37, 858–865.

Bernstein, G.A., Carroll, M.E., Thuras, P.D., Cosgrove, K.P. and Roth, M.E. (2002) Caffeine dependence in teenagers. *Drug and Alcohol Dependence*, 66, 1–6.

Bonnet, M.H. and Arand, D.L. (1992) Caffeine use as a model of acute and chronic insomnia. *Sleep*, 15, 526–536.

Carney, J.M. (1982) Effects of caffeine, theophylline and theobromine on scheduled controlled responding in rats. *British Journal of Pharmacology*, 75, 451–454.

Carroll, M.E., Hagen, E.W., Asencio, M. and Brauer, L.H. (1989) Behavioral dependence on caffeine and phencyclidine in rhesus monkeys: interactive effects. *Pharmacology, Biochemistry and Behavior*, 31, 927–932.

Chait, L.D., Uhlenhuth, E.H. and Johanson, C.E. (1987) Reinforcing and subjective effects of several anorectics in normal human volunteers. *Journal of Pharmacology and Experimental Therapeutics*, 242, 777–783.

Colton, T., Gosselin, R.E. and Smith, R.P. (1968) The tolerance of coffee drinkers to caffeine. *Clinical Pharmacology and Therapeutics*, 9, 31–39.

Couturier, E.G.M., Laman, D.M., van Duijn, M.A.J. and van Duijn, H. (1997) Influence of caffeine and caffeine withdrawal on headache and cerebral blood flow velocities. *Cephalalgia*, 17, 188–190.

Curatolo, P.W. and Robertson, D. (1983) The health consequences of caffeine. *Annals of Internal Medicine*, 98, 641–653.

D'Amicis, A. and Viani, R. (1993) The consumption of coffee, in *Caffeine, Coffee and Health*, Garattini, S., Ed., Raven Press, New York, pp. 1–16.

Debry, G. (1994) *Coffee and Health*, John Libbey, London.

Denaro, C.P., Brown, C.R., Jacob, P., III and Benowitz, N.L. (1991) Effects of caffeine with repeated dosing. *European Journal of Clinical Pharmacology*, 40, 273–278.

Denaro, C.P., Brown, C.R., Wilson, M., Jacob, P., III and Benowitz, N.L. (1990) Dose-dependency of caffeine metabolism. *Clinical Pharmacology and Therapeutics*, 31, 358–369.

Deneau, G., Yanagita, T. and Seevers, M.H. (1969) Self-administration of psychoactive substances in the monkey: a measure of psychological dependence. *Psychopharmacologia*, 16, 30–48.

Dews, P.B., O'Brien, C.P. and Bergman, J. (2002) Caffeine: behavioral effects of withdrawal and related issues. *Food and Chemical Toxicology*, 40, 1257–1261.

Di Chiara, G. (1998) A motivational learning hypothesis of the role of mesolimbic dopamine in compulsive drug use. *Journal of Psychopharmacology*, 12, 54–67.

Di Chiara, G. (1999) Drug-addiction as dopamine-dependent associative learning disorder. *European Journal of Pharmacology*, 375, 13–30.

Dixon, A.K., Gubitz, A.K., Sirinathsinghji, D.J., Richardson, P.J. and Freeman, T.C. (1996) Tissue distribution of adenosine receptor mRNAs in the rat. *British Journal of Pharmacology*, 118, 1461–1468.

Dreisbach, R.H. and Pfeiffer, C. (1943) Caffeine withdrawal headache. *Journal of Laboratory and Clinical Medicine*, 28, 1212–1219.

Dworkin, S.I., Vrana, S.L., Broadbent, J. and Robinson, J.H. (1993) Comparing the reinforcing effects of nicotine, caffeine, methylphenidate and caffeine. *Medical and Chemical Research*, 2, 593–602.

Eddy, N.B. and Downs, A.W. (1928) Tolerance and cross-tolerance in the human subject to the diuretic effect of caffeine, theobromine and theophylline. *Journal of Pharmacology and Experimental Therapeutics*, 33, 167–174.

Ellison, C.R., Singer, M.R., Moore, L.L., Nguyen, U.S.D.T., Garrahie, E. and Maror, J.K. (1995) Current caffeine intake in young children: amount and sources. *Journal of the American Dietetic Association*, 95, 802–804.

Evans, S.M. and Griffiths, R.R. (1992) Caffeine tolerance and choice in humans. *Psychopharmacology*, 108, 51–59.

Evans, S.M., Critchfield, T.S. and Griffiths, R.R. (1994) Caffeine reinforcement demonstrated in a majority of moderate caffeine users. *Behavioral Pharmacology*, 5, 231–238.

Finn, I.B. and Holtzman, S.G. (1986) Tolerance to caffeine-induced stimulation of locomotor activity in rats. *Journal of Pharmacology and Experimental Therapeutics*, 238, 542–546.

Finn, I.B. and Holtzman, S.G. (1987) Pharmacological specificity of tolerance to caffeine-induced stimulation of locomotor activity. *Psychopharmacology*, 93, 428–434.

Finn, I.B. and Holtzman, S.G. (1988) Tolerance and cross-tolerance to theophylline-induced stimulation of locomotor activity in rats. *Pharmacology, Biochemistry and Behavior*, 35, 477–479.

Fredholm, B.B., Bättig, K., Holmen, J., Nehlig, A. and Zvartau, E. (1999) Actions of caffeine on the brain with special reference to the factors that contribute to its widespread use. *Pharmacological Reviews*, 51, 83–133.

Garrett, B.E. and Holtzman, S.G. (1994) Caffeine cross-tolerance to selective dopamine D_1 and D_2 receptor agonists but not to their synergistic interaction. *European Journal of Pharmacology*, 262, 65–75.

Georgiev, V., Johansson, B. and Fredholm, B.B. (1993) Long-term caffeine treatment leads to a decreased susceptibility to NMDA-induced clonic seizures in mice without changes in adenosine A_1 receptor number. *Brain Research*, 612, 271–277.

Gilliland, K. and Bullock, W. (1984) Caffeine: a potential drug of abuse. *Advances in Alcohol and Substances of Abuse*, 3, 53–73.

Goldstein, A. and Kaizer, S. (1969a) Psychotropic effects of caffeine in man. III. A questionnaire survey of coffee drinking and its effects in a group of housewifes. *Clinical Pharmacology and Therapeutics*, 10, 477–488.

Goldstein, A., Kaizer, S. and Whitby, O. (1969b) Psychotropic effects of caffeine in man. IV. Quantitative and qualitative differences associated with habituation to coffee. *Clinical Pharmacology and Therapeutics*, 10, 489–497.

Goldstein, A. and Kalant, H. (1990) Drug policy: striking the right balance. *Science*, 249, 1513–1521.

Goldstein, A., Warren, R. and Kaizer, S. (1965) Psychotropic effects of caffeine in man. I. Interindividual differences in sensitivity to caffeine-induced wakefulness. *Journal of Pharmacology and Experimental Therapeutics*, 149, 156–159.

Griffiths, R.R., Bigelow, G.E. and Liebson, I.A. (1986) Human coffee drinking: reinforcing and physical dependence producing effects of caffeine. *Journal of Pharmacology and Experimental Therapeutics*, 239, 416–425.

Griffiths, R.R., Bigelow, G.E. and Liebson, I.A. (1989) Reinforcing effects of caffeine in coffee and capsules. *Journal of Experimental and Analytical Behavior*, 52, 127–140.

Griffiths, R.R., Bigelow, G.E., Liebson, I.A., O'Keffe, M., O'Leary, D. and Russ, N. (1986) Human coffee drinking: manipulation of concentration and caffeine dose. *Journal of Experimental and Analytical Behavior*, 45, 133–148.

Griffiths, R.R., Brady, J.V. and Bradford, L. D. (1979) Predicting the abuse liability of drugs with animal drug self-administration procedures: psychomotor stimulants and hallucinogens, in Advances in *Behavioral Pharmacology*, Vol. 2, Thompson, D. and Dews, P.B., Eds., Academic Press, New York, pp. 163–208.

Griffiths, R.R., Evans, S.M., Heishman, S.J., Preston, K.L., Sannerud, C.A., Wolf, B. and Woodson, P.P. (1990) Low-dose caffeine physical dependence in humans. *Journal of Pharmacology and Experimental Therapeutics*, 255, 1123–1132.

Griffiths, R.R. and Mumford, G.K. (1995) Caffeine: a drug of abuse?, in *Psychopharmacology: The Fourth Generation of Progress*, Bloom, F.E. and Kupfer, D.J., Eds., Raven Press, New York, pp. 1699–1713.

Griffiths, R.R. and Mumford, G.K. (1996) Caffeine reinforcement, discrimination, tolerance and physical dependence in laboratory animals and humans, in *Handbook of Experimental Pharmacology*, Vol. 118, Schuster, C.R. and Kuhar, M.J., Eds., Springer Verlag, Heidelberg, pp. 315–341.

Griffiths, R.R. and Woodson, P.P. (1988a) Caffeine physical dependence: a review of human and laboratory animal studies. *Psychopharmacology*, 94, 437–451.

Griffiths, R.R. and Woodson, P.P. (1988b) Reinforcing effects of caffeine in humans. *Journal of Pharmacology and Experimental Therapeutics*, 246, 21–28.

Griffiths, R.R. and Woodson, P.P. (1988c) Reinforcing properties of caffeine: studies in humans and laboratory animals. *Pharmacology, Biochemistry and Behavior*, 29, 419–427.

Heimer, L., Zahm, D.S., Churchill, L., Kalivas, P.W. and Wohltmann, C. (1991) Specificity in the projection patterns of accumbal core and shell in the rat. *Neuroscience*, 41, 89–125.

Heishman, S.J. and Henningfield, J.E. (1992) Stimulus functions of caffeine in humans: relation to dependence potential. *Neuroscience and Biobehavioral Reviews*, 16, 273–287.

Heishman, S.J. and Henningfield, J.E. (1994) Is caffeine a drug of dependence?: criteria and comparisons. *Pharmacopsychoecologia*, 7, 127 136.

Hirsch, K. (1984) Central nervous system pharmacology of the dietary methylxanthines, in *The Methylxanthine Beverages and Food: Chemistry, Consumption, and Health Effects*, Spiller, G.A., Ed., Alan Liss, New York, pp. 235–301.

Höfer, L. and Bättig, K. (1994a) Cardiovascular, behavioral, and subjective effects of caffeine under field conditions. *Pharmacology, Biochemistry and Behavior*, 48, 899–908.

Höfer, L. and Bättig, K. (1994b) Psychophysiological effects of switching to caffeine tablets or decaffeinated coffee under field conditions. *Pharmacopsychoecologia*, 7, 169–177.

Holtzman, S.G. (1983) Complete, reversible, drug-specific tolerance to stimulation of locomotor activity by caffeine. *Life Sciences*, 33, 779–787.

Holtzman, S.G. (1990) Caffeine as a model drug of abuse. *Trends in Pharmacological Sciences*, 11, 355–356.

Holtzman, S.G. and Finn, I.B. (1988) Tolerance to the behavioral effects of caffeine in rats. *Pharmacology, Biochemistry and Behavior*, 29, 411–418.

Holtzman, S.G., Mante, S. and Minneman, K.P. (1991) Role of adenosine receptors in caffeine tolerance. *Journal of Pharmacology and Experimental Therapeutics*, 256, 62–68.

Hughes, J.R., Higgins, S.T., Bickel, W.K., Hunt, W.K., Fenwick, J.W., Gulliver, S.B. and Mireault, G.C. (1991) Caffeine self-administration, withdrawal, and adverse effects among coffee drinkers. *Archives of General Psychiatry*, 48, 611–617.

Hughes, J.R., Hunt, W.K., Higgins, S.T., Bickel, W.K., Fenwick, J.W. and Pepper, S.L. (1992) Effect of dose on the ability of caffeine to serve as a reinforcer in humans. *Behavioral Pharmacolology*, 3, 211–218.

Hughes, J.R., Oliveto, A.H., Bickel, W.K., Higgins, S.T. and Badger, G.J. (1993) Caffeine self-administration and withdrawal: incidence, individual differences and interrelationships. *Drug and Alcohol Dependence*, 32, 239–246.

Hughes, J.R., Oliveto, A.H., Helzer, J.E., Bickel, W.K. and Higgins, S.T. (1993) Indications of caffeine dependence in a population-based sample, in *Problems of Drug Dependence 1992*, NIDA Research Monographs No. 132, NIH Publication No. 93-3505, Harris, L.S., Ed., U.S. Government Printing Office, Washington, DC, p. 194.

Hughes, J.R., Oliveto, A.H., Helzer, J.E., Higgins, S.T. and Bickel, W.K. (1992) Should caffeine abuse, dependence, or withdrawal be added to DSM-IV and ICD-10? *American Journal of Psychiatry*, 149, 33–40.

James, J.E. (1991) *Caffeine and Health*, Academic Press, London.

James, J.E., Paull, I., Cameron-Traub, E., Miners, J.O., Lelo, A. and Birkett, D.J. (1988) Biochemical validation of self-reported caffeine consumption during caffeine fading. *Journal of Behavioral Medicine*, 11, 15–30.

Johanson, C.E., Kilgore, K. and Uhlenhuth, E.H. (1983) Assessment of dependence potential of drugs in humans using multiple indices. *Psychopharmacology*, 81, 144–149.

Kessler, R.C., McGonagh, K.A., Zhao, S., Nelson, C.B., Hughes, M., Eshleman, S. and Wittchen, H.U. (1994) Lifetime and 12-month prevalence of DSM-III-R psychiatric disorders in the United States: results from the National Comorbidity Survey. *Archives of General Psychiatry*, 51, 8–19.

Koob, G.F. (1992) Drugs of abuse: anatomy, pharmacology and function of reward pathways. *Trends in Pharmacological Sciences*, 13, 177–184.

Kuzmin, A., Johansson, B., Semenova, S. and Fredholm, B.B. (2000) Differences in the effects of chronic and acute caffeine on self-administration of cocaine in mice. *European Journal of Neuroscience*, 12, 3026–3032.

Kuzmin, A., Johansson, B., Zvartau, E.E. and Fredholm, B.B. (1999) Caffeine, acting on adenosine A_1 receptors, prevents the extinction of cocaine-seeking behavior in mice. *Journal of Pharmacology and Experimental Therapeutics*, 290, 535–542.

Lane, J.D. (1997) Effects of brief caffeinated-beverage deprivation on mood, symptoms, and psychomotor performance. *Pharmacology, Biochemistry and Behavior*, 58, 203–208.

Lau, C.E. and Falk, J.L. (1995) Dose-dependent surmountability of locomotor activity in caffeine tolerance. *Pharmacology, Biochemistry and Behavior*, 52, 139–143.

Levy, M. and Zylber-Katz, E. (1983) Caffeine metabolism and coffee attributed sleep disturbances. *Clinical Pharmacology and Therapeutics*, 33, 770–775.

Lorist, M.M., Snel, J. and Kok, A. (1994) Influence of caffeine on information processing stages in well rested and fatigued subjects. *Psychopharmacology*, 113, 411–421.

Matthew, R.J. and Wilson, W.H. (1985) Caffeine consumption, withdrawal and cerebral blood flow. *Headache*, 25, 305–309.

McGowan, J.D., Altman, R.E. and Kanto, W.P. (1988) Neonatal withdrawal symptoms after chronic maternal ingestion of caffeine. *South Medical Journal*, 81, 1092–1094.

Mumford, G.K., Neill, D.B. and Holtzman, S.G. (1988) Caffeine elevates reinforcement threshold for electrical brain stimulation: tolerance and withdrawal changes. *Brain Research*, 459, 163–167.

Nawrot, P., Jordan, S., Eastwood, J., Rotstein, J., Hugenholtz, A. and Feeley, M. (2003) Effects of caffeine on human health. *Food Additive Contaminants*, 20, 1–30.

Nehlig, A. (1999) Are we dependent on coffee and caffeine?: a review on human and animal data. *Neuroscience and Biobehavioral Reviews*, 23, 563–576.

Nehlig, A. and Boyet, S. (2000) Dose-response study of caffeine effects on cerebral functional activity with a specific focus on dependence. *Brain Research*, 858, 71–77.

Nehlig, A., Daval, J.L., Boyet, S. and Vert, P. (1986) Comparative effects of acute and chronic administration of caffeine on local cerebral glucose utilization in the conscious rat. *European Journal of Pharmacology*, 129, 93–103.

Nehlig, A., Daval, J.L. and Debry, G. (1992) Caffeine and the central nervous system: mechanisms of action, biochemical, metabolic and psychostimulant effects. *Brain Research Reviews*, 17, 139–170.

Nehlig, A. and Debry, G. (1994) Effects of coffee on the central nervous system, in *Coffee and Health*, Debry, G., Ed., John Libbey, London, pp. 157–249.

Nehlig, A., Lucignani, G., Kadekaro, M., Porrino, L.J. and Sokoloff, L. (1984) Effects of acute administration of caffeine on local cerebral glucose utilization in the rat. *European Journal of Pharmacology*, 101, 91–100.

Oberstar, J.V., Bernstein, G.A. and Thuras, P.D. (2002) Caffeine use and dependence in adolescents: one-year follow-up. *Journal of Child and Adolescent Psychopharmacology*, 12, 127–135.

Okada, M., Kiryu, K., Kawata, Y., Mizuno, K., Wada, K., Tasaki, H. and Kaneko, S. (1997) Determination of the effects of caffeine and carbamazepine on striatal dopamine release by *in vivo* microdialysis. *European Journal of Pharmacology*, 321, 181–188.

Okada, M., Mizuno, K. and Kaneko, S. (1996) Adenosine A1 and A2 receptors modulate extracellular dopamine levels in rat striatum. *Neuroscience Letters*, 212, 53–56.

Oliveto, A.H., Hughes, J.R., Higgins, S.T., Bickel, W.K., Pepper, S.L., Shea, P.J. and Fenwick, J.W. (1992) Forced-choice versus free-choice procedures: caffeine self-administration in humans. *Psychopharmacology*, 109, 85–91.

Oliveto, A.H., Hughes, J.R., Pepper, S.L., Bickel, W.K. and Higgins, S.T. (1991) Low doses of caffeine can serve as reinforcers in humans, in *Problems of Drug Dependence, 1990*, NIDA Research Monograph No. 105, Harris, L.S., Ed., U.S. Government Printing Office, Washington, DC, p. 442.

Orzi, F., Morelli, M., Fieschi, C. and Pontieri, F.E. (1993) Metabolic mapping of the pharmacological and toxicological effects of dopaminergic drugs in experimental animals. *Cerebrovascular and Brain Metabolism Reviews*, 5, 95–121.

Palmer, T.M. and Stiles, G.L. (1995) Adenosine receptors. *Neuropharmacology*, 34, 683–694.

Pontieri, F.E., Tanda, G. and Di Chiara, G. (1995) Intravenous cocaine, morphine, and amphetamine preferentially increase extracellular dopamine in the "shell" as compared with the "core" of the rat nucleus accumbens. *Proceedings of the National Academy of Sciences U.S.A.*, 92, 12304–12308.

Pontieri, F.E., Tanda, G., Orzi, F. and Di Chiara, G. (1996) Effects of nicotine on the nucleus accumbens and similarity to those of addictive drugs. *Nature*, 382, 255–257.

Porrino, L.J. (1993) Functional consequences of acute cocaine treatment depend on route of administration. *Psychopharmacology*, 112, 343–351.

Porrino, L.J., Domer, F.R., Crane, A.M. and Sokoloff, L. (1988) Selective alterations in cerebral metabolism within the mesocorticolimbic dopaminergic system produced by acute cocaine administration in rats. *Neuropsychopharmacology*, 1, 109–118.

Porrino, L.J., Lucignani, G., Dow-Edwards, D. and Sokoloff, L. (1984) Correlation of dose-dependent effects of acute amphetamine administration on behavior and local cerebral metabolism in rats. *Brain Research*, 307, 311–320.

Robertson, D., Wade, D., Workman, R., Woosley, R.L. and Oates, J.A. (1981) Tolerance to the humoral and hemodynamic effects of caffeine in man. *Journal of Clinical Investigation*, 67, 1111–1117.

Rogers, P.J. and Richardson, N.J. (1993) Why do we like drinks that contain caffeine? *Trends in Food Science and Technology*, 4, 108–111.

Rogers, P.J., Richardson, N.J. and Dernoncourt, C. (1995) Caffeine use: is there a net benefit for mood and psychomotor performance? *Neuropsychobiology*, 31, 195–199.

Satoh, H. and Tanaka, T. (1997) Comparative pharmacology between habitual and non-habitual coffee-drinkers: a practical class exercise in pharmacology. *European Journal of Clinical Pharmacology*, 52, 239–240.
Schenk, S., Worley, C.M., McNamara, C. and Valadez, A. (1996) Acute and repeated exposure to caffeine: effects on reinstatement of extinguished cocaine-taking behavior in rats. *Psychopharmacology*, 126, 17–23.
Schuster, C.R., Woods, J.H. and Seevers, M.H. (1969) Self-administration of central stimulants by the monkey, in *Abuse of Central Stimulants*, Sjoqvist, F. and Tottie, M., Eds., Raven Press, New York, pp. 339–347.
SECED (1980) Rapport descriptif de la consommation de café, Paris, France.
Self, D.W. and Nestler, E.J. (1995) Molecular mechanisms of drug reinforcement and addiction. *Annual Review of Neuroscience*, 18, 463–495.
Shi, J., Benowitz, N.L., Denaro, C.P. and Sheiner, L.B. (1993) Pharmacokinetic-pharmacodynamic modeling of caffeine: tolerance to pressor effects. *Clinical Pharmacology and Therapeutics*, 53, 6–14.
Shuh, K.J. and Griffiths, R.R. (1997) Caffeine reinforcement: the role of withdrawal. *Psychopharmacology*, 130, 320–326.
Silverman, K. and Griffiths, R.R. (1992) Low-dose discrimination and self-reported mood effects in normal volunteers. *Journal of Experimental and Analytical Behavior*, 57, 91–107.
Silverman, K., Evans, S.M., Strain, E.C. and Griffiths, R.R. (1992) Withdrawal syndrome after the double-blind cessation of caffeine consumption. *New England Journal of Medicine*, 327, 1109–1114.
Sinton, C.M. and Petitjean, F. (1989) The influence of chronic caffeine administration on sleep parameters in the cat. *Pharmacology, Biochemistry and Behavior*, 32, 459–462.
Smith, A.P. (1996) Caffeine dependence: an alternative view. *Nature Medicine*, 2, 494.
Smith, A.P., Brockman, P., Flynn, R., Maben, A. and Thomas, M. (1993) Investigation of the effects of coffee on alertness and performance during the day and night. *Neuropsychobiology*, 27, 217–223.
Snel, J. (1993) Coffee and caffeine: sleep and wakefulness, in *Coffee, Caffeine and Health*, Garattini, S., Ed., Raven Press, New York, pp. 255–290.
Solinas, M., Ferré, S., You, Z.B., Karcz-Kubicha, M., Popoli, P. and Goldberg, S.R. (2002) Caffeine induces dopamine and glutamate release in the shell of the nucleus accumbens. *The Journal of Neuroscience*, 22, 6321–6324.
Stein, E.A. and Fuller, S.A. (1992) Selective effects of cocaine on regional cerebral blood flow in the rat. *Journal of Pharmacology and Experimental Therapeutics*, 262, 327–334.
Stern, K.N., Chait, L.D. and Johanson, C.E. (1989) Reinforcing and subjective effects of caffeine in normal human volunteers. *Psychopharmacology*, 98, 81–88.
Strain, E.C. and Griffiths, R.R. (1995) Caffeine dependence: fact or fiction? *Journal of the Royal Society of Medicine*, 88, 437–440.
Strain, E.C., Mumford, G.K., Silverman, K. and Griffiths, R.R. (1994) Caffeine dependence syndrome: evidence from case histories and experimental evaluations. *Journal of the American Medical Association*, 272, 1043–1048.
Svenningsson, P., Nomikos, G.G. and Fredholm, B.B. (1999) The stimulatory action and the development of tolerance to caffeine is associated with alterations in gene expression in specific brain regions. *The Journal of Neuroscience*, 19, 4011–4022.
Tanda, G., Pontieri, F.E. and Di Chiara, G. (1997) Cannabinnoid and heroin activation of mesolimbic dopamine transmission by a common μ_1 opioid receptor mechanism. *Science*, 276, 2048–2050.
Tiffin, P., Ashton, H., Marsh, R. and Kamali, F. (1995) Pharmacokinetic and pharmacodynamic responses to caffeine in poor and normal sleepers. *Psychopharmacology*, 121, 494–502.
Viani, R. (1993) Composition of coffee, in *Caffeine, Coffee and Health*, Garattini, S., Ed., Raven Press, New York, pp. 17–41.
Viani, R. (1996) Caffeine consumption, in *Proceedings of the Caffeine Workshop*, Thai FDA and ILSI, Bangkok, Thailand.
Vitiello, M. and Woods, S.C. (1977) Caffeine: preferential consumption by rats. *Pharmacology, Biochemistry and Behavior*, 3, 147–149.
von Borstel, R.W., Wurtman, R.J. and Conlay, L.A. (1983) Chronic caffeine consumption potentiates the hypotensive action of circulating adenosine. *Life Sciences*, 32, 1151–1158.
Watson, J.M., Deary, I. and Kerr, D. (2002) Central and peripheral effects of sustained caffeine use: tolerance is incomplete. *British Journal of Clinical Pharmacology*, 54, 400–406.

Watson, J.M., Lunt, M.J., Morris, S., Weiss, M.J., Hussey, D. and Kerr, D. (2000) Reversal of caffeine withdrawal by ingestion of a soft beverage. *Pharmacology, Biochemistry and Behavior*, 66, 15–18.

World Health Organization (1994) The ICD-10 classification of mental and behavioural disorders, World Health Organization, Geneva.

Worley, C.M., Valadez, A. and Schenk, S. (1994) Reinstatement of extinguished cocaine-taking behavior by cocaine and caffeine. *Pharmacology, Biochemistry and Behavior*, 48, 217–221.

Yeomans, M.R., Ripley, T., Davies, L.H., Rusted, J.M. and Rogers, P.J. (2002) Effects of caffeine on performance and mood depend on the level of caffeine abstinence. *Psychopharmacology*, 164, 241–249.

10 Caffeine and Parkinson's Disease

Michael A. Schwarzschild and Alberto Ascherio

CONTENTS

Introduction 147
The Epidemiology of Caffeine and Parkinson's Disease..147
Study Design 148
Case-Control Studies...148
Cohort Studies...149
Neuroprotection by Caffeine and More Specific Adenosine A_{2A} Antagonists in Animal
Models of Parkinson's Disease 152
Neuroprotection by Caffeine in a PD Model 152
Neuroprotection by Specific A_{2A} Antagonists in PD Models 153
A Broader Neuroprotective Role for A_{2A} Antagonists 155
Mechanisms of Protection by A_{2A} Antagonist in PD Models 156
Significance for Parkinson's Disease..159
References 159

INTRODUCTION

In Parkinson's Disease (PD) the dopamine-containing brain cells (dopaminergic neurons) that originate in the substantia nigra and innervate the striatum gradually degenerate. Dopamine released from these nigrostriatal neurons serves to enhance normal motor activity. Accordingly, the progressive loss of dopamine that accompanies the loss of these neurons leads to an insidious and eventually profound slowing of movements (bradykinesia) except for the classic tremor at rest in many PD patients. Why central dopaminergic neurons selectively degenerate in typical PD remains a mystery. The recently intensified pursuit of environmental as well as genetic clues has yielded new insights into the causes and potential for improved treatment of this disabling disease. This chapter reviews the remarkably convergent lines of epidemiological and laboratory evidence that caffeine can protect dopaminergic neurons and may help prevent the development of this disease. The possibility that caffeine also produces a symptomatic motor benefit in PD has not been substantiated and is reviewed elsewhere (Fredholm et al., 1999; Schwarzschild et al., 2002).

THE EPIDEMIOLOGY OF CAFFEINE AND PARKINSON'S DISEASE

Whereas in individuals under the age of 50 years, PD is often associated with inherited genetic variants of the disease, over 90% of PD cases are diagnosed after age 50 and typically occur without a clear family history. The predominance of nongenetic determinants is further supported by the fact that only 11% of the monozygotic twins of individuals diagnosed with PD after the age of 50 also have the disease, a rate similar to that found among dizygotic twins (Tanner et al., 1999). Epidemiologists have long been intrigued by the observation that cigarette smokers rarely develop PD, but in spite of extensive investigation it remains unclear whether nicotine or some other component of cigarette smoke reduces the risk of PD, or whether people predisposed to PD have an early aversion to smoking (Morens et al., 1995). More recently, coffee and caffeine consumption have emerged as powerful predictors of risk of PD. In this section we will review the epidemiological evidence supporting this association and discuss the contribution of epidemiological studies in clarifying the potential biological mechanisms.

STUDY DESIGN

To identify risk factors, epidemiologists are searching for clues in the diet, lifestyle, exposure to occupational or environmental chemicals, and other potentially significant aspects in the life of people with PD. Two approaches are commonly pursued in this research. The first is called a case-control study and entails identifying a group of individuals with PD (cases) and a comparable group of individuals of the same age and sex who do not have PD (controls). Information is then elicited from both groups using standardized techniques and focusing on lifestyle and events during the years preceding the diagnosis of PD for the cases, and a comparable period for the controls. A potential limitation of this research, as described in more detail below, is that the recall of events that occurred many years in the past is often inaccurate, and, more troublesome, the quality of information elicited from cases and controls may not be comparable. Cases are typically more motivated than healthy controls and may thus provide more accurate histories; on the other hand, beliefs about the causes of the disease, changes in lifestyle due to the disease itself, and cognitive impairment may affect recall. This problem is compounded by the difficulty in identifying and recruiting an appropriate control group, in part because many healthy people are unwilling to take part in the research. The second approach, known as a cohort study, consists of collecting information on the factors of interest in large numbers of healthy individuals that are then followed for many years to determine whether PD occurs more commonly among people with or without certain characteristics. A strength of this approach is that the information of interest is collected before the onset of PD and is thus not biased by the presence of the disease. Whereas a case-control study usually includes a few hundred people and can be completed in 2 to 3 years, a cohort study typically includes many thousands of people and a follow-up of 10 or more years. Not surprisingly, therefore, prospective studies are few. Below we will summarize the results of case-control and cohort studies on caffeine and PD.

CASE-CONTROL STUDIES

Questions on coffee and tea consumption and risk of PD were included in some of the early case-control studies on smoking (Nefzger et al., 1968; Baumann et al., 1980) or environmental factors (Zayed et al., 1990) and risk of PD, but published results were not adjusted for smoking and thus are difficult to interpret. The first smoking-adjusted analysis of coffee consumption and PD was reported by Jiménez-Jiménez et al. (1992), who studied 128 patients with PD (60 women) attending a movement disorder clinic in Madrid and 256 controls attending the emergency room of the same hospital for non-neurological disorders. The data supported a 30% lower risk of PD among coffee drinkers (coffee drinking was reported as yes/no only) as compared with nondrinkers in both men and women, but the results were not significant and received little attention. More detailed analyses on coffee consumption and risk of PD were later conducted in case-control studies in Germany (Hellenbrand et al., 1997) and Sweden (Fall et al., 1999). Hellenbrand et al. (1996) compared the dietary habits of 342 PD patients (118 women) recruited from nine German clinics with those of 342 controls from the same neighborhood or region. Only cases diagnosed in 1987 or later and 65 years of age or less were included. The odds ratio, which is an estimate of the relative risk, comparing subjects in the highest to those in the lowest quartile of coffee intake was 0.27 (95% CI: 0.14 to 0.52); no inverse association was found with tea. The authors also reported a strong inverse association between niacin intake and risk of PD and suggested that the observed association between coffee consumption and risk of PD could be due to its high niacin content. Niacin is required for the synthesis of the cofactor nicotinamide adenine dinucleotide (NADH) and for the function of glutathione reductase, and the authors speculated that a defective or suboptimal functioning of this and other enzymes requiring NADH might be relevant to the development of PD. However, a significant inverse association between coffee intake and risk of PD persisted after adjusting for niacin, in addition to smoking and energy intake (Hellenbrand et al., 1996); further, an inverse association between niacin and risk of PD has not been confirmed in prospective studies (see below). In a separate investigation including 113 individuals with PD and 263 controls, Fall et al. (1999) reported a highly significant inverse association between risk of PD and both coffee and tea consumption. The multivariate odds ratio comparing drinkers of five or more cups of coffee per day to nondrinkers was 0.14 (95% CI: 0.03 to 0.60). Although there are exceptions (Checkoway et al., 2002), overall the results of case-control investigations favored the existence of an inverse association between coffee or caffeine intake and risk of PD (Hernán et al., 2002).

COHORT STUDIES

The weaknesses of case-control studies for the investigation of the dietary etiology of chronic diseases are well known (Willett, 1998) and include the difficulty of recruiting an appropriate control group, the generally inaccurate assessment of diet during the relevant period that for PD may be several years before the diagnosis, and selective differences in recall of dietary habits between cases and controls. Thus, although the results of case-control studies suggest a potential protective effect of caffeine, recall or selection bias cannot be confidently excluded. These problems have been overcome in prospective investigations that took advantage of large cohorts of individuals without PD who reported their coffee and caffeine consumption and were then followed for many years. The relation between coffee or caffeine intake and risk of PD has been reported in four cohorts, the Honolulu Heart Program (HHP) (Ross et al., 2000), the Health Professionals Follow-up Study (HPFS) (Ascherio et al., 2001), the Nurses' Health Study (NHS) (Ascherio et al., 2001), and the Leisure World Cohort (LWC) (Paganini-Hill, 2001) (Table 10.1). In the latter, however, participants who already had PD at the beginning of the follow-up were not excluded. Results from a smaller cohort study (53 cases of PD) were published only as an abstract. Findings in the HHP, HPFS, and NHS cohorts are presented below.

TABLE 10.1
Prospective Cohort Studies on Coffee/Caffeine and Risk of Parkinson's Disease

Study	Participants	Place	Length of Follow-Up	Number of Cases
HHP	8,004 men of Japanese ancestry, aged 45–68 yr	Hawaii	30 yr	102
HPFS	47,351 men, aged 40–75 yr	U.S.	10 yr	157
NHS	88,565 women, aged 30–55 yr	U.S.	20 yr	131
LWC	13,979 residents of retirement community (2/3 women)	California	17 yr	395[a]

Note: HHP = Honolulu Heart Program; HPFS = Health Professionals Follow-up Study; NHS = Nurses' Health Study; LWC = Leisure World Cohort.

[a] Members of the LWC with Parkinson's disease at the beginning of the study were not excluded.

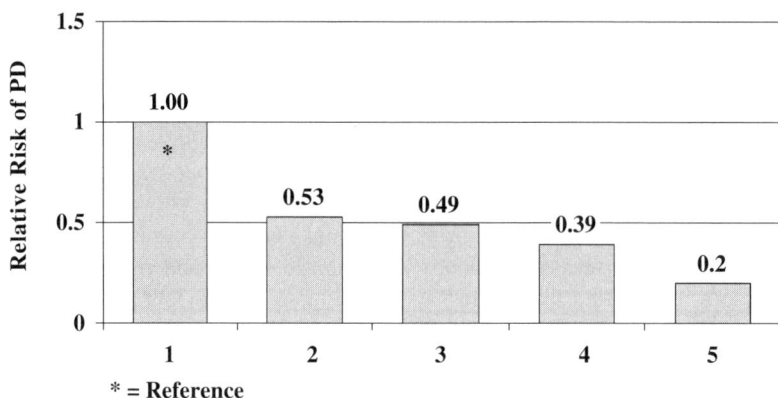

FIGURE 10.1 Age- and smoking-adjusted relative risk (RR) of Parkinson's disease by quintiles of caffeine intake, Honolulu Heart Program.

The HHP comprised over 8000 Japanese-American men who completed a 24-h diet recall at enrollment (1965 to 1968) and a food frequency questionnaire 6 years later and were followed for up to 30 years (Ross et al., 2000). In this cohort, the age- and smoking-adjusted risk of PD was five times higher among men who reported no coffee consumption at baseline compared to men who reported a consumption of 28 ounces (approximately 3.5 cups) of coffee or more (Figure 10.1). This association was also present among men who never smoked, and it cannot therefore be the result of residual confounding by cigarette smoking. Further, men with higher caffeine consumption at baseline still had a lower risk of PD during the second half of the follow-up (more than 15 years later). This last result argues against the possibility that the lower caffeine consumption among men who developed PD resulted from early, unrecognized symptoms of the disease. Analyses adjusted for niacin, alcohol, and other nutrients suggested that caffeine was responsible for this association. However, the size of the study was insufficient to examine the association between caffeine from sources other than coffee and risk of PD among noncoffee drinkers, and no information was collected on consumption of decaffeinated coffee. Thus, the possibility that compounds other than caffeine contributed to the observed inverse association could not be excluded.

The HPFS and NHS cohorts included data on both caffeinated and decaffeinated coffee and multiple assessments of caffeine consumption during the follow-up period (Ascherio et al., 2001). The HPFS was established in 1986 and comprised 51,529 male health professionals (dentists, optometrists, pharmacists, podiatrists, and veterinarians), aged 40 to 75 years. The NHS cohort was established in 1976 and compared 121,700 women who were registered nurses residing in 11 large states, aged 30 to 55 years. Participants in both cohorts provided detailed information about their medical history, lifestyle practices, and diet (Colditz et al., 1997). Follow-up questionnaires are mailed to participants every 2 years to update information on potential risk factors for chronic diseases and to ascertain whether major medical events have occurred. The caffeine analyses were based on 157 new cases of PD during 10 years of follow-up in men and 131 during 16 years of follow-up in women; most diagnoses were confirmed by a neurologist (85%) or by review of the medical records (5%), whereas the remaining 10% were confirmed by an internist or general physician. Intriguingly, we found an inverse association between caffeine intake and risk of PD in men, but not in women (Figure 10.2). Men consuming caffeine from either coffee or noncoffee sources had a lower risk of PD than noncaffeine consumers. A 50% reduction in risk of PD was observed among men consuming an amount of caffeine corresponding to one cup of coffee per day compared with men consuming no caffeine. A similar result was found in the Honolulu cohort. Further, in the HPFS cohort, consumption of tea and other caffeinated beverages among men who were not regular coffee drinkers (consumption < 1 cup/day) was also inversely associated with risk

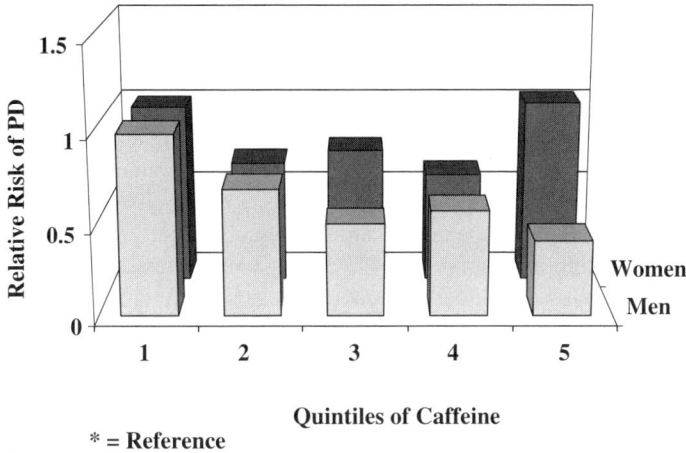

FIGURE 10.2 Age- and smoking-adjusted relative risk (RR) of Parkinson's disease by quintiles of caffeine intake, Nurses' Health Study and Health Professionals Follow-up Study.

of PD, whereas no association was found with consumption of decaffeinated coffee. These findings established that the reduction in risk of PD in men correlated directly with consumption of caffeine and not other coffee components. Among women, neither coffee nor caffeine from any source was significantly associated with risk of PD.

The finding that the risk of PD among caffeine consumers is lower than among nonconsumers in two large cohorts of men can only be explained if either caffeine consumption reduces the risk of PD or if some common genetic or environmental factor predisposes to both caffeine avoidance and PD (Schwarzschild et al., 2002). Specific explanations that have been proposed include the existence of a premorbid parkinsonian personality characterized by reduced novelty seeking (Paulson and Dadmehr, 1991) or an underlying preclinical olfactory deficit that prevents the rewarding effects related to the smell of coffee or tobacco (Benedetti et al., 2000). This is the same question raised by the results on cigarette smoking, and unfortunately it is difficult to reach a conclusive answer. A few further clues can, however, be inferred from epidemiological studies. At least for cigarette smoking, the underlying common predisposing factor is unlikely to be genetic, as an inverse association between cigarette smoking and PD risk has also been observed within monozygotic twin pairs (Tanner et al., 2002). Further, again for smoking, any predisposing factor must already be present during adolescence, as we found in the HPFS cohort that smoking at age 19 is associated with a significantly lower risk of PD after the age of 50 years (unpublished data). Similar evidence for caffeine is not available, although the observation that caffeine intake at the beginning of the study predicted risk of PD between 15 and 30 years later in the Honolulu cohort suggests that any common predisposing factor must precede the diagnosis of PD by at least 15 years (Ross et al., 2000). To further assess the temporal relationship between caffeine consumption and PD, we have examined the relative consumption of caffeine among men with PD in the HPFS according to the time of diagnosis as compared with caffeine consumption of men of the same age without PD. Although caffeine intake tends to decline with age, we found no evidence that this decline was steeper among men with PD, a result that suggests that the progressive degeneration of nigrostriatal dopaminergic neurons that accompanies PD has little effect on caffeine consumption.

A further clue to the mechanisms underlying the lower risk of PD among caffeine drinkers compared with nondrinkers may come from studies in women. The fact that we found no association between coffee or caffeine intake and risk of PD in our large prospective study of women, despite use of repeated and validated measures of consumption over a follow-up of 16 years (Ascherio et al., 2001), is unlikely to be due to chance. Further, a similar gender difference in the relationship of caffeine to PD has also been found in a case-control study that relied on prospectively collected information on coffee consumption (Benedetti et al., 2000). These findings suggest that caffeine may have different effects in men and women, perhaps because of hormonal differences. Caffeine is largely metabolized by the CYP1A2 isoenzyme of the P450 family (Fredholm et al., 1999), which also metabolizes estrogen (Pollock et al., 1999); by competing for the same enzyme, exogenous estrogen in oral contraceptives (Abernethy and Todd, 1985) or postmenopausal hormones (Pollock et al., 1999) inhibit caffeine metabolism. To address this possibility, we have therefore examined the interaction between use of postmenopausal hormones, caffeine consumption, and risk of PD among participants in the Nurses' Health Study. Overall, use of postmenopausal hormones was not associated with risk of PD. However, we found that among hormone users, women consuming six or more cups of coffee per day had a fourfold higher risk of PD (RR: 3.92, 95% CI: 1.49 to 10.34; $p = .006$) than women who never drink coffee; in contrast, among women who never used postmenopausal hormones, coffee drinkers had a lower risk of PD than nondrinkers. If caffeine avoidance and increased risk of PD were both caused by a premorbid personality or an olfactory deficit occurring more than 15 years before the diagnosis of PD, it would remain unclear how use of postmenopausal hormones would modify this association. Thus, albeit indirectly, this interaction supports a biological effect of caffeine on risk of PD. Independent confirmation in epidemiological studies and animal experiments exploring the mechanistic basis of this finding will be important. Meanwhile, this possible interaction should be considered in the planning and interpretation of trials of estrogen supplementation or caffeine use in women with PD, particularly because our results suggest that estrogen supplementation could be harmful among women consuming high amounts of caffeine.

In summary, the strength and consistency of the association between caffeine consumption and risk of PD, the lack of a convincing and specific alternative hypothesis, and the evidence of an interaction between caffeine and postmenopausal hormones suggest that caffeine consumption reduces the risk of PD in men. Further research is needed to clarify the possible mechanisms and the effects of caffeine among women.

NEUROPROTECTION BY CAFFEINE AND MORE SPECIFIC ADENOSINE A_{2A} ANTAGONISTS IN ANIMAL MODELS OF PARKINSON'S DISEASE

The recent epidemiological studies described above have established an association between the common consumption of coffee or other caffeinated beverages and a reduced risk of developing PD later in life. Despite their strength, these epidemiological investigations cannot conclusively answer the fundamental question: Does caffeine help prevent PD, or does PD or its causes help prevent the habitual use of caffeine? Although this question of causality is difficult to address in humans, animal models can offer useful clues to the answer. Here we review the evidence that caffeine and more specific antagonists of the adenosine A_{2A} receptor are in fact capable of protecting dopaminergic and other neurons from degeneration and death.

NEUROPROTECTION BY CAFFEINE IN A PD MODEL

The effect caffeine on the demise of nigrostriatal dopaminergic neurons has recently been investigated in the well-established 1-methyl-4-phenyl-1,2,3,6-tetrahydropyridine (MPTP) model of PD (Chen et al., 2001; Oztas et al., 2002; Xu et al., 2002a). Mice exposed to the dopamine neuron-specific toxin MPTP develop biochemical and anatomical lesions of the nigrostriatal system that parallel characteristic features of PD (Gerlach and Riederer, 1996). Caffeine, at doses in mice (5 to 30 mg/kg) comparable to those of typical human exposure, dose-dependently reverses the loss of striatal dopamine triggered by MPTP (Chen et al., 2001). Caffeine similarly attenuated the toxin-induced loss of dihydroxyphenylacetic acid (DOPAC), dopamine's major central nervous system metabolite, suggesting that caffeine is not simply altering dopamine metabolism in the remaining nigrostriatal nerve terminals. In addition to these biochemical markers of dopaminergic nigrostriatal function, the density of dopamine transporter (DAT) binding sites was measured as an anatomical marker of nigrostriatal innervation. Again, MPTP toxicity was diminished in the presence of caffeine, which significantly attenuated the MPTP-induced loss of striatal DAT (^{3}H-mazindol) binding sites. Caffeine's protective influence on the dopaminergic innervation of the striatum can be directly attributed to its ability to prevent the death of dopaminergic (DA) neurons originating in the substantia nigra. Stereological analysis of nigral dopaminergic (tyrosine hydroxylase-immunoreactive) neurons showed their MPTP-induced loss could be prevented by caffeine pretreatment (Oztas et al., 2002). The protective effect of caffeine was observed with different MPTP exposure paradigms (single and multiple doses) and in different mouse strains (C57Bl/6 and 129-Steel) (Chen et al., 2001).

Caffeine (1,3,7-trimethylxanthine) is metabolized by demethylation, initially to the dimethylxanthines theophylline, theobromine, and paraxanthine (1,3-, 3,7-, and 1,7- dimethylxanthine, respectively); the latter is the major dimethyl metabolite of caffeine in humans (Benowitz et al., 1995). Preliminary studies in mice demonstrate that theophylline and paraxanthine, which like caffeine are nonspecific adenosine receptor antagonists at low micromolar concentrations (Fredholm et al., 1999), can also attenuate MPTP toxicity (Xu et al., 2002b). In humans, the serum half-life of caffeine (which is typically ingested once to several times per day) is approximately 4 h (Morgan et al., 1982; Kaplan et al., 1997). Moreover, > 80% of caffeine is metabolized to paraxanthine (Benowitz et al., 1995). Thus, the findings of a protective effect of paraxanthine as well as caffeine in mice suggest that precise temporal pairing between caffeine and putative dopaminergic neurotoxin exposures in humans would not be critical for caffeine to reduce the risk of developing PD (if in fact caffeine were protective in humans).

Although the typical human pattern of frequent caffeine exposure indirectly supports the plausibility of neuroprotection against PD, it also raises the possibility that tolerance would develop to this protective action of caffeine. A characteristic feature of caffeine's psychomotor stimulant effect is that it diminishes after repeated exposure (i.e., that it shows tolerance) (Holtzman et al., 1991; Kaplan et al., 1993). To investigate the possibility that the neuroprotective effect of caffeine is also affected by prior exposure, its motor stimulant and neuroprotective effects were assessed in mice treated daily with caffeine or saline for over a week (Xu et al., 2002a). Repeated daily caffeine administration, under conditions that produced substantial locomotor tolerance, did not attenuate the protective effect of caffeine on MPTP-induced dopaminergic toxicity. Together these protective effects of caffeine and its metabolites support a causal basis for the inverse relationship between caffeine consumption and the risk of subsequently developing PD.

NEUROPROTECTION BY SPECIFIC A_{2A} ANTAGONISTS IN PD MODELS

The protective effect of caffeine in a mouse model of PD provides a compelling clue to the pathophysiology as well as the epidemiology of PD. Insight into how caffeine protects dopaminergic neurons may also lead to improved PD therapeutics aimed at slowing the underlying neurodegenerative process. A first step in pursuing this "caffeinated" clue has been the consideration of which of caffeine's known molecular targets may mediate its protective effect. Pharmacological studies indicate that the effects of caffeine on the central nervous system are mediated primarily by its antagonistic actions at the A_1 and A_{2A} subtypes of adenosine receptor (Fredholm et al., 1999). A_{2A} adenosine receptors may be particularly relevant because their expression in the brain is largely restricted to the striatum (see Figure 10.3) (Svenningsson et al., 1999), the major target of the dopaminergic neurons that degenerate in PD. Furthermore, their blockade or inactivation has been known to protect against excitotoxic and ischemic neuronal injury (see below).

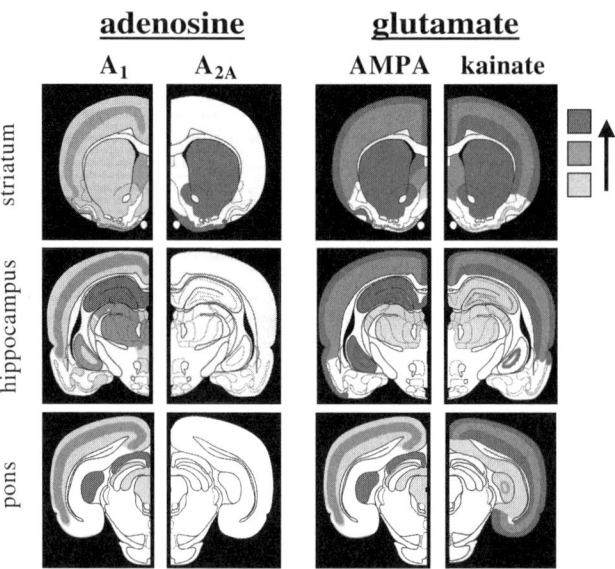

FIGURE 10.3 Brain expression patterns for subtypes of receptors for two neurotransmitters known to modulate dopaminergic neuron function. Composite distributions of specific radioligand binding to subtypes of adenosine and glutamate receptors are shown in coronal sections from the rostral, mid, and caudal rat brain (containing striatum, hippocampus, and pons, respectively). (Adapted with permission from M. Tohyama and K. Takatsuji, *Atlas of Neuroactive Substances and Their Receptors in the Rat*, Oxford University Press, Oxford, 1998). Increasing density of radioligand-binding to the receptors is indicated by increasing darkness along the gray scale on the right. Most receptors that modulate dopaminergic transmission in the striatum are widely distributed throughout the brain, whereas the A_{2A} subtype of adenosine receptor is largely restricted in its expression to the striatum and the underlying olfactory tubercle.

Accordingly, relatively specific A_{2A} as well as A_1 receptor antagonists were tested for their ability to mimic caffeine's attenuation of MPTP toxicity. MPTP-induced nigrostriatal lesions were attenuated by pretreatment with all A_{2A} antagonists tested, including both xanthine-based compounds [8-(3-chlorostyryl)caffeine (CSC) (Chen et al., 2002), 3,7-dimethyl-1-propargylxanthine (DMPX) (Chen et al., 2001), KW-6002 ((E)-1,3-diethyl-8-(3,4-dimethoxystyryl)-7-methyl-3,7-dihydro-1H-purine-2,6-dione) (Chen et al., 2001; Ikeda et al., 2002)] and those with nonxanthine structures [SCH 58261 (7-(2-phenylethyl)-5-amino-2-(2-furyl)-pyrazolo-[4,3-e]-1,2,4-triazolo[1,5-c]pyrimidine) (Chen et al., 2001) and ZM241385 (4-(2-[7-amino-2-{2-furyl}{1,2,4}triazolo{2,3-a}{1,3,5,}triazin-5-yl amino]ethyl)phenol)] (Neal Castagnoli, Jr., personal communication). The specificity of CSC with respect to its neuroprotective effect in the MPTP model has recently been called into question with the serendipitous finding that it possesses dual independent actions of high-potency inhibition of monoamine oxidase (MAO) B, as well as antagonism of the A_{2A} receptor (Chen et al., 2002; Petzer et al., 2003). Though none of the other xanthine- or nonxanthine-based A_{2A} antagonists above possesses comparable (if any) MAO B activity, the unexpected incomplete specificity of CSC even at low (nanomolar) concentrations highlights the pitfalls of adenosine pharmacology.

To circumvent such pharmacological limitations and definitively address the question of whether A_{2A} receptor blockade mimics the neuroprotective effect of caffeine, mice lacking functional A_{2A} receptors in which the gene for the receptor had been "knocked out" (A_{2A} KO mice) (Ledent et al., 1997; Chen et al., 1999) were assessed for their susceptibility to MPTP toxicity. MPTP-induced losses of striatal dopamine and dopamine transporter were significantly attenuated in A_{2A} KO mice compared to their wild-type littermates (Chen et al., 2001). These complimentary genetic and pharmacological approaches clearly demonstrate that A_{2A} receptor inactivation, like caffeine, reduces

MPTP toxicity. By contrast, multiple concentrations of the A_1 receptor antagonist 8-cyclopentyl-1,3-dipropylxanthine (CPX) showed no evidence of neuroprotection against the dopaminergic toxicity induced by multiple concentrations of MPTP (Chen et al., 2001). Recently, the neuroprotective effect of A_{2A} receptor blockade against dopaminergic neuron injury has been extended to another species and to another model of PD. The A_{2A} antagonist KW-6002 was found to prevent nigral dopaminergic neuron loss induced by 6-hydroxydopamine in rats (Ikeda et al., 2002) as well as MPTP toxicity in mice. Together these data suggest that caffeine can protect against dopaminergic neuron injury and death through its antagonistic action at the adenosine A_{2A} receptor.

A BROADER NEUROPROTECTIVE ROLE FOR A_{2A} ANTAGONISTS

These findings implicate endogenous adenosine acting on the A_{2A} receptor in the pathophysiology of nigrostriatal neuron lesions. However, this role for the A_{2A} receptor clearly extends beyond its targeting of central dopaminergic pathways to other populations of CNS neurons. For example, recent studies have demonstrated that the A_{2A} receptor contributes to the death of striatal medium spiny neurons in rodent models of Huntington's disease (HD). At very low doses, the A_{2A} antagonist SCH58261 attenuates striatal lesions induced by local infusion of the excitotoxin quinolinate (Popoli et al., 2002). Moreover, preliminary findings in A_{2A} receptor KO mice show that loss of striatal neurons induced by systemically administered 3-nitroproprionic acid (a complex II inhibitor and relatively specific striatal neuron toxin) is markedly reduced in the absence of the A_{2A} receptor or in the presence of the A_{2A} antagonist CSC (Fink et al., 2002). Of note, it is the subset of GABAergic striatal output neurons expressing high levels of A_{2A} receptor (i.e., those that project to the lateral globus pallidus) that degenerates earliest in HD (Glass et al., 2000) and whose absence may account for the involuntary choreic movements characteristic of this disorder.

In addition to the protection against striatal and nigral neuron loss offered by A_{2A} antagonists, their ability to protect neuronal populations outside the basal ganglia is well documented. For example, local injection of an A_{2A} antagonist can prevent the excitotoxic death of neurons in hippocampal cortex produced by the ionotropic glutamate receptor agonists kainate and quinolinate (Jones et al., 1998; Stone et al., 2001). Wider cortical damage in a variety of ischemic stroke models can also be attenuated by A_{2A} receptor blockers administered at the time of cerebral blood flow disruption (Gao and Phillis, 1994; Phillis, 1995; Von Lubitz et al., 1995; Monopoli et al., 1998). Similarly, transient focal ischemia produces substantially less brain damage in the cortex as well as the striatum of adult A_{2A} receptor KO mice compared to their wild-type littermates (Chen et al., 1999). Interestingly, a finding that focal ischemic brain injury in rats is dramatically attenuated by treatment with low doses of caffeine together with ethanol (Strong et al., 2000) has led to a therapeutic trial of this adenosine antagonist-CNS depressant combination in humans suffering acute stroke (Piriyawat et al., 2003).

It should be noted, however, that A_{2A} antagonists are not universally protective. In fact, outside the CNS the A_{2A} receptor may generally serve to attenuate ischemic and inflammatory tissue damage (Ohta and Sitkovsky, 2001; Okusa, 2002), such that A_{2A} agonists (rather than antagonists) have also emerged as promising therapeutic candidates. For example, ischemic cardiac and renal damage can be attenuated by A_{2A} agonists, effects of which are reversed by A_{2A} receptor blockade (Lozza et al., 1997; Belardinelli et al., 1998; Okusa et al., 2000). Even within the CNS under some circumstances, A_{2A} receptor stimulation can confer neuroprotection. Administration of an A_{2A} agonist at the time of spinal cord ischemia and reperfusion significantly reduces resultant neuronal damage (Cassada et al., 2001). The basis for protective vs. pathological effects of A_{2A} receptor activation likely relates to the variety of cellular and molecular couplings of the A_{2A} receptor, and thus the safe and effective development of A_{2A} receptor agents will rely on efforts to clarify the mechanisms of their actions.

TABLE 10.2
How A_{2A} Antagonists Might Protect in Parkinson's Disease Models

Mechanism	Pros	Cons
1. Global ↓ glutamate release (↓ direct excitotoxicity)	A_{2A}Rs generally ↑ glutamate release Explains protection of multiple neuronal populations	Relies on presumed low level of A_{2A}Rs on excitatory neurons
2. Local ↓ GABA release (↓ indirect excitotoxicity from Gpe→STN→SNc)	Dense striatopallidal A_{2A}Rs A_{2A}Rs ↑ GPe GABA release STN → SNc excitotoxicity data	Does not explain protection at other CNS sites
3. Glial cell modulation (e.g., ↑ glutamate buffering)	A_{2A} agonists ↓ glut uptake in cultured CNS glia; modulate NOS and other activities	Relies on presumed low level of A_{2A}Rs on glial cells Based only on *in vitro* studies
4. Direct DA neuron protection	Possible vesicular mechanism	? A_{2A}Rs on DA neuron A_{2A} antagonists not protective in cell cultures
5. Altered toxin metabolism		MPTP metabolites and MAO B unaffected by A_{2A} antagonists Protection in multiple toxin model

Note: See text for details and Figure 10.4 for schematic representation of mechanisms 1 to 4 as well as abbreviations.

MECHANISMS OF PROTECTION BY A_{2A} ANTAGONIST IN PD MODELS

How the A_{2A} receptor or its blockade influences the death of dopaminergic neurons remains uncertain (see Table 10.2 and Figure 10.4). An intuitive explanation that the high levels of striatal A_{2A} receptors (Jarvis and Williams, 1989) (see Figure 10.3) directly trigger the demise of the dopaminergic neurons innervating the striatum belies the cellular anatomy of the A_{2A} receptor within the basal ganglia. The vast majority of these receptors are expressed on GABAergic striatopallidal output neurons (Schiffmann et al., 1991; Fink et al., 1992), which are postsynaptic to the dopaminergic neurons that degenerate in PD (Figure 10.4). By contrast, there is little evidence for appreciable expression of A_{2A} receptors on the dopaminergic nigrostriatal neurons themselves (Fink et al., 1992; Dixon et al., 1996; Hettinger et al., 1998). Thus, A_{2A} receptors on nondopaminergic neurons (or even on non-neuronal cells) may indirectly influence the viability of the dopaminergic nigral neurons.

How might the blockade of postsynaptic A_{2A} receptor on GABAergic striatopallidal neurons improve the survival of presynaptic dopaminergic neurons? The shortest path back to the nigrostriatal dopaminergic neurons may be taken by the retrograde transsynaptic elaboration of a protective factor (Appel, 1981). Although specific neurotrophic factors in striatal neurons have been hypothesized to maintain the integrity of innervating dopaminergic neurons (Siegel and Chauhan, 2000), there is no evidence that striatal A_{2A} receptor stimulation inhibits this hypothetical retrograde neurotrophic influence (or, conversely, that A_{2A} antagonists enhance it).

Perhaps more realistic, even if more circuitous, is the possibility that striatal A_{2A} receptor blockade leads to improved dopaminergic neuron survival through a polysynaptic feedback loop involving the A_{2A} receptor-laden striatopallidal neurons (see Figure 10.4, site 2). A_{2A} receptor stimulation of GABAergic striatopallidal neurons increases extracellular GABA in the globus pallidus (Mayfield et al., 1996). The increased pallidal level of inhibitory transmitter in turn may reduce the activity the GABAergic projection from the globus pallidus to the subthalamic nucleus (STN), leading to disinhibition of its glutamatergic projections (Shindou et al., 2001). One of these

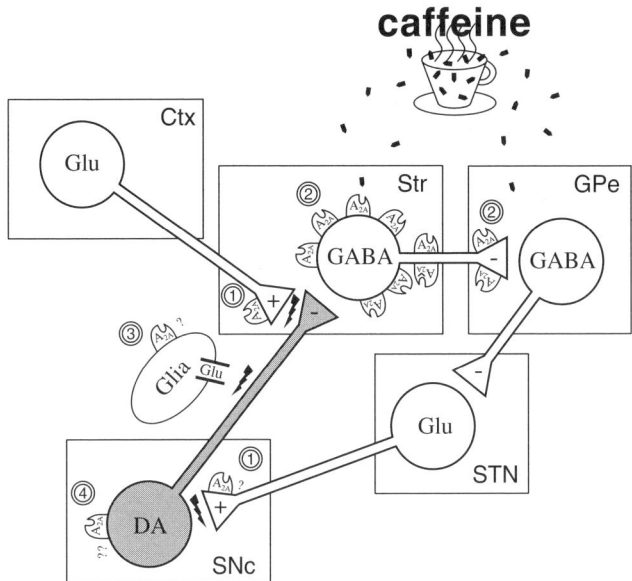

FIGURE 10.4 Sites of adenosine A_{2A} receptors whose blockade by caffeine could protect dopaminergic nigrostriatal neurons. As detailed in the text and indicated in this simplified schematic, A_{2A} receptors on the nerve terminals of excitatory glutamatergic (Glu) neurons (1) represent a widely distributed mechanism for enhanced excitotoxicity (lightning symbol, ~) that may converge on dopaminergic (DA) nigrostriatal neurons. Blockade of A_{2A} receptor-facilitated glutamate release may account for reduced excitotoxic injury to cortical and striatal neurons (not shown) as well as nigral neurons. A_{2A} receptors are densely expressed on GABAergic striatopallidal neurons (2) where their facilitative effect on GABA release in the external globus pallidus (GPe) could indirectly lead to excitotoxic stimulation of dopaminergic nigral neurons via disinhibition of the glutamatergic projection from subthalamic nucleus (STN) to the substantia nigra *pars compacta* (SNc). Glial cells (e.g., astrocytes) can also express A_{2A} receptors (3) and actively regulate the environment of neurons throughout the central nervous system. Blockade of these receptors by caffeine may protect neighboring dopaminergic neurons, possibly by activating (disinhibiting) the glial glutamate transporter, leading to lower extracellular levels of excitatory amino acids. The unsubstantiated possibility of low levels of A_{2A} receptors on dopaminergic nigrostriatal neurons (4) would allow for a direct protective effect of caffeine on dopaminergic neurons. Question marks (?) reflect the uncertainty over the presence of A_{2A} receptors on certain neuronal or glial cells *in vivo*. Ctx and Str refer to cortex and striatum, respectively.

activated STN outputs projects to the substantia nigra *pars compacta*, where its enhanced release of glutamate may exert an excitotoxic effect on the dopaminergic nigrostriatal neurons (Blandini et al., 2000; Greenamyre, 2001). Increased excitatory tone applied to the nigral dopaminergic nigral neurons in combination with their metabolic deficits induced in the MPTP model (and possibly in PD) could contribute to the cumulative injury of dopaminergic neurons (Albin and Greenamyre, 1992). Indeed, experimental blockade or reversal of STN excitatory activity has been shown to attenuate the death of dopaminergic nigral neurons induced by 6-hydroxydopamine (Piallat et al., 1996; Luo et al., 2002). Thus, striatal A_{2A} receptor stimulation could exacerbate an STN-mediated excitotoxic component of dopaminergic nigral neuron degeneration; conversely, A_{2A} antagonists may modify the circuit to slow the degenerative process.

Although this circuitry model of dopaminergic neuron protection by A_{2A} antagonists incorporates a critical role for the prominent striatal A_{2A} receptor (Figure 10.4), it does not easily explain their protective effects on nondopaminergic neurons residing at other locations in the central nervous system. Since evidence for a broader neuroprotective effect now extends from hippocampal to frontal cortex, and from nigra to striatum, alternative hypotheses that involve A_{2A} receptor modulation of a generalized central nervous system process have become more compelling. One such mechanism is the well-established facilitation of glutamate release by A_{2A} receptor stimulation (Sebastiao and Ribeiro, 1996) (see Figure 10.4, site 1), which has been consistently observed in cortex, basal ganglia, and brainstem (O'Regan et al., 1992; Castillo-Melendez et al., 1994; Popoli et al., 1995). This phenomenon likely involves A_{2A} receptors located on glutamatergic nerve terminals; it can be observed in cortical synaptosomes (Marchi et al., 2002) (in which an indirect effect of striatal A_{2A} receptors is not plausible) as well as in intact brain. Recent ultrastructural analysis of A_{2A} receptor distribution has strengthened the evidence for its presynaptic location on glutamatergic nerve terminals (Hettinger et al., 2001). Whereas A_{2A} agonists generally enhance release or overflow of glutamate, A_{2A} antagonists have been found to attenuate glutamate release or overflow triggered by depolarization, ischemia, or the glutamate receptor agonist quinolinate in an excitotoxin model of HD (Corsi et al., 2000; Pintor et al., 2001; Popoli et al., 2002; Melani et al., 2003). Thus, A_{2A} antagonist attenuation of local excitatory amino acid release throughout the central nervous system may alleviate an excitoxic component common to most models of neurotoxicity and neurodegeneration. It remains to be seen whether MPTP-induced elevations in striatal or nigral neurotransmitters are attenuated by A_{2A} antagonists.

Recently, adenosinergic modulation of glial cell function has emerged as another widely distributed central nervous system mechanism by which A_{2A} antagonists might lessen neuronal cell death (see Figure 10.4, site #3). Stimulation of A_{2A} receptors present on cultured astrocytic glial cells from cortex or brainstem was found to enhance glutamate efflux, whereas A_{2A} blockade reduced levels of extracellular glutamate (Li et al., 2001; Nishizaki et al., 2002). Genetic and pharmacological approaches suggested that A_{2A} receptor regulation of a specific glial glutamate transporter (GLT-1) may account for this effect. Earlier studies suggested that A_{2A} receptors could modulate other glial functions (such as nitric oxide synthase and cyclooxygenase activities) that may play an important role in the survival of their neuronal neighbors (Fiebich et al., 1996; Brodie et al., 1998). Thus, ubiquitous glial elements in the central nervous sytem may also host A_{2A} receptor involvement in multiple models of neurodegeneration.

Other candidate mechanisms for dopaminergic neuron protection by A_{2A} antagonists have been suggested that are unique to the toxin models of PD in which the protection has been demonstrated. For example, reduced cAMP in dopaminergic neurons leading to increased vesicular sequestration of MPP$^+$ (the active toxin metabolite of MPTP) has been proposed as an explanation for how A_{2A} antagonists attenuate neurotoxicity in the MPTP model of PD (Ikeda et al., 2002). However, this proposal is primarily based on MPP$^+$ uptake studies in a pheochromocytoma cell line, relies on the uncertain presence of A_{2A} receptors on dopaminergic neurons, and does not explain A_{2A} antagonist protection of nondopaminergic neurons. Nevertheless, the possibility of a simple, direct cellular survival effect of a small number of (as yet unsubstantiated) A_{2A} receptors on the dopaminergic neurons themselves (see Figure 10.4, site 4) has not been ruled out.

Another important consideration is the possibility that A_{2A} receptor blockade might protect against MPTP toxicity simply by limiting MPTP access to the central nervous sytem or its conversion by MAO B to the active toxin MPP$^+$. Before it was discovered that CSC possesses potent MAO B inhibitory activity independent of its A_{2A} antagonist properties (Chen et al., 2002; Petzer et al., 2003), its attenuation of MPTP metabolism in striatum had suggested that A_{2A} blockade could reduce MPTP toxicity by inhibiting MAO B activity (Chen et al., 2000). However, caffeine and genetic inactivation of the A_{2A} receptor did not significantly alter striatal MPTP levels *in vivo*, nor did they substantially alter MAO B activity *in vitro* (Chen et al., 2001, 2002; Petzer et al., 2003). Moreover, the A_{2A} antagonist KW-6002 did not appreciably alter the brain levels or kinetics of MPP$^+$ following systemic MPTP administration (Ikeda et al., 2002). Thus, inhibition of MPTP metabolism or MAO B activity does not explain the neuroprotective quality of caffeine and more specific antagonists of the A_{2A} receptor in models of PD.

SIGNIFICANCE FOR PARKINSON'S DISEASE

The demonstration that caffeine and more specific A_{2A} antagonists protect dopaminergic nigrostriatal neurons in multiple animal models of PD has pathophysiological, epidemiological, and therapeutic significance for PD.

Understanding the neurobiology of the A_{2A} and other adenosine receptors will provide insight into the role of endogenous adenosine in basal ganglia biology as well as PD pathophysiology.

Establishing the ability of caffeine to protect dopaminergic neurons in PD models and identifying a plausible mechanism of action greatly strengthens (but does not prove) the hypothesis that a neuroprotective effect of caffeine is the basis for its inverse epidemiological association with risk of PD.

With several A_{2A} antagonists emerging as promising therapeutic candidates based on their motor-enhancing symptomatic effects, the prospects of an additional neuroprotective benefit may considerably enhance their therapeutic potential.

REFERENCES

Abernethy, D.R. and Todd, E.L. (1985) Impairment of caffeine clearance by chronic use of low-dose oestrogen-containing oral contraceptives. *European Journal of Clinical Pharmacology,* 28, 425–428.

Albin, R.L. and Greenamyre, J.T. (1992) Alternative excitotoxic hypotheses. *Neurology,* 42, 733–738.

Appel, S.H. (1981) A unifying hypothesis for the cause of amyotrophic lateral sclerosis, parkinsonism, and Alzheimer's disease. *Annals of Neurology,* 10, 499–505.

Ascherio, A., Zhang, S.M., Hernán, M.A., Kawachi, I., Colditz, G.A., Speizer, F.E. et al. (2001) Prospective study of caffeine consumption and risk of Parkinson's disease in men and women. *Annals of Neurology,* 50, 56–63.

Baumann, R.J., Jameson, D., McKean, H.E., Haack, D.G. and Weisberg, L.M. (1980) Cigarette smoking and Parkinson's disease: I. A comparison of cases with matched neighbors. *Neurology,* 30, 839–843.

Belardinelli, L., Shryock, J.C., Snowdy, S., Zhang, Y., Monopoli, A., Lozza, G. et al. (1998) The A2A adenosine receptor mediates coronary vasodilation. *Journal of Pharmacology and Experimental Therapeutics,* 284, 1066–1073.

Benedetti, M.D., Bower, J.H., Maraganore, D.M., McDonnell, S.K., Peterson, B.J., Ahlskog, J.E. et al. (2000) Smoking, alcohol, and coffee consumption preceding Parkinson's disease: a case-control study. *Neurology,* 55, 1350–1358.

Benowitz, N.L., Jacob, P., III, Mayan, H. and Denaro, C. (1995) Sympathomimetic effects of paraxanthine and caffeine in humans. *Clinical Pharmacology and Therapeutics,* 58, 684–691.

Blandini, F., Nappi, G., Tassorelli, C. and Martignoni, E. (2000) Functional changes of the basal ganglia circuitry in Parkinson's disease. *Progress in Neurobiology,* 62, 63–88.

Brodie, C., Blumberg, P. and Jacobson, K. (1998) Activation of the A_{2A} adenosine receptor inhibits nitric oxide production in glial cells. *FEBS Letters,* 429, 139–142.

Cassada, D.C., Tribble, C.G., Laubach, V.E., Nguyen, B.N., Rieger, J.M., Linden, J. et al. (2001) An adenosine A2A agonist, ATL-146e, reduces paralysis and apoptosis during rabbit spinal cord reperfusion. *Journal of Vascular Surgery,* 34, 482–488.

Castillo-Melendez, M., Krstew, E., Lawrence, A.J. and Jarrott, B. (1994) Presynaptic adenosine A2a receptors on soma and central terminals of rat vagal afferent neurons. *Brain Research,* 652, 137–144.

Checkoway, H., Powers, K., Smith-Weller, T., Franklin, G.M., Longstreth, W.T. and Swanson, P.D. (2002) Parkinson's disease risks associated with cigarette smoking, alcohol consumption, and caffeine intake. *American Journal of Epidemiology,* 155, 732–738.

Chen, J.F., Huang, Z., Ma, J., Zhu, J., Moratalla, R., Standaert, D. et al. (1999) A(2A) adenosine receptor deficiency attenuates brain injury induced by transient focal ischemia in mice. *The Journal of Neuroscience,* 19, 9192–9200.

Chen, J.F., Staal, R., Xu, K., Beilstein, M., Sonsalla, P.K. and Schwarzschild, M.A. (2000) Novel neuroprotection by adenosine A2A receptor inactivation in an MPTP model of Parkinson's disease. *Movement Disorders,* 15, 37, Abstr. P304.

Chen, J.F., Steyn, S., Staal, R., Petzer, J.P., Xu, K., Van Der Schyf, C.J. et al. (2002) 8-(3-chlorostyryl)caffeine may attenuate MPTP neurotoxicity through dual actions of monoamine oxidase inhibition and A2A receptor antagonism. *Journal of Biological Chemistry,* 277, 36040–36044.

Chen, J.F., Xu, K., Petzer, J.P., Staal, R., Xu, Y.H., Beilstein, M. et al. (2001) Neuroprotection by caffeine and A(2A) adenosine receptor inactivation in a model of Parkinson's disease. *The Journal of Neuroscience,* 21, RC143.

Colditz, G.A., Manson, J.E. and Hankinson, S.E. (1997) The Nurses' Health Study: 20-year contribution to the understanding of health among women. *Journal of Women's Health,* 6, 49–62.

Corsi, C., Melani, A., Bianchi, L. and Pedata, F. (2000) Striatal A2A adenosine receptor antagonism differentially modifies striatal glutamate outflow *in vivo* in young and aged rats. *Neuroreport,* 11, 2591–2595.

Dixon, A.K., Gubitz, A.K., Sirinathsinghji, D.J., Richardson, P.J. and Freeman, T.C. (1996) Tissue distribution of adenosine receptor mRNAs in the rat. *British Journal of Pharmacology,* 118, 1461–1468.

Fall, P., Frederikson, M., Axelson, O. and Granérus, A. (1999) Nutritional and occupational factors influencing the risk of Parkinson's disease: a case-control study in southeastern Sweden. *Movement Disorders,* 14, 28–37.

Fiebich, B.L., Biber, K., Lieb, K., van Calker, D., Berger, M., Bauer, J. et al. (1996) Cyclooxygenase-2 expression in rat microglia is induced by adenosine A2a-receptors. *Glia,* 18, 152–160.

Fink, J.S., Kalda, A., Dedeoglu, A., Schwarzschild, M.A., Chen, J.F. and Ferrante, R.J. (2002) Genetic inactivation or pharmacological antagonism of the adenosine A_{2A} receptor attenuates 3-nitroproprionic acid-induced striatal damage. *Society for Neuroscience Abstracts,* 18.

Fink, J.S., Weaver, D.R., Rivkees, S.A., Peterfreund, R.A., Pollack, A.E., Adler, E.M. et al. (1992) Molecular cloning of the rat A2 adenosine receptor: selective co-expression with D2 dopamine receptors in rat striatum. *Molecular Brain Research,* 14, 186–195.

Fredholm, B.B., Bättig, K., Holmén, J., Nehlig, A. and Zvartau, E.E. (1999) Actions of caffeine in the brain with special reference to factors that contribute to its widespread use. *Pharmacological Reviews,* 51, 83–133.

Gao, Y. and Phillis, J.W. (1994) CGS 15943, an adenosine A2 receptor antagonist, reduces cerebral ischemic injury in the Mongolian gerbil. *Life Sciences,* 55, PL61–PL65.

Gerlach, M. and Riederer, P. (1996) Animal models of Parkinson's disease: an empirical comparison with the phenomenology of the disease in man. *Journal of Neural Transmission,* 103, 987–1041.

Glass, M., Dragunow, M. and Faull, R.L. (2000) The pattern of neurodegeneration in Huntington's disease: a comparative study of cannabinoid, dopamine, adenosine and GABA(A) receptor alterations in the human basal ganglia in Huntington's disease. *Neuroscience,* 97, 505–519.

Greenamyre, J.T. (2001) Glutamatergic influences on the basal ganglia. *Clinical Neuropharmacology,* 24, 65–70.

Hellenbrand, W., Seidler, A., Boeing, H., Robra, B.P., Vieregge, P., Nischan, P. et al. (1996) Diet and Parkinson's disease. I: A possible role for the past intake of specific foods and food groups: results from a self-administered food-frequency questionnaire in a case-control study. *Neurology,* 47, 636–643.

Hellenbrand, W., Seidler, A., Robra, B.-P., Vieregge, P., Oertel, W.H., Joerg, J. et al. (1997) Smoking and Parkinson's disease: a case control study in Germany. *International Journal of Epidemiology,* 26, 328–339.

Hernán, M.A., Takkouche, B., Caamaño-Isorna, F. and Gestal-Otero, J.J. (2002) A meta-analysis of coffee drinking, cigarette smoking, and the risk of Parkinson's disease. *Annals of Neurology,* 52, 276–284.

Hettinger, B.D., Lee, A., Linden, J. and Rosin, D.L. (1998) Adenosine A2A receptors in rat striatum: prevalent dendritic localization post-synaptic to catecholaminergic terminals. *Society for Neuroscience Abstracts,* 2056.

Hettinger, B.D., Lee, A., Linden, J. and Rosin, D.L. (2001) Ultrastructural localization of adenosine A2A receptors suggests multiple cellular sites for modulation of GABAergic neurons in rat striatum. *Journal of Comparative Neurology,* 431, 331–346.

Holtzman, S.G., Mante, S. and Minneman, K.P. (1991) Role of adenosine receptors in caffeine tolerance. *Journal of Pharmacology and Experimental Therapeutics,* 256, 62–68.

Ikeda, K., Kurokawa, M., Aoyama, S. and Yoshihisa, K. (2002) Neuroprotection by adenosine A_{2A} receptor blockade in experimental models of Parkinson's disease. *Journal of Neurochemistry,* 80, 262–270.

Jarvis, M.F. and Williams, M. (1989) Direct autoradiographic localization of adenosine A2 receptors in the rat brain using the A2-selective agonist, [3H]CGS 21680. *European Journal of Pharmacology,* 168, 243–246.

Jiménez-Jiménez, F.J., Mateo, D. and Giménez-Roldan, S. (1992) Premorbid smoking, alcohol consumption, and coffee drinking habits in Parkinson's disease: a case-control study. *Movement Disorders,* 7, 339–344.

Jones, P., Smith, R. and Stone, T. (1998) Protection against hippocampal kainate excitotoxicity by intracerebral administration of an adenosine A_{2A} receptor antagonist. *Brain Research,* 800, 328–335.

Kaplan, G.B., Greenblatt, D.J., Ehrenberg, B.L., Goddard, J.E., Cotreau, M.M., Harmatz, J.S. et al. (1997) Dose-dependent pharmacokinetics and psychomotor effects of caffeine in humans. *Journal of Clinical Pharmacology,* 37, 693–703.

Kaplan, G.B., Greenblatt, D.J., Kent, M.A. and Cotreau-Bibbo, M.M. (1993) Caffeine treatment and withdrawal in mice: relationships between dosage, concentrations, locomotor activity and A1 adenosine receptor binding. *Journal of Pharmacology and Experimental Therapeutics,* 266, 1563–1572.

Ledent, C., Vaugeois, J.M., Schiffmann, S.N., Pedrazzini, T., El Yacoubi, M., Vanderhaeghen, J.J. et al. (1997) Aggressiveness, hypoalgesia and high blood pressure in mice lacking the adenosine A2a receptor [see comments]. *Nature,* 388, 674–678.

Li, X.X., Nomura, T., Aihara, H. and Nishizaki, T. (2001) Adenosine enhances glial glutamate efflux via A2a adenosine receptors. *Life Sciences,* 68, 1343–1350.

Lozza, G., Conti, A., Ongini, E. and Monopoli, A. (1997) Cardioprotective effects of adenosine A1 and A2A receptor agonists in the isolated rat heart. *Pharmacological Research,* 35, 57–64.

Luo, J., Kaplitt, M.G., Fitzsimons, H.L., Zuzga, D.S., Liu, Y., Oshinsky, M.L. et al. (2002) Subthalamic GAD gene therapy in a Parkinson's disease rat model. *Science,* 298, 425–429.

Marchi, M., Raiteri, L., Risso, F., Vallarino, A., Bonfanti, A., Monopoli, A. et al. (2002) Effects of adenosine A1 and A2A receptor activation on the evoked release of glutamate from rat cerebrocortical synaptosomes. *British Journal of Pharmacology,* 136, 434–440.

Mayfield, R.D., Larson, G., Orona, R.A. and Zahniser, N.R. (1996) Opposing actions of adenosine A2a and dopamine D2 receptor activation on GABA release in the basal ganglia: Evidence for an A2a/D2 receptor interaction in globus pallidus. *Synapse,* 22, 132–138.

Melani, A., Pantoni, L., Bordoni, F., Gianfriddo, M., Bianchi, L., Vannucchi, M.G. et al. (2003) The selective A(2A) receptor antagonist SCH 58261 reduces striatal transmitter outflow, turning behavior and ischemic brain damage induced by permanent focal ischemia in the rat. *Brain Research,* 959, 243–250.

Monopoli, A., Lozza, G., Forlani, A., Mattavelli, A. and Ongini, E. (1998) Blockade of adenosine A2A receptors by SCH 58261 results in neuroprotective effects in cerebral ischaemia in rats. *Neuroreport,* 9, 3955–3959.

Morens, D.M., Grandinetti, A., Reed, D., White, L.R. and Ross, G.W. (1995) Cigarette smoking and protection from Parkinson's disease: false association or etiologic clue (review). *Neurology,* 45, 1041–1051.

Morgan, K.J., Stults, V.J. and Zabik, M.E. (1982) Amount and dietary sources of caffeine and saccharin intake by individuals ages 5 to 18 years. *Regulatory Toxicology and Pharmacology,* 2, 296–307.

Nefzger, M.D., Quadfasel, F.A. and Karl, V.C. (1968) A retrospective study of smoking in Parkinson's disease. *American Journal of Epidemiology,* 88, 149–158.

Nishizaki, T., Nagai, K., Nomura, T., Tada, H., Kanno, T., Tozaki, H. et al. (2002) A new neuromodulatory pathway with a glial contribution mediated via A(2a) adenosine receptors. *Glia,* 39, 133–147.

Ohta, A. and Sitkovsky, M. (2001) Role of G-protein-coupled adenosine receptors in downregulation of inflammation and protection from tissue damage. *Nature,* 414, 916–920.

Okusa, M.D. (2002) A(2A) adenosine receptor: a novel therapeutic target in renal disease *American Journal of Physiology Renal Physiology,* 282, F10–F18.

Okusa, M.D., Linden, J., Huang, L., Rieger, J.M., Macdonald, T.L. and Huynh, L.P. (2000) A(2A) adenosine receptor-mediated inhibition of renal injury and neutrophil adhesion. *American Journal of Physiology Renal Physiology,* 279, F809–F818.

O'Regan, M.H., Simpson, R.E., Perkins, L.M. and Phillis, J.W. (1992) The selective A2 adenosine receptor agonist CGS 21680 enhances excitatory transmitter amino acid release from the ischemic rat cerebral cortex. *Neuroscience Letters,* 138, 169–172.

Oztas, E., Kalda, A., Xue, K., Irrizary, M.C., Schwarzschild, M.A. and Chen, J. (2002) Caffeine attenuates MPTP-induced loss of dopaminergic neurons in substantia nigra in mice. *Society for Neuroscience Abstracts*, 487.6.

Paganini-Hill, A. (2001) Risk factors for Parkinson's disease: the leisure world cohort study. *Neuroepidemiology*, 20, 118–124.

Paulson, G. and Dadmehr, N. (1991) Is there a premorbid personality typical for Parkinson's disease? *Neurology*, 41, 73–76.

Petzer, J.P., Steyn, S., Castagnoli, K.P., Chen, J.F., Schwarzschild, M.A., Van Der Schyf, C.J. et al. (2003) Inhibition of monoamine oxidase B by selective adenosine A2A receptor antagonists. *Bioorganic and Medical Chemistry*, 11, 1299–1310.

Phillis, J.W. (1995) The effects of selective A1 and A2a adenosine receptor antagonists on cerebral ischemic injury in the gerbil. *Brain Research*, 705, 79–84.

Piallat, B., Benazzouz, A. and Benabid, A.L. (1996) Subthalamic nucleus lesion in rats prevents dopaminergic nigral neuron degeneration after striatal 6-OHDA injection: behavioural and immunohistochemical studies. *European Journal of Neuroscience*, 8, 1408–1414.

Pintor, A., Quarta, D., Pezzola, A., Reggio, R. and Popoli, P. (2001) SCH 58261 (an adenosine A(2A) receptor antagonist) reduces, only at low doses, K(+)-evoked glutamate release in the striatum. *European Journal of Pharmacology*, 421, 177–180.

Piriyawat, P., Labiche, L.A., Burgin, W.S., Aronowski, J.A. and Grotta, J.C. (2003) Pilot dose-escalation study of caffeine plus ethanol (caffeinol) in acute ischemic stroke. *Stroke*, 34, 1242–1245.

Pollock, B.G., Wylie, M., Stack, J.A., Sorisio, D.A., Thompson, D.S., Kirshner, M.A. et al. (1999) Inhibition of caffeine metabolism by estrogen replacement therapy in postmenopausal women. *Journal of Clinical Pharmacology*, 39, 936–940.

Popoli, P., Betto, P., Reggio, R. and Ricciarello, G. (1995) Adenosine A2A receptor stimulation enhances striatal extracellular glutamate levels in rats. *European Journal of Pharmacology*, 287, 215–217.

Popoli, P., Pintor, A., Domenici, M.R., Frank, C., Tebano, M.T., Pezzola, A. et al. (2002) Blockade of striatal adenosine A2A receptor reduces, through a presynaptic mechanism, quinolinic acid-induced excitotoxicity: possible relevance to neuroprotective interventions in neurodegenerative diseases of the striatum. *Journal of Neuroscience*, 22, 1967–1975.

Ross, G.W., Abbott, R.D., Petrovich, H., Morens, D.M., Grandineti, A., Tung, K. et al. (2000) Association of coffee and caffeine intake with the risk of Parkinson disease. *Journal of the American Medical Association*, 283, 2674–2679.

Schiffmann, S.N., Jacobs, O. and Vanderhaeghen, J.J. (1991) Striatal restricted adenosine A2 receptor (RDC8) is expressed by enkephalin but not by substance P neurons: an *in situ* hybridization histochemistry study. *Journal of Neurochemistry*, 57, 1062–1067.

Schwarzschild, M.A., Chen, J. and Ascherio, A. (2002) Caffeinated clues and the promise of adenosine A_{2A} antagonists in Parkinson's disease. *Neurology*, 58, 1154–1160.

Sebastiao, A.M. and Ribeiro, J.A. (1996) Adenosine A2 receptor-mediated excitatory actions on the nervous system. *Progress of Neurobiology*, 48, 167–189.

Shindou, T., Mori, A., Kase, H. and Ichimura, M. (2001) Adenosine A(2A) receptor enhances GABA(A)-mediated IPSCs in the rat globus pallidus. *Journal of Physiology*, 532, 423–434.

Siegel, G.J. and Chauhan, N.B. (2000) Neurotrophic factors in Alzheimer's and Parkinson's disease brain. *Brain Research Reviews*, 33, 199–227.

Stone, T.W., Jones, P.A. and Smith, R.A. (2001) Neuroprotection by A_{2A} receptor antagonist. *Drug Development Research*, 52, 323–330.

Strong, R., Grotta, J.C. and Aronowski, J. (2000) Combination of low dose ethanol and caffeine protects brain from damage produced by focal ischemia in rats. *Neuropharmacology*, 39, 515–522.

Svenningsson, P., Le Moine, C., Fisone, G. and Fredholm, B.B. (1999) Distribution, biochemistry and function of striatal adenosine A2A receptors. *Progress in Neurobiology*, 59, 355–396.

Tanner, C.M., Goldman, S.M., Aston, D.A., Ottman, R., Ellenberg, J., Mayeux, R. et al. (2002) Smoking and Parkinson's disease in twins. *Neurology*, 58, 581–588.

Tanner, C.M., Ottman, R., Goldman, S.M., Ellenberg, J., Chan, P., Mayeux, R. et al. (1999) Parkinson's disease in twins: an etiologic study. *Journal of the American Medical Association*, 281, 341–346.

Tohyama, M. and Takatsuji, K. (eds.) (1998) *Atlas of Neuroactive Substances and Their Receptors in the Rat*. Oxford University Press, Oxford, U.K., pp. 136–139 and 160–161.

Von Lubitz, D.K., Lin, R.C. and Jacobson, K.A. (1995) Cerebral ischemia in gerbils: effects of acute and chronic treatment with adenosine A2A receptor agonist and antagonist. *European Journal of Pharmacology,* 287, 295–302.

Willett, W.C. (1998) *Nutritional Epidemiology, Second Edition,* Oxford University Press, New York.

Xu, K., Xu, Y.H., Chen, J.F. and Schwarzschild, M.A. (2002a) Caffeine's neuroprotection against MPTP toxicity shows no tolerance to chronic caffeine administration in mice. *Neuroscience Letters,* 322, 13–16.

Xu, K., Xu, Y.H., Chen, J.F. and Schwarzschild, M.A. (2002b) Poster presented at the Annual Meeting of the Society for Neuroscience, Vol. 28, Abstr. 487.5.

Zayed, J., Ducic, S., Campanella, G., Panisset, J., André, P., Masson, H. et al. (1990) Facteurs environnementaux dans l'étiologie de la maladie de Parkinson. *Canadian Journal of Neurological Sciences,* 17, 286–291.

11 Caffeine in Ischemia and Seizures: Paradoxical Effects of Long-Term Exposure

Astrid Nehlig and Bertil B. Fredholm

CONTENTS

Introduction 165
Caffeine and Ischemic Brain Damage 165
Caffeine and Epilepsy 167
Acute and Chronic Caffeine Consumption Have Opposite Effects in Ischemia and
Epilepsy 168
Mechanisms Underlying the Protective Effects of Chronic Caffeine Exposure in Ischemia
and Epilepsy 168
Conclusion 169
References 169

INTRODUCTION

Human caffeine use is typically chronic. However, many of the studies that explore the actions of caffeine are acute studies, both in animals and humans. It is therefore important to realize that there are, in several instances, important differences between the effects of acute and long-term treatment (see Jacobson et al., 1996). We will discuss such differences in relation to two potentially important outcomes: damage following ischemia and seizure susceptibility.

CAFFEINE AND ISCHEMIC BRAIN DAMAGE

As discussed repeatedly elsewhere and extensively documented in several review articles, there is excellent evidence that adenosine is an endogenous neuroprotective agent (Dragunow and Faull, 1988; Marangos et al., 1990; Marangos and Miller, 1991; Rudolphi et al., 1992a,b; Fredholm, 1996; Dunwiddie and Masino, 2001; Von Lubitz, 2001). One would therefore expect that antagonism of adenosine actions (e.g., by caffeine) would cause aggravation of brain damage following events such as stroke. Indeed, and as summarized in the above review articles, there is evidence that acute administration of caffeine or theophylline can cause an aggravation of neuronal damage in several experimental model systems.

Given that caffeine (as well as its metabolites theophylline and paraxanthine) probably exerts most of its effects by blocking adenosine A_1 and A_{2A} (with a possible contribution of A_{2B}, but not A_3) receptors, we should consider the evidence that the different receptors may be responsible for the effects of caffeine on neuronal damage. All the four adenosine receptors can potentially influence processes critical in the development of postischemic damage.

There is some evidence that stimulation of adenosine A_3 receptors may affect damage (Von Lubitz et al., 1995), but the agonist used in that study is quite nonselective (Klotz et al., 1998) and the conclusions should probably be regarded as provisional. Furthermore, as discussed previously (see Chapter 1 of this book), we do not need to consider A_3 receptors when discussing effects of caffeine. Thus, we only have to consider A_1, A_{2A}, and A_{2B} receptors.

There is very strong evidence that A_1 receptors affect excitatory neurotransmission and resultant increases in intracellular calcium. Adenosine A_1 receptors provide an important inhibition of excitatory neurotransmission by reducing transmitter release (Fredholm and Dunwiddie, 1988; Dunwiddie and Fredholm, 1997; Dunwiddie and Masino, 2001). Indeed, experiments in A_1 KO mice show that all the effects of adenosine on excitatory neurotransmission depend absolutely on A_1 receptors (Johansson et al., 2001; Masino et al., 2002). In addition, activation of A_1 receptors leads to stimulation of K^+ conductance and hence to hyperpolarization, leading to decreased neuronal firing and decreased NMDA receptor activation (Dunwiddie, 1985; Dunwiddie and Masino, 2001). Finally, adenosine A_1 receptor activation leads to decreased calcium entry via N-, P-, Q-type calcium channels. It cannot be excluded that adenosine A_1 receptors on other cell types such as glial cells and leukocytes (Neely et al., 1997) may also play a role in modulating the extent of neuronal damage after an ischemic insult.

Neuronal adenosine A_{2A} receptors are virtually confined to the striatopallidal neurons, and levels in other neurons are very low (Svenningsson et al., 1999). There is no good evidence that A_{2B} receptors are highly expressed in adult neurons. This means that adenosine cannot be expected to cause any significant activation of the extrastriatal neuronal adenosine A_{2A} (or A_{2B}) receptors and, hence, that these receptors are unlikely to be a major target for caffeine action. By contrast, A_2 receptors are found on blood vessels and can mediate vasodilation (Phillis, 1989; Torregrosa et al., 1990; Weaver, 1993). Adenosine A_{2B} receptors may be even more important than A_{2A} receptors in this regard (Shin et al., 2000). Adenosine A_{2A} receptors are also found on astrocytes and microglial cells. In the latter cell type, it is likely that A_{2A} (and possibly A_{2B}) receptors regulate an oxidative burst, in the same way that they do in the peripheral equivalent, the macrophage (Fiebich et al., 1996; Si et al., 1996; Hasko et al., 2000), perhaps partly by affecting potassium channels (Kust et al., 1999). These effects of A_2 receptor stimulation could either enhance or reduce the extent of ischemic neuronal damage, and the relative importance may depend on the type of ischemic insult and other factors.

Whereas the effects of acute and long-term treatment with a rather selective adenosine A_1 receptor antagonist, 8-cyclopentyl-1,3-dipropylxanthine or DPCPX, are qualitatively different (Von Lubitz et al., 1994), this may not be the case with blockade of A_{2A} receptors. Thus, acute treatment with an adenosine A_{2A} receptor antagonist acutely leads to neuroprotection (Monopoli et al., 1998; Chen et al., 1999). Furthermore, complete elimination of A_{2A} receptors by genetic targeting affords substantial neuroprotection (Chen et al., 1999). The exact mechanism is unknown. It is, however, interesting that protection by A_{2A} antagonism is also observed when neuronal damage is induced by quinolinic acid and a mixture of xanthine and xanthine oxidase that produces free radicals (Behan and Stone, 2002).

The situation appears to differ between adult and newborn animals. Whereas most studies in adult animals show that theophylline, given before or immediately after the ischemic insult, causes increased damage, theophylline given before hypoxic ischemia in newborn rats showed a protective effect (Bona et al., 1997). In the same study, posttreatment had no significant effect. Interestingly, the adenosine A_1 antagonist DPCPX was completely inactive in these very young animals (Bona et al., 1997) and an adenosine A_1 agonist was similarly inactive in the neonates (Ådén et al., 2001). The reason appears to be that adenosine A_1 receptors are poorly expressed and poorly coupled to G proteins in young animals. Furthermore, in young animals, complete elimination of A_{2A} receptors led to an aggravated neuronal damage and increased postischemic behavioral sequelae (Ådén et al., 2002). Thus, the role of both A_1 and A_{2A} receptors is completely different in young vs. mature animals.

CAFFEINE AND EPILEPSY

Theophylline, aminophylline, and caffeine are currently prescribed as bronchodilators for the treatment of bronchial asthma and neonatal apnea (Bairam et al., 1987; Boushey, 1995; Rang et al., 1995). Theophylline, used in the long-term management of asthma, has been reported to cause hyperexcitability characterized by restlessness and tremor. The most serious adverse effect of theophylline is convulsive activity (O'Riordan et al., 1994), and at toxic levels, focal and generalized seizures (Stone and Javid, 1980; Dunwiddie and Worth, 1982), and even status epilepticus (Shannon, 1993; Oki et al., 1994). Likewise, caffeine has been used for a long time to lengthen seizures and improve the efficacy of the electroconvulsive therapy (ECT) in severely depressed patients (Shapira et al., 1987; Coffey et al., 1990).

In animal models, both theophylline and caffeine, in nonconvulsive doses, prolong kindled seizures in rats (Albertson et al., 1983; Dragunow, 1990), induce seizures in genetically epilepsy-prone rats (De Sarro et al., 1997), and have proconvulsant effects on seizures induced by kainic acid, pentylenetetrazol, or pilocarpine (Turski et al., 1985; Ault et al., 1987; Cutrufo et al., 1992). The sensitivity to the proconvulsant effects of methylxanthines is inversely related to age, with higher sensitivity in young animals (Yokohama et al., 1997; Bernášková and Mareš, 2000). In children with epilepsy, theophylline-induced convulsions are more frequent in children under the age of 1 year compared to children over 1 year (Miura and Kimura, 2000). A pivotal role in methylxanthine-induced seizures may be played by the balance between GABA and glutamate (Corradetti et al., 1984; Segev et al., 1988; De Sarro and De Sarro, 1991; Amabeoku, 1999). The convulsant action of methylxanthines is most likely linked to the blockade of the effects of endogenous adenosine at A_1 receptors located on glutamatergic neurons, thus allowing a larger release of the excitatory neurotransmitter glutamate (Dunwiddie, 1980; Dunwiddie et al., 1981).

It has been suggested that adenosine may provide an inhibitory tone in the mammalian nervous system (for review see Knutsen and Murray, 1997). Thus, adenosine could act as an endogenous anticonvulsant (Dragunow et al., 1985; Dragunow, 1986). During epileptic seizures, large quantities of adenosine are released by the cells surrounding the epileptic focus (Winn et al., 1979, 1980; Park et al., 1987; During and Spencer, 1992; Berman et al., 2000). The dramatic increase in adenosine levels recorded during epileptiform activity seems to contribute to the termination of ongoing seizure activity as well as to the postictal refractory period (During and Spencer, 1992). The most likely candidate of the antiepileptic effect of adenosine is the A_1 receptor because of its known inhibitory action on the release of neurotransmitters, especially excitatory transmitters whose release is increased during seizures (Fredholm and Hedqvist, 1980; Fredholm and Dunwiddie, 1988). Indeed, A_1 receptor agonists reduce the seizures induced by chemical or electrical stimuli (Barraco et al., 1984; Concas et al., 1993; Klitgaard et al., 1993; Young and Dragunow, 1994; Zhang et al., 1994; De Sarro et al., 1999; Wiesner et al., 1999). In addition, adenosine is able to inhibit calcium fluxes and to open 4-aminopyridine-sensitive potassium channels (Schubert and Lee, 1986; Schubert et al., 1986). Both of these actions would result in membrane hyperpolarization and in an increase in threshold for the activation of NMDA receptors.

The role of the A_{2A} receptor in epilepsy is more controversial, and it is still not clear whether this receptor is involved in the regulation of convulsive seizures. Indeed, A_{2A} agonists may aggravate (De Sarro et al., 1999) or antagonize seizures in rodents (Von Lubitz et al., 1993), while A_{2A} receptor antagonists have only a limited capacity to antagonize chemically induced seizures (Klitgaard et al., 1993). In fact, A_{2A} receptors are found at low density in the hippocampus and cortex (Rosin et al., 1998; Cunha et al., 1999), together with a high density of A_1 receptors (Goodman and Snyder, 1982; Fastbom et al., 1987; Ochiishi et al., 1999). In the hippocampus, which is often involved in seizure activity, the localization of A_1 and A_{2A} receptors overlaps in the pyramidal layers of the CA1, CA2, and CA3 regions, where they modulate excitability in opposite ways. For example, the activation of the A_{2A} receptor attenuates the capacity of an A_1 agonist to lower hippocampal excitability (Cunha et al., 1994). Moreover, the activation of the adenosine A_{2A} receptor can induce the release of two excitatory neurotransmitters, acetylcholine and glutamate (Sebastiao and Ribeiro, 1996; Dunwiddie and Fredholm, 1997). These two neurotransmitters are released during alcohol withdrawal seizures (Rossetti and Carboni, 1995; Imperato et al., 1998), and an A_1 agonist (Concas et al. 1994) or the deletion of the A_{2A} receptor gene in mice (El Yacoubi et al., 2001) reduces alcohol withdrawal syndrome. The effect of A_{2A} receptor ligands depends on tonic A_1 receptor activation (Lopes et al., 2002). The role of the A_3 adenosine receptor in seizures is also controversial. A somewhat selective A_3 receptor agonist, IB-MECA, has neuroprotective properties against NMDA- and pentylenetetrazol-induced seizures (Von Lubitz et al., 1995), while it is entirely ineffective in ameliorating tonic convulsions induced by electroshocks (Jacobson et al., 1996). Thus, the role of adenosine during seizures appears to be complex and to involve more than its commonly accepted endogenous anticonvulsant effect mediated by the activation of A_1 receptors.

ACUTE AND CHRONIC CAFFEINE CONSUMPTION HAVE OPPOSITE EFFECTS IN ISCHEMIA AND EPILEPSY

Given the large body of information available in the literature about the deleterious role of acute caffeine administration on stroke-induced damage (Dragunow and Faull, 1988; Marangos et al., 1990; Marangos and Miller, 1991; Rudolphi et al., 1992a,b; Fredholm, 1996; Dunwiddie and Masino, 2001; Von Lubitz, 2001), it was quite surprising when it was shown that long-term treatment with caffeine reduced the damage induced by 10 min of bilateral carotid occlusion in gerbils and rats (Rudolphi et al., 1989; Sutherland et al., 1991). Furthermore, treatment of rat dams with peroral caffeine protected their offspring against neuronal damage after hypoxic ischemia in the early postnatal period (Bona et al., 1995). This effect could be reproduced by an A_{2A} antagonist but not by an A_1 antagonist (Bona et al., 1997). These results clearly indicate that long-term caffeine use does not introduce any major risk for increased damage after an acute stroke. Indeed, there are no consistent data linking caffeine use with increased mortality due to stroke (Willett et al., 1996).

Likewise, an acute exposure to methylxanthines decreases the threshold to various convulsants (Albertson et al., 1983; Turski et al., 1985; Ault et al., 1987; Dragunow, 1990; Cutrufo et al., 1992; De Sarro et al., 1997) and worsens brain damage induced by seizures (Pinard et al., 1990). Conversely, chronic caffeine treatment leads to decreased susceptibility to seizures (Georgiev et al., 1993; Von Lubitz et al., 1993; Johansson et al., 1996) and protects the hippocampus against neuronal damage caused by lithium-pilocarpine–induced status epilepticus (Rigoulot et al., 2003). Chronic treatment with an adenosine A_1 antagonist mimics caffeine's chronic effects on seizure susceptibility (Von Lubitz et al., 1994), whereas chronic treatment with an adenosine A_1 agonist, cyclopentyladenosine, results in pronounced increase in intensity of seizure and mortality (Von Lubitz et al., 1994). Chronic treatment with the somewhat selective A_3 receptor agonist, IB-MECA, is able to protect against NMDA-induced seizures and to reduce mortality after electroshock- and pentylenetetrazol-induced seizures (Von Lubitz et al., 1995).

MECHANISMS UNDERLYING THE PROTECTIVE EFFECTS OF CHRONIC CAFFEINE EXPOSURE IN ISCHEMIA AND EPILEPSY

Intially, the mechanism underlying the neuroprotective effect of long-term caffeine treatment was assumed to be due to an up-regulation of adenosine A_1 receptors. However, later it was found that the neuroprotective and antiepileptic effects of chronic caffeine exposure occur in the complete absence of any change in the number of adenosine A_1 receptors (Georgiev et al., 1993). Moreover, these effects are most markedly observed during the ongoing treatment and not after it, as should be observed if an increased transmission through adenosine receptors had occurred as a result of the chronic exposure to caffeine (Georgiev et al., 1993). Nevertheless, the effect of chronic caffeine exposure may somehow involve an action at the A_1 receptors, since the decreased susceptibility to seizures can be obtained also by chronic exposure to an adenosine A_1 receptor antagonist (Von Lubitz et al., 1994).

Most likely, as suggested by Johansson et al. (1996), a primary effect of chronic exposure to caffeine on adenosine receptors would lead to adaptive changes in other neurotransmission systems and/or in fundamental properties related to neuronal excitability, as reflected by a reduction in *c-fos* expression after seizures occurring in animals treated with caffeine (Johansson et al., 1996). The nature of these changes is still unknown, but a reduction of the excitatory action of acetylcholine on cholinergic neurons has been shown to occur after chronic caffeine treatment (Lin and Phillis, 1990). Moreover, long-term treatment with methylxanthines alters the coupling of receptors to G-proteins (Ramkumar et al., 1988; Fastbom and Fredholm, 1990). It has also been reported recently that the psychostimulant effects of caffeine involve the phosphorylation of DARPP-32 (dopamine- and cyclic AMP-regulated phosphoprotein of 32,000 kDa) (Lindskog et al., 2002). Thus, the effects of chronic caffeine exposure in ischemia and epilepsy are most likely mediated via a complex cascade of downstream reactions initiated by blockade of adenosine receptors but probably are not critically dependent on a change in the total number of adenosine receptors. Possibly, other proteins such as G-proteins and DARPP-32 are important. Furthermore, these effects may be different in different brain regions (Svenningsson et al., 1999).

CONCLUSION

These data extend the reports that daily caffeine consumption in reasonable amounts cannot be considered of concern for human health (Nawrot et al., 2003). Indeed, such consumption may reduce and/or delay neurological diseases such as stroke, epilepsy, Parkinson's disease (see Chapter 10 of this book), and, possibly, Alzheimer's disease (Maia and de Mendonça, 2002).

REFERENCES

Åden, U., Halldner, L., Lagercrantz, H., Dalmau, I., Ledent, C. and Fredholm, B.B. (2003) Aggravated ischemic brain damage after hypoxic ischemia in immature adenosine A_{2A} knockout mice. *Stroke*, 734, 739–744.

Åden, U., Leverin, A.L., Hagberg, H. and Fredholm, B.B. (2001) Adenosine A_1 receptor agonism in the immature rat brain and heart. *European Journal of Pharmacology*, 426, 185–192.

Albertson, T.E., Joy, R.M. and Stark, L.G. (1983) Caffeine modification of kindled amygdaloid seizures. *Pharmacology, Biochemistry and Behavior*, 19, 339–344.

Amabeoku, G.J. (1999) Gamma-aminobutyric acid and glutamic acid receptors may mediate theophylline-induced seizures in mice. *General Pharmacology*, 32, 365–372.

Ault, B., Olney, M.A., Joyner, J.L., Boyer, C.E., Notrica, M.A., Soroko, F.E. et al. (1987) Proconvulsant actions of theophylline and caffeine in the hippocampus: implications for the management of temporal lobe epilepsy. *Brain Research*, 426, 93–102.

Bairam, A., Boutroy, M.J., Badonnel, Y. and Vert, P. (1987) Theophylline versus caffeine: comparative effects in treatment of idiopathic apnea in the preterm infant. *Journal of Pediatrics,* 110, 636–639.

Barraco, R.A., Swanson, T.H., Phillis, J.W. and Berman, R.F. (1984) Anticonvulsant effects of adenosine analogues on amygdaloid-kindled seizures in rats. *Neuroscience Letters*, 46, 317–322.

Behan, W.M. and Stone, T.W. (2002). Enhanced neuronal damage by co-administration of quinolinic acid and free radicals, and protection by adenosine A2A receptor antagonists. *British Journal of Pharmacology*, 135, 1435–1442.

Berman, R.F., Fredholm, B.B., Aden, U. and O'Connor, W.T. (2000) Evidence for increased dorsal hippocampal adenosine release and metabolism during pharmacologically induced seizures. *Brain Research*, 872, 44–53.

Bernášková, K. and Mareš, P. (2000) Proconvulsant effects of aminophylline on cortical epileptic after discharges varies during ontogeny. *Epilepsy Research*, 39, 183–190.

Bona, E., Åden, U., Fredholm, B.B. and Hagberg, H. (1995) The effect of long term caffeine treatment on hypoxic-ischemic brain damage in the neonate. *Pediatric Research*, 38, 312–318.

Bona, E., Ådén, U., Gilland, E., Fredholm, B.B. and Hagberg, H. (1997) Neonatal cerebral hypoxia-ischemia: the effect of adenosine receptor antagonists. *Neuropharmacology*, 36, 1327–1338.

Boushey, H.A. (1995) Bronchodilators and other agents used in asthma, in *Basic and Clinical Pharmacology*, Katzung, B.G., Ed., Appleton and Lange, Stamford, CT, pp. 305–319.

Chen, J.F., Huang, Z., Ma, J., Zhu, J., Moratalla, R., Standaert, D. et al. (1999) A(2A) adenosine receptor deficiency attenuates brain injury induced by transient focal ischemia in mice. *The Journal of Neuroscience*, 19, 9192–9200.

Coffey, C.E., Figiel, G.S., Weiner, R.D. and Saunders, W.B. (1990) Caffeine augmentation of ECT. *American Journal of Psychiatry*, 147, 579–585.

Concas, A., Cuccheddu, T., Floris, S., Mascia, M.P. and Biggio, G. (1994) 2-Chloro-N6-cyclopentyladenosine (CCPA), an adenosine A1 receptor agonist, suppresses ethanol withdrawal syndrome in rats. *Alcohol*, 29, 261–264.

Concas, A., Santoro, G., Mascia, M.P., Maciocco, E., Dazzi, L., Ongini, E. et al. (1993) Anticonvulsant doses of 2-chloro-N6-cyclopentyladenosine, an adenosine A1 receptor agonist, reduce GABAergic transmission in different areas of the mouse brain. *Journal of Pharmacology and Experimental Therapeutics*, 267, 844–851.

Corradetti, R., Lo Conte, R., Moroni, F., Passani, M.B. and Pepeu, G. (1984) Adenosine decreases aspartate and glutamate release from rat hippocampal slices. *European Journal of Pharmacology*, 104, 19–26.

Cunha, R.A., Milusheva, E., Vizi, E.S., Ribeiro, J.A. and Sebastiao, A.M. (1994) Excitatory and inhibitory effects of A1 and A2 adenosine receptor activation on the electrically evoked (^{3}H)acetylcholine release from different areas of the rat hippocampus. *Journal of Neurochemistry*, 63, 207–214.

Cunha, R.A., Constantino, M.D. and Ribeiro, J.A. (1999) G protein coupling of CGS 21680 binding sites in the rat hippocampus and cortex is different from that of adenosine A_1 and striatal A_{2A} receptors. *Naunyn Schmiedebergs Archives of Pharmacology*, 359, 295–302.

Cutrufo, C., Bortot, L., Giachetti, A. and Manzini, S. (1992) Differential effects of various xanthines on pentylenetetrazol-induced seizures in rats: an EEG and behavioural study. *European Journal of Pharmacology*, 222, 1–6.

De Sarro, A. and De Sarro, G.B. (1991) Responsiveness of genetically epilepsy-prone rats to aminophylline-induced seizures and interactions with quinolones. *Neuropharmacology*, 30, 169–176.

De Sarro, A., Grasso, S., Zappala, M., Nava, F. and De Sarro, G.B. (1997) Convulsant effects of some xanthine derivatives in genetically epilepsy-prone rats. *Naunyn-Schmiedeberg's Archives of Pharmacology*, 356, 48–55.

De Sarro, G.B., De Sarro, A., Di Paola, E.D. and Bertorelli, R. (1999) Effects of adenosine receptor agonists and antagonists on audiogenic seizure-sensible DBA/2 mice. *European Journal of Pharmacology*, 371, 137–145.

Dragunow, M. (1986) Adenosine: the brain's natural anticonvulsant? *Trends in Pharmacological Sciences*, 7, 128–130.

Dragunow, M. (1990) Adenosine receptor antagonism accounts for the seizure-prolonging effects of aminophylline. *Pharmacology, Biochemistry and Behavior*, 36, 751–755.

Dragunow, M. and Faull, R.L. (1988) Neuroprotective effects of adenosine: see comments. *Trends in Pharmacological Sciences*, 9, 193–194.

Dragunow, M., Goddard, G.V. and Laverty, R. (1985) Is adenosine an endogenous anticonvulsant? *Epilpesia*, 26, 480–487.

Dunwiddie, T.V. (1980) Endogenously released adenosine regulates excitability in the *in vitro* hippocampus. *Epilepsia*, 21, 541–548.

Dunwiddie, T.V. (1985) The physiological role of adenosine in the central nervous system. *International Review of Neurobiology*, 27, 63–139.

Dunwiddie, T.V. and Fredholm, B.B. (1997) Adenosine modulation, in *Purinergic Approaches in Experimental Therapeutics*, Jacobson, K.A. and Jarvis, M.F., Eds., Wiley-Liss, New York, pp. 359–382.

Dunwiddie, T.V. and Worth, T. (1982) Sedative and anticonvulsant effects of adenosine analogs in mouse and rat. *Journal of Pharmacology and Experimental Therapeutics*, 220, 70–76.

Dunwiddie, T.V., Hoffer, B.J. and Fredholm, B.B. (1981) Alkylxanthines elevate hippocampal excitability: evidence for a role of endogenous adenosine. *Naunyn-Schmiedeberg's Archives of Pharmacology*, 316, 326–330.

Dunwiddie, T.V. and Masino, S.A. (2001) The role and regulation of adenosine in the central nervous system. *Annual Review of Neuroscience*, 24, 31–55.

During, M.J. and Spencer, D.D. (1992) Adenosine: a potential mediator of seizure arrest and postictal refractoriness. *Annals of Neurology*, 32, 618–624.

El Yacoubi, M., Ledent, C., Parmentier, M., Daoust, M., Costentin, J. and Vaugeois, J.M. (2001) Absence of the adenosine A_{2A} receptor or its chronic blockade decrease ethanol withdrawal-induced seizures in mice. *Neuropharmacology*, 40, 424–432.

Fastbom, J. and Fredholm, B.B. (1990) Effects of long-term theophylline treatment on adenosine A_1-receptors in rat brain: autoradiographic evidence for increased receptor number and altered coupling to G-proteins. *Brain Research*, 507, 195–199.

Fastbom, J., Pazos, A. and Palacios, J.M. (1987) The distribution of adenosine A1 receptors and 5´-nucleotidase in the brain of some commonly used experimental animals. *Neuroscience*, 22, 813–826.

Fiebich, B.L., Biber, K., Lieb, K., van Calker, D., Berger, M., Bauer, J. and Gebicke-Haerter, P.J. (1996) Cyclooxygenase-2 expression in rat microglia is induced by adenosine A2a-receptors. *Glia*, 18, 152–160.

Fredholm, B.B. (1980) Are methylxanthine effects due to antagonism of endogenous adenosine? *Trends in Pharmacological Sciences*, 1, 129–132.

Fredholm, B.B. (1996) Adenosine and neuroprotection, in *Neuroprotective Agents and Cerebral Ischemia*, Green, A.R. and Cross, A.J., Eds., Academic Press, London, pp. 259–280.

Fredholm, B.B. and Hedqvist, P. (1980) Modulation of neurotransmission by purine nucleotides and nucleosides. *Biochemical Pharmacology*, 29, 1635–1643.

Fredholm, B.B. and Dunwiddie, T.V. (1988) How does adenosine inhibit transmitter release? *Trends in Pharmacological Sciences*, 9, 130–134.

Georgiev, V., Johansson, B. and Fredholm, B.B. (1993) Long-term treatment leads to a decreased susceptibility to NMDA-induced clonic seizures in mice without changes in adenosine A_1 receptor number. *Brain Research*, 612, 271–277.

Goodman, R.R. and Snyder, S.H. (1982) Autoradiographic localization of adenosine receptor in rat brain using (^{3}H)cyclohexyladenosine. *Journal of Neuroscience*, 2, 1230–1241.

Hasko, G., Kuhel, D.G., Chen, J.F., Schwarzschild, M.A., Deitch, E.A., Mabley, J.G. et al. (2000) Adenosine inhibits IL-12 and TNF-[alpha] production via adenosine A2a receptor-dependent and independent mechanisms. *FASEB Journal*, 14, 2065–2074.

Imperato, A., Dazzi, L., Carta, G., Colombo, G. and Biggio, G. (1998) Rapid increase in basal acetylcholine release in the hippocampus of freely moving rats induced by withdrawal from long-term ethanol intoxication. *Brain Research*, 784, 347–350.

Jacobson, K.A., von Lubitz, D.K.J.E., Daly, J.W. and Fredholm, B.B. (1996) Adenosine receptor ligands: differences with acute versus chronic treatment. *Trends in Pharmacological Sciences*, 17, 108–113.

Johansson, B., Georgiev, V., Kuosmanen, T. and Fredholm, B.B. (1996) Long-term treatment with some methylxanthines decreases the susceptibility to bicuculline- and pentylenetetrazol-induced seizures in mice: relationship to c-fos expression and receptor binding. *European Journal of Neuroscience*, 295, 147–154.

Johansson, B., Halldner, L., Dunwiddie, T.V., Masino, S.A., Poelchen, W., Giménez-Llort, L. et al. (2001) Hyperalgesia, anxiety, and decreased hypoxic neuroprotection in mice lacking the adenosine A_1 receptor. *Proceedings of the National Academy of Sciences of the U.S.A.*, 98, 9407–9412.

Klitgaard, H., Knutsen, L.J.S. and Thomsen, C. (1993) Contrasting effects of adenosine A1 and A2 receptor ligands in different chemoconvulsive rodent models. *European Journal of Pharmacology*, 242, 221–228.

Klotz, K.N., Hessling, J., Hegler, J., Owman, C., Kull, B., Fredholm, B.B. and Lohse, M.J. (1998) Comparative pharmacology of human adenosine receptor subtypes: characterization of stably transfected receptors in CHO cells. *Naunyn-Schmiedeberg's Archives of Pharmacology*, 357, 1–9.

Knutsen, L.J.S. and Murray, T.F. (1997) Adenosine and ATP in epilepsy, in *Purinergic Approaches in Experimental Therapeutics*, Jacobson, K.A. and Jarvis, M.F., Eds., pp. Wiley Liss, New York, 423–447.

Kust, B.M., Biber, K., van Calker, D. and Gebicke-Haerter, P.J. (1999) Regulation of K+ channel mRNA expression by stimulation of adenosine A2a-receptors in cultured rat microglia. *Glia*, 25, 120–130.

Lin, Y. and Phillis, J.W. (1990) Chronic caffeine exposure reduces the excitant action of acetylcholine on cerebral cortical neurons. *Brain Research*, 524, 316–318.

Lindskog, M., Svenningsson, P., Pozzi, L., Kim, Y., Fienberg, A.A., Bibb, J.A. et al. (2002) Involvement of DARPP phosphorylation in the stimulant action of caffeine. *Nature*, 418, 774–778.

Lopes, L.V., Cunha, R.A., Kull, B., Fredholm, B.B. and Ribeiro, J.A. (2002) Adenosine A_{2A} receptor facilitation of hippocampal synaptic transmission is dependent on the tonic A_1 receptor inhibition. *Neuroscience*, 112, 319–329.

Maia, L. and de Mendonça, A. (2002) Does caffeine intake protect from Alzheimer's disease? *European Journal of Neurology*, 9, 377–382.

Marangos, P.J. and Miller, L. (1991) Adenosine-based therapeutics in neurologic disease, in *Adenosine Nucleotides as Regulators of Cellular Function*, Phillis, J.W., Ed., CRC Press, Boca Raton, FL, pp. 413–422.

Marangos, P.J., von Lubitz, D., Daval, J.L. and Deckert, J. (1990) Adenosine: its relevance to the treatment of brain ischemia and trauma. *Progress in Clinical and Biological Research*, 361, 331–349.

Masino, S.A., Diao, L., Illes, P., Zahniser, N.R., Larson, G.A., Johansson, B. et al. (2002) Modulation of hippocampal glutamatergic transmission by ATP is dependent on adenosine A_1 receptors. *Journal of Pharmacology and Experimental Therapeutics*, 303, 356–363.

Miura, T. and Kimura, K. (2000) Theophylline-induced convulsions in children with epilepsy. *Pediatrics*, 105, 920.

Monopoli, A., Lozza, G., Forlani, A., Mattavelli, A. and Ongini, E. (1998) Blockade of adenosine A2A receptors by SCH 58261 results in neuroprotective effects in cerebral ischaemia in rats. *Neuroreport*, 9, 3955–3959.

Nawrot, P., Jordan, S., Eastwood, J., Rotstein, J., Hugenholtz, A. and Feeley, M. (2003) Effects of caffeine on human health. *Food Additive Contaminants*, 20, 1–30.

Neely, C.F., Jin, J. and Keith, I.M. (1997) A1-adenosine receptor antagonists block endotoxin-induced lung injury. *American Journal of Physiology*, 272, L353–L361.

Ochiishi, T., Chen, L., Yukawa, A., Saitoh, Y., Sekino, Y., Arai, T. et al. (1999) Cellular localization of adenosine A1 receptors in rat forebrain: immunohistochemical analysis using adenosine A1 receptor-specific monoclonal antibody. *Journal of Comparative Neurology*, 411, 301–316.

Oki, J., Yamamoto, M., Yanagawa, J., Ikeda, K., Taketazu, M. and Miyamota, A. (1994) Theophylline-induced seizures in a 6-month-old girl: serum and cerebrospinal fluid levels. *Brain & Development*, 16, 162–164.

O'Riordan, J.I., Hutchinson, J., FitzGerald, M.X. and Hutchinson, M. (1994) Amnesic syndrome after theophylline associated seizures: iatrogenic brain injury. *Journal of Neurology, Neurosurgery and Psychiatry*, 57, 643–645.

Park, T.S., Van Wylen, D.G.L., Rubio, R. and Berne, R.M. (1987) Interstitial fluid adenosine analogs on amygdala, hippocampus and caudate nucleus after kindled seizures. *Epilepsia*, 28, 658–666.

Phillis, J.W. (1989) Adenosine in the control of the cerebral circulation. *Cerebrovascular and Brain Metabolism Reviews*, 1, 26–54.

Pinard, E., Riche, D., Puiroud, S. and Seylaz, J. (1990) Theophylline reduces cerebral hyperemia and enhances brain damage induced by seizures. *Brain Research*, 511, 303–309.

Ramkumar, V., Bumgarner, J.R., Jacobson, K.A. and Stiles, G.L. (1988) Multiple components of the A1 adenosine receptor-adenylate cyclase system are regulated in rat cerebral cortex by caffeine ingestion. *Journal of Clinical Investigation*, 82, 242–247.

Rang, H.P., Dale, M.M. and Ritter, J.M. (1995) *Pharmacology*, 3rd ed., Churchill Livingstone, Edinburgh, pp. 358–361.

Rigoulot, M.A., Leroy, C., Koning, E., Ferrandon, A. and Nehlig, A. (2003) A chronic low-dose caffeine exposure protects against hippocampal damage but not against the occurrence of epilepsy in the lithium-pilocarpine model in the rat. *Epilepsia*, 44, 529–535.

Rosin, D.L., Robeva, A., Woodard, R.L., Guyenet, P.G. and Linden, J. (1998) Immunohistochemical localization of adenosine A2A receptors in the rat central nervous system. *Journal of Comparative Neurology*, 401, 163–186.

Rossetti, Z.L. and Carboni, S. (1995) Ethanol withdrawal is associated with increased extracellular glutamate in the rat striatum. *European Journal of Pharmacology*, 283, 177–183.

Rudolphi, K.A., Keil, M., Fastbom, J. and Fredholm, B.B. (1989) Ischaemic damage in gerbil hippocampus is reduced following upregulation of adenosine (A_1) receptors by caffeine treatment. *Neuroscience Letters*, 103, 275–280.

Rudolphi, K.A., Schubert, P., Parkinson, F.E. and Fredholm, B.B. (1992a) Adenosine and brain ischemia. *Cerebrovascular and Brain Metabolism Reviews*, 4, 346–369.

Rudolphi, K.A., Schubert, P., Parkinson, F.E. and Fredholm, B.B. (1992b) Neuroprotective role of adenosine in cerebral ischaemia. *Trends in Pharmacological Sciences*, 13, 439–445.

Schubert, P. and Lee, K.S. (1986) Non-synaptic modulation of repetitive firing by adenosine is antagonized by 4-aminopyridine in rat hippocampal slice. *Neuroscience Letters*, 67, 334–338.

Schubert, P., Heinemann, U. and Kolb, R. (1986) Differential effects of adenosine on pre- and postsynaptic calcium fluxes. *Brain Research*, 376, 382–386.

Sebastiao, A.M. and Ribeiro, J.A. (1996) Adenosine A2 receptor-mediated excitatory actions on the nervous system. *Progress in Neurobiology*, 48, 167–189.

Segev, S., Rehavi, M. and Rubistein, E. (1988) Quinolones, theophylline and diclofenac interactions with the γ-aminobutyric acid receptor. *Antimicrobial Agents Chemotherapy*, 32, 1624–1626.

Shannon, M. (1993) Predictors of major toxicity after theophylline overdose. *Annals of Internal Medicine*, 119, 1161–1167.

Shapira, B., Lerer, B., Gilboa, D., Drexler, H., Kugelmass, S. and Calev, A. (1987) Facilitation of ECT by caffeine pretreatment. *American Journal of Psychiatry*, 144, 1199–1202.

Shin, H.K., Shin, Y.W. and Hong, K.W. (2000) Role of adenosine A(2B) receptors in vasodilation of rat pial artery and cerebral blood flow autoregulation. *American Journal of Physiology Heart and Circulatory Physiology*, 278, H339–H344.

Si, Q.S., Nakamura, Y., Schubert, P., Rudolphi, K. and Kataoka, K. (1996) Adenosine and propentofylline inhibit the proliferation of cultured microglial cells. *Experimental Neurology*, 137, 345–349.

Stone, W.E. and Javid, M.J. (1980) Aminophylline and imidazole as convulsants. *Archives Internationales de Pharmacodynamie*, 248, 120–131.

Sutherland, G.R., Peeling, J., Lesiuk, H.J., Brownstone, R.M., Rydzy, M., Saunders, J.K. and Geiger, J.D. (1991) The effects of caffeine on ischemic neuronal injury as determined by magnetic resonance imaging and histopathology. *Neuroscience*, 42, 171–182.

Svenningsson, P., Le Moine, C., Fisone, G. and Fredholm, B.B. (1999) Distribution, biochemistry and function of striatal adenosine A2A receptors. *Progress in Neurobiology*, 59, 355–396.

Svenningsson, P., Nomikos, G.G. and Fredholm, B.B. (1999) The stimulatory action and the development of tolerance to caffeine is associated with alterations in gene expression in specific brain regions. *The Journal of Neuroscience*, 19, 4011–4022.

Torregrosa, G., Salom, J.B., Miranda, F.J., Alabadi, J.A., Alvarez, C. and Alborch, E. (1990) Adenosine A2 receptors mediate cerebral vasodilation in the conscious goat. *Blood Vessels*, 27, 24–27.

Turski, W.A., Cavalheiro, E.A., Ikonomidou, C., Mello, L.E., Bortolotto, Z.A. and Turski, L. (1985) Effects of aminophylline and 2-chloroadenosine on seizures produced by pilocarpine in rats: morphological and electroencephalographic correlates. *Brain Research*, 361, 309–323.

Von Lubitz, D.K.J.E. (2001) Adenosine in the treatment of stroke: yes, maybe, or absolutely not? *Expert Opinion on Investigational Drugs*, 10, 619–632.

Von Lubitz, D.K.J.E., Carter, M.F., Deutsch, S.I., Lin, R.C., Mastropaolo, J., Meshulam, Y. and Jacobson, K.A. (1995) The effects of adenosine A_3 receptor stimulation on seizures in mice. *European Journal of Pharmacology*, 275, 23–29.

Von Lubitz, D.K.J.E., Paul, I.A., Carter, M. and Jacobson, K.A. (1993) Effects of N^6 cyclopentyl adenosine and 8-cyclopentyl-1,3-dipropylxanthine on N-methyl-D-aspartate induced seizures in mice. *European Journal of Pharmacology*, 249, 265–270.

Von Lubitz, D.K.J.E., Paul, I.A., Ji, X.D., Carter, M.F. and Jacobson, K.A. (1994) Chronic adenosine A1 receptor agonist and antagonist: effect on receptor density and N-methyl-D-aspartate induced seizures in mice. *European Journal of Pharmacology*, 253, 95–99.

Weaver, D.R. (1993) A2a adenosine receptor gene expression in developing rat brain. *Molecular Brain Research*, 20, 313–327.

Wiesner, J.B., Ugarkar, B.G., Castellino, A.J., Barankiewicz, J., Dumas, D.P., Gruber, H.E. et al. (1999) Adenosine kinase inhibitors as a novel approach to anticonvulsant therapy. *Journal of Pharmacology and Experimental Therapeutics*, 289, 1669–1677.

Willett, W.C., Stampfer, M.J., Manson, J.E., Colditz, G.A., Rosner, B.A., Speizer, F.E. and Hennekens, C.H. (1996) Coffee consumption and coronary heart disease in women: a ten-year follow-up. *Journal of the American Medical Association*, 275, 458–462.

Winn, H.R., Welsh, J.E., Bryner, C., Rubio, R. and Berne, R.M. (1979) Brain adenosine production during the initial 60 seconds of bicuculline seizures in rats. *Acta Neurologica Scandinavica*, 72, 536–537.

Winn, H.R., Welsh, J.E., Rubio, R. and Berne, R.M. (1980) Changes in brain adenosine during bicuculline-induced seizures in rats: effects of hypoxia and altered systemic blood pressure. *Circulation Research*, 47, 868–877.

Yamamoto, K., Toyama, E., Kawakami, J., Sawada, Y. and Iga, T. (1996) Neurotoxic convulsions induced by theophylline and its metabolites in mice. *Biological Pharmacy Bulletin*, 19, 869–872.

Yokohama, H., Onodera, K., Yagi, T. and Iinuma, K. (1997) Therapeutic doses of theophylline exert proconvulsant effects in developing mice. *Brain & Development*, 19, 403–407.

Young, D. and Dragunow, M. (1994) Status epilepticus may be caused by loss of adenosine anticonvulsant mechanisms. *Neuroscience*, 58, 245–261.

Zhang, G., Franklin, P.H. and Murray, T.F. (1994) Activation of adenosine A_1 receptors underlies anticonvulsant effect of CGS 21680. *European Journal of Pharmacology*, 255, 239–243.

12 Caffeine and Headache: Relationship with the Effects of Caffeine on Cerebral Blood Flow

Astrid Nehlig

CONTENTS

Introduction 175
Antinociceptive Properties of Caffeine..176
Efficacy of Caffeine in the Treatment of Tension-Type Headaches 177
Efficacy of Caffeine in the Treatment of Migraine Headaches 178
Caffeine Withdrawal Headaches 179
Vascular Changes during Various Types of Headaches 179
Cerebral Vasoconstrictive Properties of Caffeine 180
Conclusion 180
References 181

INTRODUCTION

Caffeine is the most widely used psychoactive substance in the world. Most of the caffeine consumed comes from dietary sources such as coffee, tea, cola drinks, and chocolate. The consumption of caffeine is most often correlated with the positive effects following its ingestion, namely increased alertness, energy, and ability to concentrate (Benowitz, 1990; Fredholm et al., 1999; Nawrot et al., 2003). In addition to its dietary use, caffeine can also be found in a number of medications, mainly in over-the-counter and prescription preparations for weight loss and pain relief. The use of caffeine as an adjunctive constituent of analgesic medications can be dated back to 1875 when caffeine was first isolated and characterized structurally (Arnaud, 1987). The first clinical trials concerning the possible contribution of caffeine to the analgesic properties of aspirin and acetaminophen were only performed at the beginning of the second part of the 20th century. However, the interpretation of these studies is rendered difficult by their methodological limitations (Beaver, 1966, 1981).

More recent, well-conducted studies have focused on the association of caffeine, mainly with aspirin or acetaminophen, and the contribution of caffeine to the antipain efficacy of analgesics in migraine and tension-type headache. In 1988, the FDA classified caffeine as belonging to Category I ingredients considered to be "generally regarded as safe and effective" and to Category III ingredients considered as requiring additional research to prove efficacy as over-the-counter analgesic adjuvants (Beaver, 1981; FDA, 1988). In 1994, a review on caffeine data by the Nonprescription Drugs Advisory Committee reported that a 130-mg caffeine dose provided significant adjuvant efficacy when combined with aspirin or acetaminophen/aspirin in a variety of pain states (Dalessio, 1994). The combination of caffeine with current prescription and nonprescription drugs in alleviating migraine and tension-type headache-related symptoms and pain has been the subject of a number of studies that will be summarized below. These studies are of clinical importance since most approved prescription drugs for headache, mainly migraine, are expensive and many have therapy-limiting side effects and contraindications (Von Seggern, 1992; Solomon, 1993; Kumar and Cooney, 1995).

ANTINOCICEPTIVE PROPERTIES OF CAFFEINE

As reviewed in Chapter 1 of this book, most of the actions of caffeine reflect its action as a nonspecific antagonist at adenosine receptors (Nehlig et al., 1992; Fredholm et al., 1999). Among its numerous pharmacological effects on the body, caffeine has been reported to express intrinsic antinociceptive actions in animal models (Daly, 1993); however, they depend on the type of pain stimulus: mechanical, thermal, or electrical (Seegers et al., 1981; Person et al., 1985; Castaneda-Hernandez et al., 1994; Engelhardt et al., 1997; Diaz-Reval et al., 2001). Only rare human studies have assessed whether caffeine alone has independent analgesic properties. In two studies including 301 and 53 subjects suffering from tension-type headache, caffeine significantly relieved pain in a dose-dependent manner (Ward et al., 1991; Diamond et al., 2000). The antinociceptive actions of caffeine are considered to be linked to its antagonism at the level of adenosine receptors. In the peripheral nervous system, adenosine A_1 receptor activation produces antinociceptive actions by decreasing, while adenosine A_2 receptor activation produces pronociceptive or pain-enhancing properties by increasing cyclic AMP levels in the sensory nerve terminals. Adenosine A_3 receptor activation produces pain behaviors due to the release of histamine and serotonin from mast cells and subsequent actions on nerve cell terminals. In humans, the peripheral administration of adenosine produces pain responses resembling those generated under ischemic conditions (Sawynok, 1995, 1998). Thus, adenosine systems appear to contribute to the antinociceptive properties of caffeine. Antinociception results from the inhibition of intrinsic neurons by an increase in K^+ conductance and presynaptic inhibition of sensory nerve terminals to inhibit the release of substance P and glutamate with subsequent actions on pain perception (Sawynok and Yaksh, 1993; Sawynok, 1998). The intrinsic antinociceptive actions of caffeine have been proposed to result from actions at supraspinal sites because manipulation of central monoaminergic pathways can inhibit such actions (Sawynok and Reid, 1996) and may involve inhibition of presynaptic adenosine receptors on cholinergic nerve terminals (Ghelardini et al., 1997). In addition, during headaches, adenosine receptor-mediated vasodilation and/or irregular vascular tone contribute to pain (Dalessio, 1979; Ferrari, 1991).

A number of studies have addressed the question of the adjunctive analgesic properties of caffeine in over-the-counter and prescription pain medications for both headache and other types of pain. For example, 65 to 100 mg of caffeine potentiate the analgesic effects of 500 mg of acetaminophen or 200 mg of ibuprofen in postpartum uterine cramping (Jain et al., 1978; Laska et al., 1983, 1984; Akin et al., 1996), postpartum pain and episiotomy (Laska et al., 1983, 1984), third molar extraction (Laska et al., 1983; Forbes et al., 1991; McQuay et al., 1996), and dentoalveolar pain (Kiersch and Minic, 2002). Based on a pooled analysis of 30 clinical studies including about 10,000 patients suffering from various types of pain, including headaches, Laska et al. (1983, 1984) estimated that the combination of caffeine with current analgesics has a potency of 1.41, which means that it would take 41% more aspirin or acetaminophen alone to reach the same analgesic level as the combined medication.

EFFICACY OF CAFFEINE IN THE TREATMENT OF TENSION-TYPE HEADACHES

Tension-type headaches, as classified by the International Headache Society (1988), occur usually at a maximal frequency of 15 per month and respond to nonprescription medications. They are most often not accompanied by disabling pain and work arrest (Jensen and Paiva, 1993; Matthew, 1993). In the studies detailed below, the patients recruited experienced between four and ten tension-type headaches per month during the year preceding the study (International Headache Society, 1988). In tension-type headaches, the efficacy of the combination of caffeine with acetaminophen and aspirin has been repeatedly reported (Laska et al., 1984; Schachtel et al., 1991; Migliardi et al., 1994; Diamond et al., 2000).

In a randomized, double-blind, parallel-design trial, Schachtel et al. (1991) compared the analgesic properties of a single dose of acetaminophen (1000 mg) or aspirin (1000 mg) with caffeine (64 mg) or placebo in 302 subjects with tension-type headache. Acetaminophen and the combination of aspirin with caffeine were significantly more efficient than placebo in terms of the sum of pain intensity difference from baseline, total pain relief, and percentage of patients experiencing total pain relief. The aspirin/caffeine combination was rated superior to acetaminophen alone for the sum of pain intensity difference from baseline and percentage of patients experiencing complete pain relief.

In a series of six randomized, double-blind, two-period crossover studies, conducted under similar protocols, Migliardi et al. (1994) compared the analgesic efficacy of two combinations containing caffeine in subjects with tension-type headache. In the first four studies involving 1900 patients, tablets of acetaminophen (500 mg), aspirin (500 mg), and caffeine (130 mg) (APAP/ASA/CAF) were combined, while in two other studies involving 911 subjects, two tablets of acetaminophen (1000 mg) and caffeine (130 mg) (APAP/CAF) were combined. The patients were all involved in two treatment periods during which they took two separate medications for two different headache attacks. Acetaminophen was found superior to placebo, but the APAP/CAF and APAP/ASA/CAF combinations were significantly superior to acetaminophen alone and placebo. The caffeine adjuvant effect reached 76% for sum of pain intensity difference from baseline, 89% for percentage of pain intensity difference from baseline, and 97% for total pain relief. For peak analgesia and duration, the caffeine adjuvant effect of the two combinations compared to acetaminophen alone ranged from 63 to 85%. The pooled analgesic responses for the four studies of APAP/ASA/CAF and the two studies of APAP/CAF were virtually superimposable. The effect of caffeine in this antipain medication was totally independent of the usual dietary caffeine intake of the patients or their caffeine consumption in the 4 h preceding medication. However, the two combinations produced more stomach discomfort, nervousness, and dizziness than acetaminophen or placebo (Migliardi et al., 1994).

More recently, in a randomized, double-blind, multicenter trial, Diamond et al. (2000) compared the analgesic effect of ibuprofen alone (400 mg), ibuprofen (400 mg) plus caffeine (200 mg), caffeine alone (200 mg), or placebo in 301 subjects with tension-type headache. The combination of ibuprofen and caffeine provided significantly greater analgesic effects than ibuprofen alone, caffeine alone, or placebo. This was true for the delay between the medication and meaningful improvement in headache relief, total analgesia provided over 4 and 6 h, peak relief, and the number of patients reporting these improvements. This study also showed a similar efficacy of caffeine (200 mg) or ibuprofen (400 mg) alone (Diamond et al., 2000), which is in line with the data of an older study reporting a similar efficacy of caffeine (65 or 130 mg) or acetaminophen (650 mg) alone in the treatment of tension-type headache (Ward et al., 1991).

The analgesic potency of caffeine alone was also shown in another type of headache, postdural puncture headache. The addition of caffeine to saline during the first 90 min after spinal anesthesia reduced moderate and severe postdural puncture headaches as well as the analgesic demand for 4

d. In these cases, caffeine was found to be a simple and safe way to minimize postdural puncture headache (Ford et al., 1989; Yucel et al., 1999; Vincent and Aboff, 2001).

In conclusion, in tension-type headaches, caffeine has both analgesic properties by itself and appears to be able to potentiate the analgesic properties of common antipain drugs such as acetaminophen, aspirin, and ibuprofen.

EFFICACY OF CAFFEINE IN THE TREATMENT OF MIGRAINE HEADACHES

Migraine is a recurrent disorder that produces a wide spectrum of pain and disability (Stewart et al., 1996a; Von Korff et al., 1998). An estimated 23 to 25 million Americans, 18% of women and 6% of men, suffer from migraine attacks (Stewart et al., 1992), among whom more than 50% experience at least one episode per month (Rasmussen and Olesen, 1993). Only 5 to 15% of migraine attacks are associated with mild pain and no disability; moderate to severe pain and disability are reported in 60 to 70% of attacks, and incapacitating pain, bed rest, and total disability occur in 25 to 35% of the attacks (Lipton and Stewart, 1993; Von Korff et al., 1994; Stewart et al., 1996b). The severity of the attacks is variable in a given person (Johannes et al., 1995; Von Korff et al., 1998). Most migraine sufferers (65% of men and 57% of women) manage their headache with nonprescription medications (Celentano et al., 1992), but it is not clear how effective these medications are in the treatment of severe migraine attacks. Indeed, the optimal treatment depends both on the severity of the disease and the severity of individual attacks (Lipton et al., 1994; Michel et al., 1997; Pryse-Phillips et al., 1997; Lipton, 1998).

In a series of three double-blind, randomized, parallel-group, single-dose, placebo-controlled trials, Lipton et al. (1998) examined the efficacy of the treatment of migraine symptoms in a population of 1220 subjects by a nonprescription combination of acetaminophen (500 mg), aspirin (500 mg), and caffeine (65 mg) taken orally as a single-dose treatment. Significantly greater reductions in pain intensity were seen from 1 to 6 h after treatment in patients taking the medication compared to the placebo group. The percentage of patients free of pain at 6 h was twice as high in the medication group as in the placebo group. The other symptoms of migraine, such as nausea, photophobia, phonophobia, and functional disability, were significantly improved in the medicated compared with the placebo group. Goldstein et al. (1999) extended the analysis of the studies performed by Lipton et al. (1998) to a subgroup of 172 subjects, within the initial group of 1220, that were suffering from severe, disabling migraine symptoms. They reached conclusions similar to those in the study including the whole population of migraineurs and showed the superiority of the combination of acetaminophen, aspirin, and caffeine over the placebo in alleviating pain and associated symptoms in this group of patients more severely disabled by their migraines. Finally, the conclusions of the two previous studies were extended to menstruation-associated migraine in a group of 967 women extracted from the initial group of 1220 subjects. A similar efficacy of the combination of the medications cited above was also found in the case of menstruation-associated migraine (Silberstein et al., 1999).

However, in the three studies cited above, the objective was to analyze whether over-the-counter, rather inexpensive, medication could be effective in reducing migraine-associated pain and symptoms. The data of the three studies (Lipton et al., 1998; Goldstein et al., 1999; Silberstein et al., 1999) confirm the safety and efficacy of the acetaminophen/aspirin/caffeine combination on migraine symptoms but do not outline the specific role of caffeine in this combination, since no comparison was performed with combinations containing or not containing caffeine or with caffeine alone. In fact, the efficacy of the acetaminophen/aspirin/caffeine combination was postulated by Strong (1997) to be possibly linked to the presence of caffeine; however, this hypothesis is based on the observation of only one case, the author of the paper himself (Strong, 1997). He found that only the combination of medication including caffeine and caffeine alone (100 mg) were effective in his migraine attacks. However, a large-scale study is still missing to clearly outline the potentiation of analgesic properties afforded by the addition of caffeine to an analgesic drug or combination of drugs as well as the efficacy of caffeine alone in the relief of pain and other symptoms associated with migraine.

More recent randomized, double-blind, multicenter, parallel-group studies also tested the efficacy of ergotamine (1 or 2 mg) plus caffeine (100 or 200 mg) vs. calcium carbasalate/lysine acetylsalicylate (equivalent to 900 mg of aspirin) plus metoclopramide (10 mg), eletriptan (80 mg), or rizatriptan (10 mg) in migraine. In the four studies, the mixture of aspirin equivalents and metoclopramide (Le Jeunne et al., 1999; Titus et al., 2000) and the serotonin receptor agonists eletriptan (Diener et al., 2002) and rizatriptan (Christie et al., 2003) were superior to the combination of ergotamine and caffeine in alleviating all migraine-associated symptoms, such as intensity of pain, duration of pain relief, nausea, photophobia, phonophobia, and functional disability.

Recently, Geyde (2001) published his hypothesized treatment for migraines. He is proposing to use a combination of products that would act on a whole subset of physiological regulations that are impaired during migraine attacks. Migraine attacks include, but are not limited to, (1) falling blood levels of serotonin (Anthony and Lance, 1971, 1989); (2) vascular dilatation (Dalessio, 1979; Ferrari, 1991); (3) inflammatory response in intracranial structures (Pearce, 1993), and (4) in some, but not all migraines, an increasing release of the vasodilator histamine (Guyton, 1991), which continues to rise 24 h after onset (Anthony and Lance, 1971). The author is proposing to use the following combination: (1) tryptophan (500 mg) as a precursor of serotonin; (2) niacin (nicotinic acid, 100 mg) to facilitate tryptophan's conversion to serotonin; (3) acetylsalicylic acid (650 mg) to enhance tryptophan's conversion to serotonin and reduce prostaglandin-mediated inflammation; (4) calcium carbonate (500 mg) (Guyton, 1991), and (5) caffeine (64 mg) to reduce vasodilation (Leonard et al., 1987). This treatment was tested on 12 subjects, 9 of whom were relieved from migraine symptoms within 1 h after taking the treatment.

CAFFEINE WITHDRAWAL HEADACHES

In sensitive individuals, caffeine withdrawal typically induces headaches, among other symptoms (Griffiths et al., 1990; Nehlig et al., 1992; Silverman et al., 1992; Hughes et al., 1993). A strong positive correlation has also been described between caffeine consumption, fasting, and headaches before and after surgical procedures. For every increase in the usual daily consumption of 100 mg of caffeine (about a cup of coffee), the risk of headache immediately before and after surgery is increased by 12 and 16%, respectively, and correlates also with the duration of fasting (Fennelly et al., 1991; Nikolajsen et al., 1994). The risk of headaches is reduced in individuals who drink caffeine or receive substitutive caffeine tablets on the day of the surgery (Weber et al., 1993, 1997; Hampl et al., 1995). Therefore, it was advised by three studies that the numerous healthy patients who drink caffeine-containing beverages daily and are undergoing minor surgical procedures should be permitted to ingest preoperative caffeine (Weber et al., 1993, 1997; Nikolajsen et al., 1994).

Caffeine withdrawal symptoms disappear soon after absorption of caffeine. This effect is strongly linked to the psychological satisfaction related to the ingestion of caffeine; this is especially true for the first cup of the day. The potential reversal of caffeine withdrawal-induced headache and other symptoms by the absorption of caffeine alone has been known for over 50 years and shown repeatedly (Dreisbach and Pfeiffer, 1943; Goldstein and Kaizer, 1969; Goldstein et al., 1969; Griffiths and Woodson, 1988; Hughes et al., 1991). The occurrence of headaches on substitution of caffeinated by decaffeinated coffee predicts subsequent caffeine self-administration (Hughes et al., 1991).

VASCULAR CHANGES DURING VARIOUS TYPES OF HEADACHES

Tension-type headaches are not accompanied by vascular changes (Andersson et al., 1997; Diener, 1997; Sliwka et al., 2001). Conversely, migraine attacks and the interictal state are characterized by cerebral blood flow changes, but there is no real consensus on the nature of the cerebrovascular changes in migraine states. In the interictal state, cerebral blood flow and baseline flow velocity were reported to be higher (Facco et al., 1996; Valikovics et al., 1996; Vasudeva et al., 2003), similar (De Benedittis et al., 1999), or lower (De Benedittis et al., 1999) in migraine patients compared to the control population. There were also differences between migraines with or without aura characterized by different, even opposite, deviations in regional cerebral blood flow (Facco et al., 1996; De Benedittis et al., 1999). During migraine attacks, most often reports indicate a decrease in regional cerebral blood flow (La Spina et al., 1994, 1997; Andersson et al., 1997; Cutrer et al., 1998; De Benedittis et al., 1999; Sanchez del Rio et al., 1999) or no change in migraine with aura (Silvestrini et al., 1996). In migraine without aura, the reports indicate no change (Ferrari et al., 1995; Thomsen et al., 1995; Silvestrini et al., 1996) or a reduction in cerebrovascular reactivity and cerebral blood flow (Andersson et al., 1997; Bednarczyk et al., 1998). Moreover, sumatriptan, a serotonin receptor agonist commonly used for the treatment of migraine headache, does not change cerebral blood flow during migraine attacks (Ferrari et al., 1995; Limmroth et al., 1996).

Conversely, there is a relationship between caffeine withdrawal, the development of headaches, and changes in cerebral blood flow. The cerebral blood flow velocities are increased during withdrawal headaches, significantly decrease within 30 min after caffeine intake in all subjects, and return to baseline values after 2 h (Couturier et al., 1997). This study is in line with others suggesting that increased blood volume may be involved in caffeine withdrawal headache (Dreisbach and Pfeiffer, 1943; Von Borstel et al., 1983; Hirsch, 1984; Mathew and Wilson, 1985b).

CEREBRAL VASOCONSTRICTIVE PROPERTIES OF CAFFEINE

Methylxanthines such as caffeine or theophylline induce vasodilation, except in the central nervous system, where they raise cerebrovascular resistance; this actually contributes to a reduction in cerebral blood flow. The vasoconstrictive properties of methylxanthines have been demonstrated in humans (Wechsler et al., 1950; Shenkin, 1951; Moyer et al., 1952; Gottstein and Paulson, 1972; Magnussen and Hoedt-Rasmussen, 1977; Mathew et al., 1983; Mathew and Wilson, 1985b, 1990; Cameron et al., 1990) and in animals (Oberdörster et al., 1975; Morii et al., 1983; Grome and Stefanovich, 1985, 1986; Puiroud et al., 1988; Ko et al., 1990; Nehlig et al., 1990). The absorption of 250 mg of caffeine in humans induces a decrease in cerebral blood flow ranging from 20 to 30% (Mathew et al., 1983; Mathew and Wilson, 1985b, 1990). This decrease is independent from mood, peripheric physiological activity, and arterial partial pressure of CO_2 (Mathew et al., 1983; Mathew and Wilson, 1985a,b, 1990). Caffeine induces a regional decrease in cerebral blood flow, mainly in the areas where it increases metabolism (i.e., in monoaminergic cell groupings, in the motor and limbic systems, and in the thalamus) (Grome and Stefanovich, 1985, 1986; Nehlig et al., 1990).

CONCLUSION

At this point, the efficacy of caffeine in relieving headache induced by caffeine withdrawal, which leads to cerebral vasodilatation, appears clear and seems to reflect the central vasoconstrictive properties of the methylxanthine. In tension-type headache, there do not seem to be vascular changes related to the attack, and therefore the analgesic effect of caffeine per se or combined with other antipain medication is most likely mediated by other phenomena. It cannot be totally discounted, however, that the vasoconstrictive effect of caffeine could add to the mechanisms involved in pain relief. Finally, for migraine attacks, the literature is rather in favor of a decrease in cerebral blood flow during the attacks. The origin of pain in this pathology remains to be clearly defined; pain is attributed to the dilatation of the ipsilateral medial cerebral artery and also to the dilatation and increased pulsations of the superficial temporal artery and other extracranial arteries (Olesen, 1993). The role of caffeine in pain relief in migraine is not clearly understood and has not been fully explored since the effect of caffeine per se or the comparison of antipain drug combinations with and without caffeine is missing. However, the serotonin receptor agonists appear to be more effective than regular antimedications combined with caffeine in the treatment of migraine, which is in line with the hypothesis of several authors that changes in the diameter of cerebral arterial blood vessels are not so frequent in migraineurs and possibly epiphenomena unrelated to headache (Olesen, 1993).

REFERENCES

Akin, M.D., VanOsdell, D.J. and Balm, T.K. (1996) Multiple dose study evaluating the analgesic efficacy of ibuprofen plus caffeine compared to ibuprofen and placebo in women with primary dysmenorrhea. *Clinical Pharmacological and Therapeutics*, 569 (Abstr.), 131.

Andersen, A.R., Langemark, M. and Olesen, J. (1991) Regional cerebral blood flow in chronic tension-type headache, in *Frontiers of Headache Research 1. Migraine and Other Headaches: The Vascular Mechanisms*, Olesen, J., Ed., Raven Press, New York, pp. 319–321.

Andersson J.L., Muhr, C., Lilja, A., Valind, S., Lundberg, P.O. and Langstrom, B. (1997) Regional cerebral blood flow and oxygen metabolism during migraine with and without aura. *Cephalalgia*, 17, 570–579.

Anthony, M. and Lance, J.W. (1989) Plasma serotonin in patients with chronic headaches. *Journal of Neurology, Neurosurgery and Psychiatry*, 52, 182–184.

Anthony, M. and Lance, J.W. (1971) Histamine and serotonin in cluster headache. *Archives of Neurology*, 25, 225–231.

Arnaud, M.J. (1987) The pharmacology of caffeine. *Progress in Drug Research*, 31, 273–313.

Beaver, W.T. (1966) Mild analgesics: a review of their clinical pharmacology. *Journal of Medical Sciences*, 251, 576–599.

Beaver, W.T. (1981) Aspirin and acetaminophen as constituents of analgesic combinations. *Archives of Internal Medicine*, 141, 293–300.

Bednarczyk, E.M., Remler, B., Weikart, C., Nelson, A.D. and Reed, R.C. (1998) Global cerebral blood flow, blood volume, and oxygen metabolism in patients with migraine headache. *Neurology*, 50, 1736–1740.

Benowitz, N.L. (1990) Clinical pharmacology of caffeine. *Annual Review of Medicine*, 41, 277–288.

Cameron, O.G., Modell, J.G. and Hariharan, M. (1990) Caffeine and human cerebral blood flow: a positron emission tomography study. *Life Sciences*, 47, 1141–1146.

Castaneda-Hernandez, G., Castillo-Mendez, M.S., Lopez-Munoz, F.J., Granados-Soto, V. and Flores-Murrieta, F.J. (1994) Potentiation by caffeine of the analgesic effect of aspirin in the pain-induced functional impairment model in the rat. *Canadian Journal of Physiology and Pharmacology*, 72, 1127–1131.

Celentano, D., Stewart, W., Lipton, R. and Reed, M. (1992) Medication use and disability among migraineurs: a national probability sample survey. *Headache*, 32, 223–228.

Christie, S., Gobel, H., Mateos, V., Allen, C., Vrijens, F. and Shivaprakash, M. (2003) Crossover comparison of efficacy and preference for rizatriptan 10 mg versus ergotamine/caffeine in migraine. *European Neurology*, 49, 20–29.

Couturier, E.G.M., Laman, D.M., van Duijn, M.A.J. and van Duijn, H. (1997) Influence of caffeine and caffeine withdrawal on headache and cerebral blood flow velocities. *Cephalalgia*, 17, 188–190.

Cutrer, F.M., Sorensen, A.G., Weisskoff, R.M., Ostergaard, L., Sanchez del Rio, M., Lee, E.J. et al. (1998) Perfusion-weighted imaging defects during spontaneous migrainous aura. *Annals of Neurology*, 43, 25–31.

Dalessio, D.J. (1979) Classification and mechanism of migraine. *Headache*, 19, 114–119.

Dalessio, D.J. (1994) Caffeine as an analgesic adjuvant: review of the evidence. *Headache*, 34 (Suppl. 1), 1012.

Daly, J.E. (1993) Mechanism of action of caffeine, in *Caffeine and Health*, Garattini, S., Ed., Raven Press, New York, pp. 97–150.

De Beneditts, G., Ferrari da Passano, C., Granata, G. and Lorenzetti, A. (1999) CBF changes during headache-free periods and spontaneous/induced attacks in migraine with and without aura: a TCD and SPECT comparison study. *Journal of Neurosurgical Sciences*, 43, 141–146.

Diamond, S. (1999) Caffeine as an analgesic adjuvant in the treatment of headache. *Headache Quarterly, Current Treatment and Research*, 10, 119–125.

Diamond, S., Balm, T.K. and Freitag, F.G. (2000) Ibuprofen plus caffeine in the treatment of tension-type headache. *Clinical Pharmacology and Therapeutics*, 68, 312–319.

Diaz-Reval, M.I., Ventura-Martinez, R., Hernandez-Delgadillo, G.P., Dominguez-Ramirez, A.M. and Lopez-Munoz, F.J. (2001) Effect of caffeine on antinociceptive action of ketoprofen in rats. *Archives of Medical Research*, 32, 13–20.

Diener, H.C. (1997) Positron emission tomography studies in headache. *Headache*, 37, 622–625.

Diener, H.C., Jansen, J.P., Reches, A., Pascual, J., Pitei, D., Steiner, T.J. and The Eletriptan and Cafergot Comparative Study Group (2002) Efficacy, tolerability, and safety of oral eletriptan and ergotamine plus caffeine (Cafergot) in the acute treatment of migraine: a multicentre, randomised, double-blind, placebo-controlled comparison. *European Neurology*, 47, 99–107.

Dreisbach, R.H. and Pfeiffer, C. (1943) Caffeine withdrawal headache. *Journal of Laboratory and Clinical Medicine*, 28, 1212–1219.

Engelhardt, G., Mauz, A.B. and Pairet, M. (1997) Role of caffeine in combined analgesic drugs from the point of view of experimental pharmacology. *Arzneimittelforschung*, 47, 917–927.

Facco, E., Munari, M., Baratto, F., Behr, A.U., Dal Palu, A., Cesaro, S. et al. (1996) Regional cerebral blood flow (rCBF) in migraine during the interictal period: different rCBF patterns in patients with and without aura. *Cephalalgia*, 16, 161–168.

FDA (1988) Over-the-counter drugs: establishment of monograph for OTC internal analgesic, antipyretic and antirheumatic products. *Federal Register*, 53, 46204–46224.

Fennelly, M., Galletly, D.C. and Purdie, G.I. (1991) Is caffeine withdrawal the mechanism of postoperative headaches? *Anesthesia and Analgesia*, 72, 449–453.

Ferrari, M.D. and Subcutaneous Sumatriptan International Study Group (1991) Treatment of migraine attacks with sumatriptan. *New England Journal of Medicine*, 325, 316–321.

Ferrari, M.D., Haan, J., Blokland, J.A., Arndt, J.W., Minnee, P., Zwinderman, A.H. et al. (1995) Cerebral blood flow during migraine attacks without aura and effect of sumatriptan. *Archives of Neurology*, 52, 135–139.

Forbes, J.A., Beaver, W.T., Jones, K.F., Kehm, C.J., Smith, W.K., Gongloff, C.M. et al. (1991) Effect of caffeine on ibuprofen analgesia in postoperative oral surgery pain. *Clinical Pharmacology and Therapeutics*, 49, 674–684.

Ford, C.D., Ford, D.C. and Koenigsberg, M.D. (1989) A simple treatment of post-lumbar-puncture headache. *Journal of Emergency Medicine*, 7, 29–31.

Fredholm, B.B., Bättig, K., Holmen, J., Nehlig, A. and Zvartau, E. (1999) Actions of caffeine on the brain with special reference to the factors that contribute to its widespread use. *Pharmacological Reviews*, 51, 83–133.

Geyde, A. (2001) Hypothesized treatment for migraines using low doses of tryptophan, niacin, calcium, caffeine, and acetylsalicylic acid. *Medical Hypotheses*, 56, 91–94.

Ghelardini, C., Galeotti, N. and Bartolini, A. (1997) Caffeine induces central cholinergic analgesia. *Naunyn-Schmiedeberg's Archives of Pharmacology*, 356, 590–595.

Goldstein, A. and Kaizer, S. (1969) Psychotropic effects of caffeine in man. III. A questionnaire survey of coffee drinking and its effects in a group of housewifes. *Clinical Pharmacology and Therapeutics*, 10, 477–488.

Goldstein, A., Kaizer, S. and Whitby, O. (1969) Psychotropic effects of caffeine in man. IV. Quantitative and qualitative differences associated with habituation to coffee. *Clinical Pharmacology and Therapeutics*, 10, 489–497.

Goldstein, J., Hoffman, H.D., Armellino, J.J., Batthika, J.P., Hamelsky, S.W., Couch, J. et al. (1999) Treatment of severe, disabling migraine attacks in an over-the-counter population of migraine sufferers: results from three randomized, placebo-controlled studies of the combination of acetaminophen, aspirin, and caffeine. *Cephalalgia*, 19, 684–691.

Gottstein, U. and Paulson, O.B. (1972) The effects of intracarotid aminophylline on the cerebral circulation. *Stroke*, 3, 560–565.

Griffiths, R.R. and Woodson, P.P. (1988) Reinforcing effects of caffeine in humans. *Journal of Pharmacology and Experimental Therapeutics*, 246, 21–28.

Griffiths, R.R., Evans, S.M., Heishman, S.J., Preston, K.L., Sannerud, C.A., Wolf, B. and Woodson, P.P. (1990) Low-dose caffeine physical dependence in humans. *Journal of Pharmacology and Experimental Therapeutics*, 255, 1123–1132.

Grome, J.J. and Stefanovich, V. (1985) Differential effects of xanthine derivatives on local cerebral blood flow and glucose utilization in the conscious rat, in *Adenosine: Receptors and Modulation of Cell Function*, Stefanovich, V., Rudolphi, K. and Schubert, P., Eds., IRL Press Limited, Oxford, pp. 453–457.

Grome, J.J. and Stefanovich, V. (1986) Differential effects of methylxanthines on local cerebral blood flow and glucose utilization in the conscious rat. *Naunyn-Schmiedeberg's Archives of Pharmacology*, 333, 172–177.

Guyton, A.C. (1991) *Textbook of Medical Physiology*, 8th ed., W.B. Saunders, Toronto, Canada.

Hampl, K.F., Schneider, M.C., Ruttimann, U., Ummenhofer, U. and Drewe, J. (1995) Perioperative administration of caffeine tablets for prevention of postoperative headaches. *Canadian Journal of Anaesthesia*, 42, 789–792.

Hirsch, K. (1984) Central nervous system pharmacology of the dietary methylxanthines, in *The Methylxanthine Beverages and Food: Chemistry, Consumption, and Health Effects*, Spiller, G.A., Ed., Alan Liss, New York, pp. 235–301.

Hughes, J.R., Higgins, S.T., Bickel, W.K., Hunt, W.K., Fenwick, J.W., Gulliver, S.B. and Mireault, G.C. (1991) Caffeine self-administration, withdrawal, and adverse effects among coffee drinkers. *Archives of General Psychiatry*, 48, 611–617.

Hughes, J.R., Oliveto, A.H., Bickel, W.K., Higgins, S.T. and Badger, G.J. (1993) Caffeine self-administration and withdrawal: incidence, individual differences and interrelationships. *Drug and Alcohol Dependence*, 32, 239–246.

International Headache Society (1988) Classification and diagnostic criteria for headache disorders, cranial neuralgias and facial pain. *Cephalalgia*, 8 (Suppl. 7), 1–96.

Jain, A.K., McMahon, F.G., Ryan, J.R., Unger, D. and Richard, W. (1978) Aspirin and aspirin-caffeine in postpartum pain relief. *Clinical Pharmacology and Therapeutics*, 24, 69–75.

Jensen, R. and Paiva, T. (1993) Tension-type headache, cluster headache, and miscellaneous headaches: episodic tension-type headache, in *The Headaches*, Olesen, J., Tfelt-Hansen, P. and Welch, K.M.A., Eds., Raven Press, New York, pp. 497–502.

Johannes, C., Linet, M., Stewart, W., Celentano, D., Lipton, R. and Szklo, M. (1995) relationship of headache to phase of the menstrual cycle among young women: a daily diary study. *Neurology*, 45, 1076–1082.

Kiersch, T.A. and Minic, M.R. (2002) The onset of action and the analgesic efficacy of Saridon (a propyphenazone/paracetamol/caffeine combination) in comparison with paracetamol, ibuprofen, aspirin and placebo (pooled statistical analysis). *Current Medical Research Opinion*, 18, 18–25.

Ko, K.R., Ngai, A.C. and Winn, H.R. (1990) Role of adenosine in regulation of cerebral blood flow in sensory cortex. *American Journal of Physiology*, 259, H1703–H1708.

Kumar, K.L. and Cooney, T.G. (1995) Headaches. *Medical Clinics of North America*, 79, 264–271.

Laska, E.M., Sunshine, A., Mueller, F., Elvers, W.B. and Rubin, A. (1984) Caffeine as an analgesic adjuvant. *Journal of the American Medical Association*, 251, 1711–1718.

Laska, E.M., Sunshine, A., Zighelboim, I., Roure, C., Marrero, I., Wanderling, J. and Olson, N. (1983) Effect of caffeine on acetaminophen analgesia. *Clinical Pharmacology and Therapeutics*, 33, 498–509.

La Spina, I., Calloni, M.V. and Porazzi, D. (1994) Transcranial Doppler monitoring of a migraine with aura attack from the prodromal phase to the end. *Headache*, 34, 593–596.

La Spina, I., Vignati, A. and Porazzi, D. (1997) Basilar artery migraine: transcranial Doppler EEG and SPECT from the aura phase to the end. *Headache*, 37, 43–47.

Le Jeunne, C., Pascual Gómez, J., Pradalier, A., Titus, I., Albareda, F., Joffroy, A., Liaño, H. et al. (1999) Comparative efficacy and safety of calcium carbasalate plus metoclopramide versus ergotamine tartrate plus caffeine in the treatment of acute migraine attacks. *European Neurology*, 41, 37–43.

Leonard, T.K., Watson, R.R. and Mohs, M.E. (1987) The effects of caffeine on various body systems. *Journal of the American Dietetic Association*, 87, 1048–1053.

Limmroth, V., May, A., Auerbach, P., Wosnitza, G., Eppe, T. and Diener, H.C. (1996) Changes in cerebral blood flow velocity after treatment with sumatriptan or placebo and implications for the pathophysiology of migraine. *Journal of Neurological Sciences*, 138, 60–65.

Lipton, R.B. (1998) Disability assessment as a basis for stratified care. *Cephalalgia*, 18 (Suppl. 22), 40–46.

Lipton, R.B. and Stewart, W.F. (1993) Migraine in the United States: a review of epidemiology and health care use. *Neurology*, 43 (Suppl. 3), S6–S10.

Lipton, R.B., Amatniek, J., Ferrari, M. and Gross, M. (1994) Migraine: identifying and removing barriers to care. *Neurology*, 44 (Suppl. 4), S63–S68.

Lipton, R.B., Stewart, W.F., Ryan, R.E., Saper, J., Silberstein, S. and Sheftell, F. (1998) Efficacy and safety of acetaminophen, aspirin, and caffeine in alleviating migraine headache pain. *Archives of Neurology*, 55, 210–217.

Magnussen, I. and Hoedt-Rasmussen, K. (1977) The effect of intraarterial administered aminophylline on cerebral hemodynamics in man. *Acta Neurologica Scandinavica*, 55, 131–136.

Matthew, N.T. (1993) Tension-type headache, cluster headache, and miscellaneous headaches: acute pharmacotherapy, in *The Headaches*, Olesen, J., Tfelt-Hansen, P. and Welch, K.M.A., Eds., Raven Press, New York, pp. 531–536.

Mathew, R.J., Barr, D.L. and Weinman, M.L. (1983) Caffeine and cerebral blood flow. *British Journal of Psychiatry*, 143, 604–608.

Mathew, R.J. and Wilson, W.H. (1985a) Caffeine changes in cerebral circulation. *Stroke*, 16, 814–817.

Mathew, R.J. and Wilson, W.H. (1985b) Caffeine consumption, withdrawal and cerebral blood flow. *Headache*, 25, 305–309.

Mathew, R.J. and Wilson, W.H. (1990) Behavioral and cerebrovascular effects of caffeine in patients with anxiety disorders. *Acta Psychiatrica Scandinavica*, 82, 17–22.

McQuay, H.J., Angell, K., Carroll, D., Moore, R.A. and Juniper, R.P. (1996) Ibuprofen compared with ibuprofen plus caffeine after third molar surgery. *Pain*, 66, 247–251.

Michel, P., Dubroca, B., Dartigues, J., Hasnaoui, A.E. and Henry, P. (1997) Frequency of severe attacks in migraine sufferers from the Gazel cohort. *Cephalalgia*, 17, 863–866.

Migliardi, J.R., Armellino, J.J., Friedman, M., Gillings, D.B. and Beaver, W.T. (1994) Caffeine as an analgesic adjuvant in tension headache. *Clinical Pharmacology and Therapeutics*, 56, 576–586.

Morii, S., Winn, H.R. and Berne, R.M. (1983) Effect of theophylline, an adenosine receptor blocker, on cerebral blood flow (CBF) during rets and transient hypoxia. *Journal of Cerebral Blood Flow and Metabolism*, 3 (Suppl. 1), S480–S481.

Moyer, J.H., Tasnek, A.B., Miller, S.J., Snyder, H. and Bowman, R.O. (1952) The effect of theophylline with ethylene diamine (aminophylline) and caffeine on cerebral hemodynamics and cerebral fluid pressure in patients with hypertensive headache. *American Journal of Medical Sciences*, 224, 377–385.

Nawrot, P., Jordan, S., Eastwood, J., Rotstein, J., Hugenholtz, A. and Feeley, M. (2003) Effects of caffeine on human health. *Food Additive Contaminants*, 20, 1–30.

Nehlig, A., Daval, J.L. and Debry, G. (1992) Caffeine and the central nervous system: mechanisms of action, biochemical, metabolic and psychostimulant effects. *Brain Research Reviews*, 17, 139–170.

Nehlig, A., Pereira de Vasconcelos, A., Dumont, I. and Boyet, S. (1990) Effects of caffeine, L-phenylisopropyladenosine and their combination on local cerebral blood flow in the rat. *European Journal of Pharmacology*, 179, 271–280.

Nikolajsen, L., Larsen, K.M. and Kierkegaard, O. (1994) Effect of previous frequency of headache, duration of fasting and caffeine abstinence on perioperative headache. *British Journal of Anaesthesia*, 72, 295–297.

Oberdörster, G., Lang, R. and Zimmer, R. (1975) Influence of adenosine and lowered cerebral blood flow on the cerebrovascular effects of theophylline. *European Journal of Pharmacology*, 30, 197–204.

Olesen, J. (1993) Migraine: hemodynamics, in *The Headaches*, Olesen, J., Tfelt-Hansen, P. and Welch, K.M.A., Eds., Raven Press, New York, pp. 209–222.

Pearce, J.M.S. (1993) Sumatriptan: efficacy and contribution to migraine mechanisms. *Journal of Neurology, Neurosurgery and Psychiatry*, 55, 1103–1105.

Person, D.L., Kissin, I., Brown, P.T., Xavier, A.V., Vinik, H.R. and Bradley, E.L. (1985) Morphine-caffeine analgesic interaction in rats. *Anesthesia and Analgesia*, 64, 851–856.

Pryse-Phillips, W., Dodick, D., Edmeads, J., Gawel, M., Nelson, R., Purdy, R. et al. (1997) Guidelines for the diagnosis and management of migraine in clinical practice. *Canadian Medical Association Journal*, 156, 1276–1287.

Puiroud, S., Pinard, E. and Seylaz, J. (1988) Dynamic cerebral and systemic circulatory effects of adenosine, theophylline and dipyridamole. *Brain Research*, 453, 287–298.

Rasmussen, B.K. and Olesen, J. (1993) Migraine epidemiology. *Cephalalgia*, 13, 216–217.

Sanchez del Rio, M., Bakker, D., Wu, O., Agosti, R., Mitsikostas, D.D., Ostergaard, L. et al. (1999) Perfusion weighted imaging during migraine: spontaneous visual aura and headache. *Cephalalgia*, 19, 701–707.

Sawynok, J. (1995) Pharmacological rationale for the clinical use of caffeine. *Drugs*, 49, 37–50.

Sawynok, J. (1998) Adenosine receptor activation and nociception. *European Journal of Pharmacology*, 347, 1–11.

Sawynok, J. and Reid, A. (1996) Neurotoxin-induced lesions to central serotonergic, noradrenergic and dopaminergic systems modify caffeine-induced antinociception in the formalin test and locomotor stimulation in rats. *Journal of Pharmacology and Experimental Therapeutics*, 277, 646–653.

Sawynok, J. and Yaksh, T.L. (1993) Caffeine as an analgesic adjuvant: a review of pharmacology and mechanisms of action. *Pharmacological Reviews*, 45, 43–85.

Schachtel, B.P., Thoden, W.R., Konerman, J.P., Brown, A. and Chaing, D.S. (1991) Headache pain model for assessing and comparing the efficacy of over-the-counter analgesic agents. *Clinical Pharmacology and Therapeutics*, 50, 322–329.

Seegers, A.J., Jager, L.P., Zandberg, P. and van Noordwijk, J. (1981) The anti-inflammatory, analgesic and antipyretic activities of non-narcotic analgesic drug mixtures in rats. *Archives Internationales de Pharmacodynamie*, 251, 237–254.

Shenkin, H.A. (1951) Effects of various drugs upon cerebral circulation and metabolism in man. *Journal of Applied Physiology*, 3, 465–471.

Silberstein, S.D., Armellino, J.J., Hoffman, H.D., Battikha, J.P., Hamelsky, S.W., Stewart, W.F. and Lipton, R.B. (1999) Treatment of menstruation-associated migraine with the nonprescription combination of acetaminophen, aspirin, and caffeine: results from three randomized, placebo-controlled studies. *Clinical Therapeutics*, 21, 475–490.

Silverman, K., Evans, S.M., Strain, E.C. and Griffiths, R.R. (1992) Withdrawal syndrome after the double-blind cessation of caffeine consumption. *New England Journal of Medicine*, 327, 1109–1114.

Silvestrini, M., Matteis, M., Troisi, E., Cupini, L.M. and Bernardi, G. (1996) Cerebrovascular reactivity in migraine with and without aura. *Headache*, 36, 37–40.

Sliwka, U., Harscher, S., Diehl, R.R., van Schayck, R., Niesen, W.D. and Weiller, C. (2001) Spontaneous oscillations in cerebral blood flow velocity give evidence of different autonomic dysfunctions in various types of headache. *Headache*, 41, 157–163.

Solomon, G.D. (1993) Therapeutic advances in migraine. *Journal of Clinical Pharmacology*, 33, 200–209.

Stewart, W.F., Lipton, R.B., Celentano, D.D. and Reed, M.L. (1992) Prevalence of migraine headache in the United States: relation to age, income, race, and other sociodemographic factors. *Journal of the American Medical Association*, 267, 64–69.

Stewart, W.F., Lipton, R.B. and Liberman, J. (1996a) Variation in migraine prevalence by race. *Neurology*, 47, 52–59.

Stewart, W.F., Lipton, R.B. and Simon, D. (1996b) Work-related disability: results from the American migraine study. *Cephalalgia*, 16, 231–238.

Strong, F.C., III (1997) It may be the caffeine in extra strength Excedrin that is effective for migraine. *Journal of Pharmacy and Pharmacology*, 49, 1260.

Thomsen, L.L., Iversen, H.K. and Olesen, J. (1995) Cerebral blood flow velocities are reduced during attacks of unilateral migraine without aura. *Cephalalgia*, 15, 109–116.

Titus, F., Escamilla, C., Gomes da Costa Palmeira, M.M., Leira, R. and Pereira Monteiro, J.M. (2001) A double-blind comparison of lysine acetylsalicylate plus metoclopramide vs. ergotamine plus caffeine in migraine: effects on nausea, vomiting and headache symptoms. *Clinical Drug Investigation*, 21, 87–94.

Valikovics, A., Olah, L., Fulesdi, B., Kaposzta, Z., Ficzere, A., Bereczki, D. and Csiba, L. (1996) Cerebrovascular reactivity measured by transcranial Doppler in migraine. *Headache*, 36, 232–328.

Vasudeva, S., Claggett, A.L., Tietgen, G.E. and McGrady, A.V. (2003) Biofeedback-assisted relaxation in migraine headache: relationship to cerebral blood flow velocity in the middle cerebral artery. *Headache*, 43, 245–250.

Vincent, S. and Aboff, B. (2001) Use of caffeine in post-dural puncture headache: a case report. *Del Medical Journal*, 73, 97–100.

Von Borstel, R.W., Wurtman, R.J. and Conlay, L.A. (1983) Chronic caffeine consumption potentiates the hypotensive action of circulating adenosine. *Life Sciences*, 32, 1151–1158.

Von Korff, M., Stewart, W.F. and Lipton, R.B. (1994) Assessing migraine severity: new directions. *Neurology*, 44 (Suppl. 4), 40–46.

Von Korff, M., Stewart, W.F., Simon, D.S. and Lipton, R.B. (1998) Migraine and reduced work performance: a population-based diary study. *Neurology*, 50, 1741–1745.

Von Seggern, R.L., Parantainen, J., Gothoni, G. and Vapaatalo, H. (1992) Cost considerations in migraine treatment. 2. Acute migraine treatment. *Headache*, 36, 493–502.

Ward, N., Whitney, C., Avery, D. and Dunner, D. (1991) The analgesic effect of caffeine in headache. *Pain*, 44, 151–155.

Weber, J.G., Ereth, M.H. and Danielson, D.R. (1993) Perioperative ingestion of caffeine and postoperative headache. *Mayo Clinic Proceedings*, 68, 842–845.

Weber, J.G., Klindworth, J.T., Arnold, J.J., Danielson, D.R. and Ereth, M.K. (1997) Prophylactic intravenous administration of caffeine and recovery after ambulatory surgical procedure. *Mayo Clinic Proceedings*, 72, 621–626.

Wechsler, R.L., Kleiss, L.M. and Kety, S.S. (1950) The effects of intravenously administered aminophylline on cerebral circulation and metabolism in man. *Journal of Clinical Investigation*, 29, 28–30.

Yucel, A., Ozyalcin, S., Talu, G.K., Yucel, E.C. and Erdine, S. (1999) Intravenous administration of caffeine sodium benzoate for postdural puncture headache. *Regional Anesthesia and Pain Medicine*, 24, 51–54.

13 Cerebral Effects of Noncaffeine Constituents in Roasted Coffee

Tomas de Paulis and Peter R. Martin

CONTENTS

Abstract 187
Introduction 187
Antiaddiction Effects of Coffee 188
Antidepressant Effects of Coffee 191
Antioxidant Effects of Coffee 191
Conclusions 193
Acknowledgments 193
References 193

ABSTRACT

Caffeine exerts the dominant effects in the central nervous system after ingestion of coffee, so little attention has been paid to the actions of noncaffeine constituents in coffee. Similar to all plants, coffee contains numerous constituents. Because some of the effects of these compounds may modify or even oppose those of caffeine, the results of their presence in consumed coffee would be difficult to observe, except in decaffeinated coffee. Recently, it has been suggested that roasted coffee contains agents with anticraving properties, which may affect the ability of individuals who consume coffee to better cope with addiction. Another piece of evidence from epidemiological studies of suicide rates suggests that coffee constituents may have an antidepressant effect. Coffee also contains compounds with weak estrogenic activity, which may cause an increase in muscarinic cholinergic receptors involved in cognition. The antioxidant activity of caffeic acid and its metabolites may play a role in protecting brain cells from free radical oxidation. Further, because the brain, like other organs in the body, depends on blood flow to ensure access to oxygenated hemoglobin, the cardiovascular effects of roasted coffee may play a significant role in preventing stroke and other degenerative diseases of the brain.

INTRODUCTION

Every day, for millions of people worldwide the alluring aroma of hot coffee brewed from freshly roasted beans drags sleepers from their beds and pedestrians into cafes (Illy, 2002). For years, the conventional wisdom held that this attraction of coffee was the result of the stimulating effect of caffeine (Fredholm et al., 1999). However, underlying this seemingly simple beverage exists a profound chemical complexity, and noncaffeine constituents perhaps have effects on the central

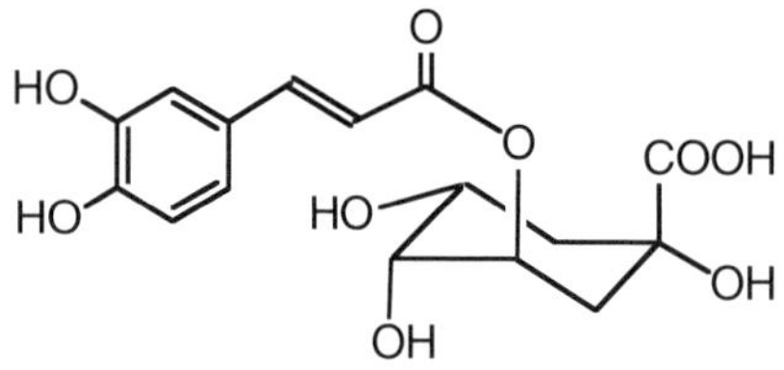

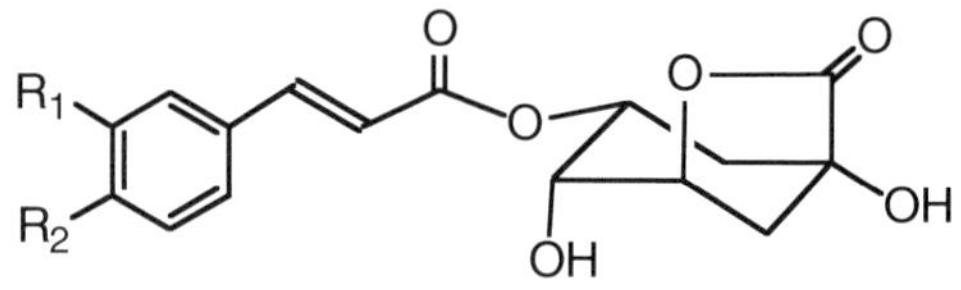

3-CAQ ($R_1=R_2=$OH), 3-caffeoyl-1,5-quinide
3-FEQ ($R_1=$OCH$_3$, $R_2=$OH), 3-feruloyl-1,5-quinide
3-COQ ($R_1=$H, $R_2=$OH), 3-coumaroyl-1,5-quinide

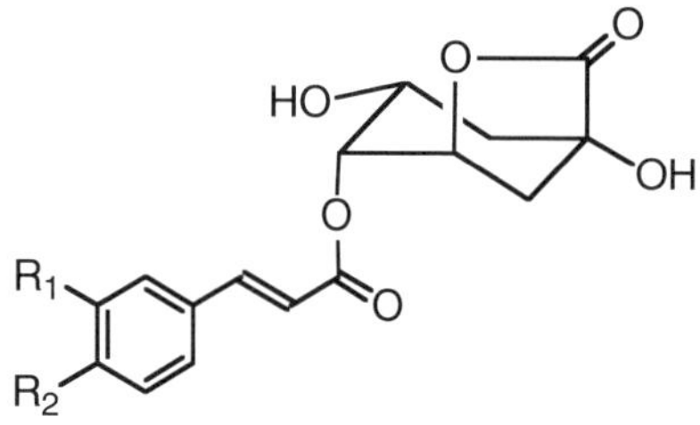

5-CAQA, 5-caffeoylquinic acid

4-CAQ ($R_1=R_2=$OH), 4-caffeoyl-1,5-quinide
4-FEQ ($R_1=$OCH$_3$, $R_2=$OH), 4-feruloyl-1,5-quinide
4-COQ ($R_1=$H, $R_2=$OH), 4-coumaroyl-1,5-quinide

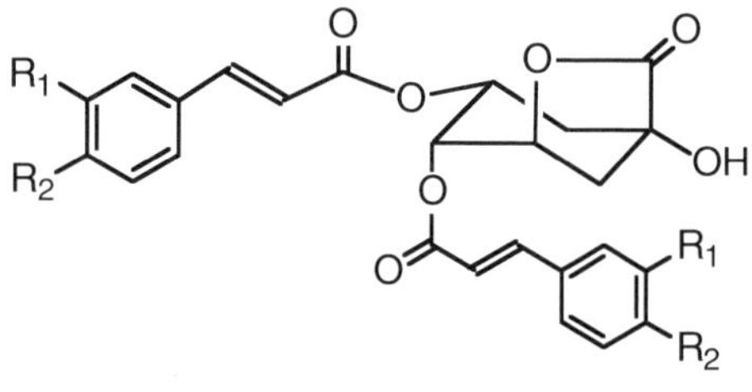

DICAQ ($R_1=R_2=$OH)
DIFEQ ($R_1=$OCH$_3$, $R_2=$OH)
DICOQ ($R_1=$H, $R_2=$OH)

FIGURE 13.1 Structures and acronyms of chlorogenic acid and lactones (quinides).

nervous system that are contrary to those of caffeine. Similar to all plant material, coffee beans contain a variety of low-molecular-weight compounds, in particular chlorogenic acids (Clarke, 1985). These constitute a mixture of mono- and di-esters of quinic acid and caffeic and ferulic acids. However, in contrast to tea or chocolate, roasted coffee is unique among beverages in that it contains quinides, formed in the roasting process from the corresponding chlorogenic acids. Figure 13.1 shows the structures of the most abundant chlorogenic acid, 5-caffeoylquinic acid (5-CAQA), and putative quinides, some of which have been identified in roasted coffee (Schrader et al., 1996), while others are expected based on the amount of their corresponding chlorogenic acids present in green coffee beans (Clarke, 1985; de Paulis et al., 2002).

ANTIADDICTION EFFECTS OF COFFEE

Accumulating clinical evidence supports the idea that opioid receptor antagonists are effective in reducing craving and relapse in patients suffering from drug or alcohol addiction and related behaviors (Garbutt et al., 1999; Kim et al., 2001). For example, the mu opioid receptor antagonists naltrexone and nalmefene have been used as an adjunct to various psychosocial treatments of alcoholism, heroin addiction, and pathological gambling. These disorders are highly prevalent in most human societies and have significant social, economic, and public health consequences in all the countries of the world. It is therefore of great interest that a weak antimorphine activity of roasted coffee has been discovered. Boublik et al. (1983) and Wynne et al. (1987) were the first to report that quinides in coffee can act as antagonists of the mu opioid receptor (i.e., they can specifically inhibit the effects of morphine and other opioid-related drugs at this receptor). The estimated total amount of quinides in an average cup of regular, instant, or decaffeinated coffee, approximately 1% of dry weight (Clarke, 1985), raises the possibility that daily coffee consumption may have beneficial effects against various addictive disorders. Although Boublik et al. (1983) reported that instant coffee was able to displace the binding of the mu opioid receptor antagonist, [^{3}H]naloxone, the active agent was never identified. However, mass spectroscopic evidence suggested an isomer of feruloylquinide (Wynne et al., 1987). Investigation of quinides in roasted coffee at the Vanderbilt Institute for Coffee Studies has demonstrated that not only does 3-feruloyl-1,5-quinide (3-FEQ) block [^{3}H]naloxone binding, but the majority of quinides present in roasted coffee, in particular 4-caffeoyl-1,5-quinide (4-CAQ) and 3,4-dicaffeoyl-1,5-quinide (DICAQ), share this antimorphine activity with potencies one order of magnitude higher than those of 3-FEQ and 4-FEQ (de Paulis et al., in press). The earlier work with instant coffee (Boublik et al., 1983; Wynne et al., 1987) did not clarify whether a single compound or a mixture of similar compounds accounted for the mu opioid receptor binding activity. The fraction containing the highest activity inhibited the binding of 1 nM [^{3}H]naloxone by 50% at 0.025 mg/ml. Assuming that this represented an equal mixture of 3-FEQ and 4-FEQ, the corresponding K_i value would be 65 μM, in agreement with the observed affinities for 3-FEQ (K_i 90 μM) and 4-FEQ (K_i 40 μM) (de Paulis et al., 2004). One reason that Boublik and Wynne failed to identify 3-CAQ and 4-CAQ in instant coffee could be their separation method (column chromatography on neutral silica gel). Under these conditions, derivatives of caffeic acid bind irreversibly to the stationary phase, the *ortho*-hydroxy groups possibly forming chelating bonds with the silicon dioxide (de Paulis et al., 2002). Thus, it is reasonable to assume that using this method of purification in these earlier studies resulted in only the less abundant (and less potent) feruloylquinides being identified.

A preliminary study of coffee extract on the inhibition of morphine-induced analgesia in mice was conducted at the Vanderbilt Institute for Coffee Studies. Decaffeinated instant coffee (Taster's Choice, Nestlé) was extracted with ethyl acetate and the solvent evaporated. The residual oil (10% of the original weight) was dissolved in 40 ml of water containing 10% tween-80 (Sigma) and diluted with saline to 10 mg/ml. Mice (N = 8) were injected intraperitoneally with a dose of 10 mg/kg body weight and tested in the hot plate test after 15 min. After receiving coffee extract, complete inhibition of the analgesic effect of morphine was noted, as measured by reduction in the time to react, e.g., jumping or licking the hind paws after the mouse was placed on a hotplate kept at 52°C (Figure 13.2). Latency time for coffee extract plus morphine was 5.4 ± 0.5 sec vs. 18.5 ± 1.5 sec for morphine alone. Coffee extract by itself was not different from control injections (7.9 ± 2.6 sec vs. 6.2 ± 0.8 sec) (McDonald, unpublished). Because the extract represented a 10-fold concentration of quinide-like constituents in instant decaffeinated coffee, the tested dose was equivalent to at least 100 mg/kg of instant coffee. These results are in contrast to a previous report by Strubelt et al. (1986) that high doses of instant, decaffeinated coffee (60 mg/kg i.v., 150 mg/kg i.p., or 2000 mg/kg p.o.) had no apparent effect on morphine-induced analgesia in mice. However, the dose of morphine used by Strubelt et al. (1986) was unusually high (10 mg/kg i.p.), the less accurate tail-flick method was used, and the analgesic activity was measured 1 h after administration of coffee. In our study of the pharmacokinetic profile of DICAQ (100 mg/kg i.p.), both brain and plasma levels in the mouse peaked after 5 to 10 min, with an apparent half-life of 2 min. No detectable amounts of DICAQ were seen in the mice's brains after 30 min (de Paulis et al., 2002).

Results of DIFEQ, DICAQ and 4-CAQ in blocking morphine effects in the hot plate test are shown in Table 13.1. Morphine (1 mg/kg i.p.) was given 15 min prior to the test drug. Testing was performed 15 min postinjection in groups of eight mice per dose. Test compounds were dissolved in 10% tween-80 (5 mg/ml). From the dose-response data, approximate effective doses (ED$_{50}$) for each compound were calculated by log-logit analysis. *In vivo* behavioral activities range from 21 mg/kg for DIFEQ to 0.2 mg/kg for 4-CAQ (McDonald, unpublished). In this preliminary study the wide dose range does not match the range of antimorphine activities of these compounds *in vitro*.

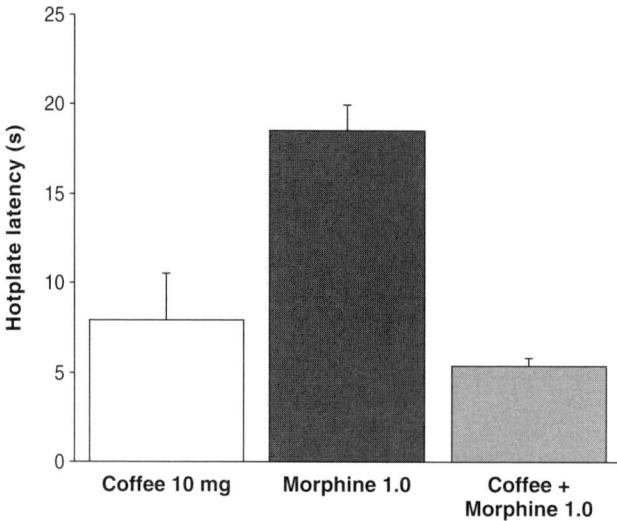

FIGURE 13.2 Ability of decaffeinated instant coffee extract (10 mg/kg i.p.) to block the effects of morphine (1 mg/kg i.p.) in the hot plate test in mice 15 min after injection. The bars represent mean ± SEM (N = 8).

This discrepancy could be the result of different pharmacokinetic properties, or suggest that inhibition of morphine-induced analgesia (antinociception) in the mouse is mediated by a mechanism other than simple blockade of mu opioid receptors. One such possible mechanism of action could be the recently discovered inhibition of the adenosine transporter by roasted coffee (de Paulis et al., 2002).

The fact that dicinnamoylquinides, such as DIFEQ and DICAQ in coffee, as well as coffee itself, are inhibitors of the human adenosine transporter was evidenced by their ability to displace the binding of [^{3}H] 4-nitrobenzyl-6-thioinosine (de Paulis et al., 2002). This effect on adenosine may confound the interpretation of the hot plate test results. When the presynaptic reuptake transporter is blocked, more adenosine becomes available for neurotransmission. The resulting activation of adenosine receptors in the spinal cord causes antinociception by inhibiting the release

TABLE 13.1
Preliminary Antimorphine Activities of Substituted Cinnamoyl-1,5-Quinides

Compound	Acronym	In vitro (K_i, μM)[a]	In vivo (ED_{50}, mg/kg i.p.)[b]
3,4-Diferuloylquinide	DIFEQ	11.9	20.5
3,4-Dicaffeoylquinide	DICAQ	15.1	1.6
3-Caffeoylquinide	3-CAQ	105	NT[c]
4-Caffeoylquinide	4-CAQ	4.4	0.20
Naloxone		0.0037	0.07[d]

[a] Calculated from the concentration that inhibits 50% of [^{3}H]naloxone binding in HEK-MOR cell homogenates.
[b] Dose that reduces morphine-induced latency by 50% in mice.
[c] Not tested.
[d] Subcutaneous administration; data taken from McGilliard and Takemori (1978).

of substance P in sensory nerve terminals (Cahill et al., 1996; Sawynok et al., 1998). Depending on the time of administration, the adenosine uptake inhibitors dilazep and dipyridamole block morphine-induced analgesia when given 10 min before morphine, but they enhance analgesia when given 5 min after morphine (Keil and Delander, 1995; Sadig-Lindell et al., 2001). Thus, quinides from coffee consumption could have the paradoxical effect of both preventing and increasing pain, in contrast to caffeine, which has no or little modulating effect on pain (Karlsten et al., 1992; Koegh and Chaloner, 2002). If confirmed in human subjects, this dual analgesic effect of these compounds could make them particularly appropriate in addiction treatment, because the combined effects on pain are less likely to lead to dependence liability than is a pure analgesic agent that is acting as an agonist at the mu opioid receptor.

ANTIDEPRESSANT EFFECTS OF COFFEE

Epidemiological studies have demonstrated that there is a negative correlation between suicide rates and coffee consumption. Individuals who drank up to three cups of coffee per day had a significantly lower risk of committing suicide than those who did not drink coffee (Klatsky et al., 1993). Similar results were found in a 10-year study of female registered nurses (Kawachi et al., 1996). A study on the effects of coffee in Brazilian schoolchildren found that having three cups of coffee with milk significantly decreased their feelings of sadness (Flores et al., 2000). Thus, coffee seems to exert an antidepressant-like action in humans. It can be argued that this effect of coffee possibly is caused by its content of caffeine, since caffeine blocks the inhibitory effect of adenosine on dopamine nerves in the brain (Fredholm et al., 1999). However, caffeine by itself has no euphoric activity (Nehlig and Boyet, 2000), and caffeine in subjects with normal mood did not show any effect on mood (Rodgers and Dernoncourt, 1998). Only direct clinical studies of coffee with and without caffeine in depressed patients will be able to clarify this controversial issue.

Another aspect of the potential role of coffee and anxiety symptoms of depression relates to the fact that some derivatives of chlorogenic acid (i.e., dicinnamoylquinides) formed in the roasting process are inhibitors of the adenosine transporter in the brain (de Paulis et al., 2002). Because inhibition of the adenosine transporter leads to an increase in adenosine neurotransmission, the result is a decrease in NMDA glutamate levels in the brain. This adenosine–glutamate regulation is disturbed after alcohol consumption and in particular during periods of alcohol withdrawal (Kaplan et al., 1999). Caffeine, being an antagonist of adenosine receptors, would, of course, be expected to have the opposite effect, thereby overwhelming any proadenosine effect of the quinides. However, chronic coffee consumption leads, paradoxically, to elevated adenosine levels (Conlay et al., 1997), and chronic administration of high doses of caffeine in mice results in profound increases in neurotransmitter receptors in the brain, in particular the GABA receptor complex that is the target for the benzodiazepine anxiolytic agents (Shi et al., 1993).

ANTIOXIDANT EFFECTS OF COFFEE

One of the fundamental questions regarding potential health effects of coffee on brain functions is whether brewed or instant roasted coffee shares the antioxidant capacity of other plant-derived beverages, such as fruit juices, tea, or wine (Richelle et al., 2001; Schilter et al., 2001). Antioxidant mechanisms may be essential for protection against a variety of disorders in which reactive oxygen species produced by the body during metabolism have been implicated in disease causation, including arteriosclerosis, degenerative brain disorders (Pratico, 2002), cancer, complications of drug and alcohol addiction, and others (Heitzer et al., 2001). One of the major constituents of coffee, caffeic acid, is an effective antioxidant agent (Raneva et al., 2001). It has been suggested that the antioxidant activity of caffeoyl derivatives emanates from the ability of caffeic acid to form an iron complex, which prevents the production of hydroxyl radicals (Kono et al., 1998; Sestili et al., 2002). It was recently demonstrated that the roasting process has no diminishing effect on the antioxidant activity of the remaining caffeoyl derivatives (Charurin et al., 2002). Because coffee contains a higher percentage of caffeic acid derivatives than other beverages (Clifford, 2000; Mattila and Kumpilainen, 2002), the antioxidant effects of its major human metabolite, dihydrocaffeic acid (DCA) (Rechner et al., 2001; Moridani et al., 2001), was investigated at the Vanderbilt Institute for Coffee Studies.

All blood vessels in the human body, including the brain, contain specialized endothelial cells, which can be cultured and harvested (Bauer et al., 1992). When such cells were treated overnight with 100 μM DCA, loaded with 450 μM cis-parinaric acid for 1 h, then subjected to oxidation with 40 mM of a free-radical initiator (AAPH) for 2 h, the fluorescence in DCA-treated cells was about 50% greater than in control cells (0.75 ± 0.09 for DCA vs. 0.55 ± 0.12 for control) in the two experiments performed so far (May, unpublished). Preloading with *alpha*-tocopherol had about the same protective effect, further supporting the antioxidant effect of DCA. These results show clearly that endothelial cells actively take up DCA against a concentration gradient, that DCA protects against loss of membrane-bound *alpha*-tocopherol during the oxidant stress, and that DCA also protects membrane fatty acids from oxidation in cells not loaded with *alpha*-tocopherol. Together with the results that DCA protects against menadione-induced oxidant stress in the cytosol, these results clearly establish an antioxidant role for DCA within endothelial cells. Although the DCA concentrations used in these studies exceed those expected *in vivo* following oral absorption (25 to 50 nM at 30 to 60 min after ingestion of 1 g of decaffeinated instant coffee), the ability of these cells to concentrate DCA by two- to fourfold within the cells is likely to produce intracellular concentrations effective in generating the antioxidant protection.

Other neuroprotective effects of coffee may be related to its estrogenic effects. In studies of dopaminergic neurons, estrogen appears to reduce the production of free radicals and protect against oxidative stress (Pita et al., 2002). In a Canadian study on health and aging in 4615 participants who were 65 years old or older in 1991, coffee consumption was associated with reduced risk of suffering from Alzheimer's disease 5 years later (Lindsay et al., 2002). Both coffee from regular whole beans and instant coffee, but not tea or cocoa, contain constituents of unknown structure that exhibit estrogenic activities equivalent to 0.11 to 0.17 mg of genistein (4′,5,7-trihydroxyisoflavone) per 10 g of original amount of coffee (Kitts, 1987). This effect cannot be caused by the known diterpenes, cafestol and kahweol, because these are removed in connection with the preparation of instant coffee (Gross et al., 1987). Recently, both known subtypes of estrogen receptors have been identified in the brain (Goldstein and Sites, 2002). In cultured cells, estrogen protects brain endothelial cells from ischemia (Galea et al., 2002). In elderly women, estrogen has been found to prolong cognitive function (Fillit, 2002) and prevent structural signs of aging of the brain (Cook et al., 2002). In addition, there may be a synergistic effect between the estrogenic activity of coffee and the antiparkinson effect of caffeine. In an epidemiological study of 78,000 healthy women who had reached menopause at the end of the study, estrogen therapy in women who had reported low caffeine consumption was associated with Parkinson's disease. This population had a significantly lower risk of developing Parkinson's disease than women with low coffee consumption and no hormone therapy (Ascherio et al., 2002). However, for women who consumed six cups of coffee or more per day, the trend was reversed. The possible association between Parkinson's disease and estrogen and coffee, with and without caffeine, needs further studies.

A third mechanism of neuroprotection of the brain relates to the recent discovery that roasted coffee inhibits the adenosine transporter (de Paulis et al., 2002). After having a stroke, the brain contains high levels of endogenous proinflammatory agents, such as the tumor necrosis factor-*alpha*, resulting in neuroinflammation (Arvin et al., 1995). Activation of adenosine A_{2A} receptors in brain, which is the expected outcome of adenosine transporter blockade, results in antiinflammatory activity. Further, activation of these receptors has been shown to decrease cell death following both intracerebral hemorrhage and focal ischemic stroke (Mayne et al., 2001).

CONCLUSIONS

Because caffeine exerts such a dominant effect on the brain, the potentially beneficial effect of noncaffeine constituents in roasted coffee is poorly understood. Recent studies of decaffeinated coffee have demonstrated that roasted coffee contains constituents that exhibit antimorphine, proadenosine, and antioxidant activities in the brain. From epidemiological studies there seems to emerge a series of beneficial effects of coffee on brain function, which may be unrelated to the effects of caffeine. Even though many of these effects may be weak and vary from individual to individual (Nardini et al., 2002), the fact that coffee is consumed in such large quantities worldwide makes the elucidation of the mechanisms of action of these constituents important.

ACKNOWLEDGMENTS

This work was sponsored by the Vanderbilt Institute for Coffee Studies. The permission to include unpublished results from the laboratories of James M. May and Michael P. McDonald is gratefully acknowledged. Support from the National Coffee Association of USA, Kraft Foods, Sara Lee, Starbucks, Nestlé, the International Coffee Organization, and a consortium of coffee-producing countries (Brazil, Colombia, Guatemala, and Japan) is gratefully acknowledged.

REFERENCES

Arvin, B., Neville, L.F., Barone, F.C. and Feuerstein, G.Z. (1995) Brain injury and inflammation: a putative role of TNF alpha. *Annals of the New York Academy of Sciences,* 765, 62–71.

Ascherio, A., Chen, H., Colditz, G. and Speizer, F. (2002) Caffeine, postmenopausal estrogen, and risk of Parkinson's disease. *Movement Disorders,* 17 (Abstr.), 1103.

Bauer, J., Margolis, M., Schreiner, C., Edgell, C.J., Azizkhan, J., LazarowskI, E. and Juliano, R.L. (1992) In vitro model of angiogenesis using a human endothelium-derived permanent cell line: contributions of induced gene expression, G-proteins, and integrins. *Journal of Cell Physiology,* 153, 437–449.

Boublik, J.H., Quinn, M.J., Clements, J.A., Herington, A.C., Wynne, K.N. and Funder, J.W. (1983) Coffee contains potent opiate receptor binding activity. *Nature,* 301, 246–248.

Cahill, C.M., White, T.D. and Sawynok, J. (1996) Synergy between mu/delta-opioid receptors mediates adenosine release from spinal cord synaptosomes. *European Journal of Pharmacology,* 298, 45–49.

Charurin, P., Ames, J.M. and Dolores del Castillo, M. (2002) Antioxidant activity of coffee model systems. *Journal of Agricultural and Food Chemistry,* 50, 3751–3756.

Clarke, R.J. (1985) Chlorogenic acids, in *Coffee, Volume 1: Chemistry,* Clarke, R.J. and Macrae, R., Eds., Elsevier Science, New York, pp 153–202.

Clifford, M.N. (2000) Chlorogenic acids and other cinnamates: nature, occurrence, diatary burden, absorption and metabolism. *Journal of Science and Food Agriculture,* 80, 1033–1043.

Conlay, L.A., Conant, J.A., deBros, F. and Wurtman, R. (1997) Caffeine alters plasma adenosine levels. *Nature,* 389, 136.

Cook, I.A., Morgan, M.L., Dunkin, J.J., David, S., Witte, E., Lufkin, R., Abrams, M., Rosenberg, S. and Leuchter, A.F. (2002) Estrogen replacement therapy is associated with less progression of subclinical structural brain disease in normal elderly women: a pilot study. *International Journal of Geriatric Psychiatry,* 17, 610–618.

de Paulis, T., Schmidt, D.E., Bruchey, A.K., Kirby, M.T., McDonald, M.P., Commers, P., Lovinger, D.M. and Martin, P.R. (2002) Dicinnamoylquinides in roasted coffee inhibit the human adenosine transporter. *European Journal of Pharmacology,* 242, 213–221.

de Paulis, T., Commers, P., Farah, A., Zhao, J., McDonald, M.P., Galici, R. and Martin, P.R. (in press) 4-Caffeoyl-1,5-quinide in roasted coffee inhibits [^{3}H]naloxone binding and reverses antinociceptive effects of morphine in mice. *Psychopharmacology.*

Fillit, H.M. (2002) The role of hormone replacement therapy in the prevention of Alzheimer's disease. *Archives of Internal Medicine,* 162, 1934–1942.

Flores, G.B., Andrade, F. and Lima, D.R. (2000) Can coffee help fighting the drug problem?: preliminary results of a Brazilian youth drug study. *Acta Pharmacologica Sinica,* 21, 1057–1216.

Fredholm, B.B., Battig, K., Holmen, J., Nehlig, A. and Zvartau, E.E. (1999) Actions of caffeine in the brain with special reference to factors that contribute to its widespread use. *Pharmacological Reviews,* 51, 83–133.

Galea, E., Santizo, R., Feinstein, D.L., Adamsom, P., Greenwood, J., Koenig, H.M. and Pelligrino, D.A. (2002) Estrogen inhibits NF-*kappa*-B-dependent inflammation in brain endothelium without interfering with I-*kappa*-B degradation. *Neuroreport,* 13, 1469–1472.

Garbutt, J.C., West, S.L., Carey, T.S., Lohr, K.N. and Crews, F.T. (1999) Pharmacological treatment of alcohol dependence: a review of the evidence. *Journal of the American Medical Association,* 281, 1318–1325.

Gross, G., Jaccaud, E. and Huggett, A.C. (1987) Analysis of the content of the diterpenes cafestol and kahweol in coffee brews. *Food and Chemical Toxicology,* 35, 547–554.

Heitzer, T., Schlinzig, T., Krohn, K., Meinertz, T. and Münzel, T. (2001) Endothelial dysfunction, oxidative stress, and risk of cardiovascular events in patients with coronary artery disease. *Circulation,* 104, 2673–2678.

Illy, E. (2002) The complexity of coffee. *Scientific American,* June, 86–91.

Kaplan, G.B., Bharmal, N.H., Leite-Morris, K.A. and Adams, W.R. (1999) Role of adenosine A_1 and A_{2A} receptors in the alcohol withdrawal syndrome. *Alcohol,* 19, 157–162.

Karlsten, R., Gordh, T. and Post, C. (1992) Local antinociceptive and hyperalgesic effects in the formalin test after peripheral administration of adenosine analogues in mice. *Pharmacology and Toxicology,* 70, 434–438.

Kawachi, I., Willett, W.C., Colditz, G.A., Stampfer, M.J. and Spiieizer, F.E. (1996) A prospective study of coffee drinking and suicide in women. *Archives of Internal Medicine,* 156, 521–525.

Keil, G.J. and Delander, G.E. (1995) Time-dependent antinociceptive interactions between opioids and nucleoside transport inhibitors. *Journal of Pharmacology and Experimental Therapeutics,* 274, 1387–1392.

Kim, S.W., Grant, J.E., Adson, D.E. and Shin, Y.C. (2001) Double-blind naltrexone and placebo comparison study in the treatment of pathological gambling. *Biological Psychiatry,* 49, 914–921.

Kitts, D.D. (1987) Studies on the estrogenic activity of coffee extract. *Journal of Toxicology and Environmental Health,* 20, 37–49.

Klatsky, A.L., Armstrong, M.A. and Friedman, G.D. (1993) Coffee, tea, and mortality. *Annals of Epidemiology,* 3, 375–381.

Koegh, E. and Chaloner, N. (2002) The modulating effect of anxiety sensitivity on caffeine-induced hypoalgesia in healthy women. *Psychopharmacology,* 164, 429–431.

Kono, Y., Kashine, S., Yoneyama, T., Sakamoto, Y., Matsui, Y. and Shibata, H. (1998) Iron chelation by chlorogenic acid as a natural antioxidant. *Bioscience, Biotechnology and Biochemistry,* 62, 22–27.

Lindsay, J., Laurin, D., Verreault, R., Hebert, R., Helliwell, B., Hill, G.B. and McDowell, I. (2002) Risk factors for Alzheimer's disease: a prospective analysis from the Canadian Study of Health and Aging. *American Journal of Epidemiology,* 156, 445–453.

Mattila, P. and Kumpilainen, J. (2002) Determination of free and total phenolic acids in plant-derived foods by HPLC and diode-array detection. *Journal of Agricultural and Food Chemistry,* 50, 3660–3667.

Mayne, M., Fotheringham, J., Yan, H.-J., Power, C., Del Bigio, M.R., Peeling, J. and Geiger, J.D. (2001) Adenosine A_{2A} receptor activation reduces proinflammatory events and decreases cell death following intracerebral hemorrhage. *Annals of Neurology,* 49, 727–735.

McGilliard, K.L. and Takemori, A.E. (1978) Antagonism by naloxone of narcotic-induced respiratory depression and analgesia. *Journal of Pharmacology and Experimental Therapeutics,* 207, 494–503.

Moridani, M.Y., Scobie, H., Jamshidzadeh, A., Salehi, P. and O'Brien, P.J. (2001) Caffeic acid, chlorogenic acid, and dihydrocaffeic acid metabolism: glutathione conjugate formation. *Drug Metabolism and Disposition,* 29, 1432–1439.

Nardini, M., Cirillo, E., Natella, F. and Scaccini, C. (2002) Absorption of phenolic acids in humans after coffee consumptions. *Journal of Agricultural and Food Chemistry,* 50, 5735–5741.

Nehlig, A. and Boyet, S. (2000) Dose-response study of caffeine effects on cerebral functional activity with specific focus on dependence. *Brain Research,* 858, 71–77.

Pita, I., Wojna, V., Rodriguez, K. and Serrano, C. (2002) Estrogen effect on Parkinson's disease severity. *Movement Disorders,* 17 (Abstr.), 1111.

Pratico, D. (2002) Oxidative imbalance and lipid peroxidation in Alzheimer's disease. *Drug Development Research,* 56, 446–451.

Raneva, V., Shimasaki, H., Ishida, I., Ueta, N. and Niki, E. (2001) Antioxidantive activity of 3,4-dihydroxyphenylacetic acid and caffeic acid in rat plasma. *Lipids,* 36, 1111–1116.

Rechner, A.R., Spencer, J.P.E., Kuhnle, G., Hahn, U. and Rice-Evans, C.A. (2001) Novel biomarkers of the metabolism of caffeic acid derivatives in vivo. *Free Radical Biology and Medicine,* 30, 1213–1222.

Richelle, M., Tavazzi, I. and Offord, E. (2001) Comparison of the anti-oxidant activity of commonly consumed polyphenolic beverages (coffee, cocoa, and tea) prepared per cup serving. *Journal of Agricultural and Food Chemistry,* 49, 3438–3442.

Rodgers, P.J. and Dernoncourt, C. (1998) Regular caffeine consumption: a balance of adverse and beneficial effects for mood and psychomotor performance. *Pharmacology, Biochemistry and Behavior,* 59, 1039–1045.

Sadig-Lindell, B., Sylven, C., Hagerman, I., Berglund, M., Terenius, L., Franzen, O. and Eriksson, B.E. (2001) Oscillation of pain intensity during adenosine infusion. Relationship to *beta*-endorphin and sympathetic tone. *Neuroreport,* 12, 1571–1575.

Sawynok, J. (1998) Adenosine receptor activation and nociception. *European Journal of Pharmacology,* 347, 1–11.

Schilter, B., Cavin, C., Tritscher, A. and Constable, A. (2001) Health effects and safety consideration, in *Coffee: Recent Developments*, Clarke, R.J. and Vitzthum, O.G., Eds., Blackwell Science, London, pp. 165–183.

Schrader, K., Kiehne, A., Engelhardt, U.H. and Maier, H.G. (1996) Determination of chlorogenic acids with lactones in roasted coffee. *Journal of Science and Food Agriculture,* 71, 392–398.

Sestili, P., Diamantini, G., Bedini, A., Cerioni, L., Tommasini, I., Tarzia, G. and Cantoni, O. (2002) Plant-derived phenolic compounds prevent the DNA single-strand breakage and cytotoxicity induced by *tert*-butylhydroperoxide via an iron-chelating mechanism. *Biochemical Journal,* 364, 121–128.

Shi, D., Nikodijevic, O., Jacobson, K.A. and Daly, J.W. (1993) Chronic caffeine alters the density of adenosine, adrenergic, cholinergic, GABA, and serotonin receptors and calcium channels in mouse brain. *Cellular and Molecular Neurobiology,* 13, 247–261.

Strubelt, O., Kaschube, M. and Zetler, G. (1986) Failure of coffee to inhibit the pharmacodynamic activity of morphine in vivo. *Experientia,* 42, 35–37.

Wynne, K.N., Familari, M., Boublik, J.H., Drummer, O.H., Rae, I.D. and Funder, J.W. (1987) Isolation of opiate receptor ligands in coffee. *Clinical and Experimental Pharmacology and Physiology,* 14, 785–790.

14 Can Tea Consumption Protect against Stroke?

Astrid Nehlig

CONTENTS

Abstract 197
Tea Consumption and Composition 197
General Protective Properties of Tea 198
Tea and Ischemia 198
Mechanisms of Tea-Induced Protection against Ischemia 199
Conclusion and Future Directions 201
References 201

ABSTRACT

Antioxidant properties of tea that lead to health-protective effects have been reported, mainly in relation to various types of cancer, coronary heart disease, and inflammation (Mukhtar and Ahmad, 2000). Only a few experimental data are available about the possible neuroprotective effects of tea consumption in human stroke or animal models of ischemia-reperfusion. The purpose of this chapter is to analyze the available data and the mechanisms by which tea may protect brain areas from stroke and to briefly compare the data on tea with those on coffee and stroke.

TEA CONSUMPTION AND COMPOSITION

The tea plant, *Camellia sinensis*, also known as *Thea sinensis* L., originates from southeast Asia but is currently cultivated in more than 30 countries around the world. Tea, the water extract of the dry leaves of *Camellia sinensis*, is consumed worldwide, although in greatly different quantities; it is generally accepted that, after water, tea is the most consumed beverage in the world, with a per capita consumption of approximately 120 ml/d (Katiyar and Mukhtar, 1996). Of the total amount of tea produced and consumed in the world, 78% is black, 20% is green, and less than 2% is oolong tea. Black tea is mostly consumed in Western countries and in some Asian countries, while green tea is consumed primarily in Japan, China, India, and a few countries in North Africa and the Middle East. This variety of tea has become progressively more popular and is consumed in increasing quantities in Western countries. Oolong tea production and consumption are limited to southeastern China and Taiwan (Katiyar and Mukhtar, 1996).

The different types of tea undergo different manufacturing processes. To produce green tea, freshly harvested leaves are rapidly steamed or pan-fried to inactivate enzymes, thereby preventing fermentation and producing a dry, stable product. Epicatechins are the main compounds in green tea, accounting for its characteristic color and flavor.

For the production of black and oolong teas, the fresh leaves are allowed to wither until their moisture content is reduced to about 55% of the original leaf weight, which results in the concentration of polyphenols in the leaves. The withered leaves are then rolled and crushed, thus initiating the "fermentation" of polyphenols via polyphenol oxidase-catalyzed oxidative polymerization. Oolong tea is prepared by firing the leaves shortly after rolling to terminate the oxidation and dry the leaves. Normal oolong tea is considered to be about half less fermented than black tea.

The composition of tea leaves depends on a variety of factors, including climate, season, horticulture practices, and the type and age of the plant. The chemical composition of green tea is similar to that of the leaf. Green tea contains polyphenolic compounds that include flavanols, flavandiols, flavonoids, and phenolic acids that account for 25 to 30% of the solids in water extracts of green tea leaves. Most of the polyphenols in green tea are (–)-epicatechin, (–)-epicatechin-3-gallate, (–)-epigallocatechin, (–)-epigallocatechin-3-gallate (ECGC), and (+)-catechin. In black tea, the major polyphenols are bisflavanols, theaflavins, and thearubigins that are formed from catechins during the process of polymerization. Theaflavins (about 1 to 2% of the solid in water extracts of black tea leaves) include theaflavin, theaflavin-3-gallate, and theaflavin-3,3'-digallate, and these substances contribute to the typical color and flavor of black tea. A substantial proportion of the solids in water extracts of black tea leaves represent thearubigins that have a wide range of molecular weights and are poorly characterized (Graham, 1992). Thus, it has been assumed that there are significant differences in antioxidant properties between green and black teas.

GENERAL PROTECTIVE PROPERTIES OF TEA

Tea has been consumed by some human populations for many generations and, in some parts of the world, has been considered to have health-promoting properties (Weisburger et al., 1997). Extensive laboratory research and the epidemiological findings of the past 20 years reported that polyphenolic compounds present in tea may reduce the risk of a variety of illnesses. Mainly, green tea catechins display pharmacological properties such as anticarcinogenic activity (Bu-Abbas et al., 1994; Katiyar and Mukhtar, 1996; Xu et al., 1996; Dreosti et al., 1997; Kohlmeier et al., 1997; Weisburger et al., 1997; Mukhtar and Ahmad, 2000), antioxidant activity (Serafini et al., 1996; Uchida et al., 1992), and anti-inflammatory activity (Katiyar et al., 1995a,b; Katiyar and Mukhtar, 1996) and appear able to prevent cardiovascular disease (Yamaguchi et al., 1991; Stensvold et al., 1992; Uchida et al., 1995; Tijburg et al., 1997), stroke (Weisburger, 1996), osteoporosis (Fujita, 1994), liver disease (Imai and Nakachi, 1995), and bacterial and viral infections (Nakayama et al., 1990; Horiba et al., 1991). Drinking green tea daily would contribute to maintaining plasma levels of catechins sufficient to exert antioxidant activities against oxidative modification of lipoproteins in circulating blood (Nagakawa et al., 1997). Japanese epidemiologists reported that among patients with the highest consumption of green tea (10 or more cups per day), a significant decrease in the risk of gastric cancer was obvious. Moreover, one cup of green tea infusion contains 100 to 200 mg of polyphenolic compounds, which leads to the possible daily consumption of about 1 g of (–)-epigallocatechin-3-gallate in green tea (Kono et al., 1988).

TEA AND ISCHEMIA

In ischemia, the mode of neuronal death is considered to be a continuum between apoptosis and necrosis: ischemic neurons appear cytologically necrotic while exhibiting many biochemical features of apoptosis. Ischemia-induced cell death is active, energy-dependent, and the result of a cascade of detrimental events that include disturbance of calcium homeostasis leading to increased excitotoxicity, dysfunction of the endoplasmic reticulum and mitochondria, elevation of oxidative stress causing DNA damage, lipid peroxidation, alteration of proapoptotic gene expression, and activation of caspases and endonucleases leading to the final degradation of the genome. The purpose of the present chapter is not to review the whole cascade of the molecular events occurring in ischemia/reperfusion-induced neuronal death, which has been described in detail elsewhere (Chan, 2001; Graham and Chen, 2001; Hou and MacManus, 2002), but rather to focus on the steps of the ischemic cascade that have been considered as targets for the potential protective action of tea and tea extracts.

Ischemia/reperfusion-mediated brain injury results, at least partly, from the oxidation of cellular macromolecules (Siesjö, 1993; Kawase et al., 1999). Indeed, because of the high consumption of oxygen by the brain, the high concentrations of polyunsaturated fatty acids and transition metals, and the low concentration of antioxidants, the brain is very vulnerable to reperfusion-induced reactive oxygen species that lead to oxidative damage to lipids and DNA. Within the molecules most harmful to the brain, eicosanoids (prostaglandins and leukotrienes), thromboxane A_2, malondialdehyde, and oxygen radicals are considered as mediators of ischemia/reperfusion-induced brain injury (Matsuo et al., 1996; Islekel et al., 1999). Eicosanoids and lipid peroxides are involved in ischemia/reperfusion-induced brain damage because of their ability to alter membrane permeability, induce brain edema, and ultimately lead to neuronal death (Watanabe et al., 1994).

At present, only a limited number of groups have studied the potential neuroprotective effects of tea and tea constituents on ischemia/reperfusion-induced ischemic neuronal damage. In global ischemia, the consumption of 0.5% green tea extracts by rats subjected to a 5-min ligation of the two common carotid arteries followed by 48 h of recirculation allowed a 37% reduction of the infarcted volume compared to animals drinking regular water (Hong et al., 2000). Similarly, 41 and 60% reductions in the infarcted volume were recorded in Mongolian gerbils subjected to a 3-week 0.5 or 2% green tea extract regimen prior to a 60-min focal ischemia induced by the occlusion of the middle cerebral artery, followed by 48 h of blood recirculation (Hong et al., 2001). A similar protection from cerebral infarction induced by global ischemia consecutive to a 3-min bilateral ligation of the common carotid arteries followed by 5 d of recirculation was found in gerbils treated immediately after the induction of the ischemia with 25 or 50 mg/kg of the green tea polyphenol, (–)-epigallocatechin-3-gallate (EGCG) (Lee et al., 2000). Likewise, a long-term administration of 0.5% EGCG to spontaneously hypertensive rats decreased the incidence of stroke and prolonged the rats' life span without affecting blood pressure (Uchida et al., 1995). The protective potency of EGCG on memory function following transient global ischemia induced by the bilateral ligation of the common carotid arteries is, however, weaker than that of two other tea polyphenols, (–)-epicatechin and (+)catechin (Matsuoka et al., 1995).

MECHANISMS OF TEA-INDUCED PROTECTION AGAINST ISCHEMIA

Epidemiological studies suggest that the consumption of tea polyphenols (also called flavonoids) may be associated with reduced risk of coronary heart disease, stroke, and cancer-related deaths (Weisburger et al., 1997; Mukhtar and Ahmad, 2000). Tea polyphenols are rapidly absorbed into the circulation following oral ingestion and are found predominantly after a single administration in the plasma, colon and small intestine, liver, lungs, pancreas, mammary glands, and skin, also in brain, kidneys, and reproductive organs (Nagakawa and Miyazawa, 1997; Suganuma et al., 1998). Moreover, a second administration of the polyphenol EGCG enhances tissue levels in blood, brain, liver, pancreas, bladder, and bones four to six times above those observed after a single administration (Suganuma et al., 1998). These results suggest that frequent consumption of green tea enables the body to maintain a high organic level of tea polyphenols, and these experimental data are in accordance with the report that the antioxidant potential is increased in humans by the consumption of green and black teas (Langley-Evans, 2000; Sung et al., 2000). The antioxidant potential of tea measured *in vitro* compares to that of beverages from fruits or vegetables. From calculations of the antioxidant levels of green tea, it appears that a consumption of 150 ml of tea could make a significant contribution to the total daily antioxidant capacity intake (Prior and Cao, 1999). In humans, the antioxidant capacity of plasma is significantly and progressively increased after taking green or black tea in amounts of 300 and 450 ml (Langley-Evans, 2000; Sung et al., 2000). However, the antioxidant potential of black tea appears to be totally negated by the simultaneous consumption of milk in tea (Langley-Evans, 2000). The protective role of different green tea extracts against oxidative damage relates to their polyphenol composition, directly to their amounts of EGCG and (–)-epigallocatechin (Toschi et al., 2000). Tea polyphenols inhibit the liver enzyme xanthine oxidase, which produces reactive oxygen species, and thus act at a quite early level in the oxidative cascade by inhibiting production rather than only by neutralizing already-formed reactive oxygen species (Aucamp et al., 1997; Kondo et al., 1999).

Reactive oxygen species are largely involved in the pathogenesis of ischemia/reperfusion brain injury. These reactive oxygen species lead to oxidative damage to lipids and DNA. Oxygen radicals, eicosanoids (i.e., prostaglandins and leukotrienes) that result from the metabolism of arachidonic acid by lipoxygenase and cyclooxygenase, thromboxane A_2, and malondialdehyde are considered as mediators of ischemia/reperfusion-induced brain injury by altering membrane permeability, inducing brain edema, and ultimately leading to neuronal death (Matsuo et al., 1996; Islekel et al., 1999). A chronic treatment with green tea before experimental global or focal ischemia has been shown to reduce ischemia/reperfusion-induced elevation of reactive oxygen species such as the hydrogen peroxide level; this confirms that green tea can act in ischemia either by preventing or scavenging oxygen-free radicals, as shown in other *in vivo* or *in vitro* systems (Lee et al., 1995; Hiramaoto et al., 1996; Wei et al., 1999). This effect may reflect the green tea extract-induced increase in the level of the antioxidant enzyme catalase in liver (Khan et al., 1992; Lee et al., 1995). Thromboxane A_2 and platelet-activating factor (PAF) are the mediators of neutrophil activation during cerebral ischemia/reperfusion (Matsuo et al., 1996) that leads to an additional deleterious generation of oxygen radicals during ischemia/reperfusion injury (Matsuo et al., 1995). Green tea polyphenols exhibit antiplatelet activity, which results in antithrombotic activities that may add to the neuroprotective properties of green tea polyphenols (Kang et al., 1999, 2001).

Likewise, the production of eicosanoids (leukotriene C_4 and prostaglandin E_2), thromboxane A_2, lipid peroxidation products (malonaldehyde and 4-hydroxynonenal), and 8-hydroxydeoxyguanosine, a form of oxidative DNA damage is largely increased after ischemia/reperfusion (Hong et al., 2000, 2001). Chronic exposure to green tea prior to the ischemic insult or the acute administration of the green tea polyphenol EGCG reduces the production of the damaging compounds cited above (Lee et al., 2000). The mechanisms underlying these effects are unknown. Green tea extracts reduce the activities of the enzymes phospholipase A_2 and cyclooxygenase in rat platelets that lead to the enhanced synthesis of eicosanoids (Yang et al., 1999). Alternately, green tea could also increase the degradation pathway of eicosanoids (Hong et al., 2000). The green tea-mediated protection against damage to lipids and DNA leads to an attenuation of ischemia/reperfusion-induced brain apoptosis and cell death (Matsuoka et al., 1995; Lee et al., 2000; Hong et al., 2001). Similar protective effects of green tea extracts or polyphenols were observed in other models of oxidative stress-induced neuronal cell death, such as *in vitro* cell models of Parkinson's disease (Levites et al., 2001, 2002b; Nie et al., 2002). The neuroprotective mechanisms of tea polyphenols against oxidative stress-induced cell death include the stimulation of protein kinase C and modulation of cell survival/cell cycle genes (Levites et al., 2002a).

Another factor involved in ischemia/reperfusion-induced cell damage is the production of nitric oxide (NO). Brain ischemia leads to a rise in NO_3^-/NO_2^- concentrations mainly produced by inducible nitric oxide synthase (iNOS), which forms large amounts of NO in macrophages several hours after an insult. These high concentrations of NO produce large amounts of reactive oxygen species that have been shown to cause deamination of DNA deoxynucleotides and bases. The large energy depletion leading to energy failure that is caused by the activation of the enzyme poly(ADP-ribose) polymerase that repairs strand breakage in DNA appears as a major factor involved in ischemia/reperfusion-induced cell death (Eliasson et al., 1997; Szabo and Dawson, 1998). EGCG from green tea has been shown to serve as an iNOS inhibitor (Chan et al., 1997). EGCG is able to block the early events mediated by iNOS activation by inhibiting the binding of the transcription factor, nuclear factor-kappa B, to the iNOS promoter, thereby inhibiting iNOS transcription (Lin and Lin, 1997). EGCG is also able to directly act as an antioxidant and scavenge reactive oxygen radicals produced by ischemia/reperfusion (Lin and Lin, 1997; Nagai et al., 2002).

CONCLUSION AND FUTURE DIRECTIONS

At this point, only a few experimental studies have concentrated on the potential neuroprotective effects of green tea extracts on neuronal damage induced by global or focal ischemia/reperfusion. As reported, the studies have all been focused on the antioxidant properties of green tea polyphenols, mainly EGCG. By contrast, studies of the neuroprotective effects of coffee in global or focal ischemia/reperfusion-induced damage are fully devoted to the caffeine contained in coffee (for details see Chapter 11 in this book), although coffee also contains polyphenols and antioxidant compounds such as caffeic acid (for more details, see Chapter 13 in this book). This would in fact be particularly relevant, since the total antioxidant capacity of coffee appears to be higher than that of tea (Natella et al., 2002). Thus, further studies should examine the potential cumulative neuroprotective effect of polyphenols and caffeine contained in coffee and tea against injuries such as ischemia/reperfusion.

It will also be important to assess whether the amounts of green tea extracts or EGCG given in experimental studies are really representative of the human situation. A recent phase I clinical trial was initiated after allowance from the U.S. Food and Drug Administration in order to examine the safety and efficacy of consuming the equivalent of at least 10 cups (2.4 l) of green tea per day in cancer patients. The results are not known at this point (Mukhtar and Ahmad, 2000). This type of study should be extended to stroke and other neurodegenerative diseases such as Parkinson's disease, for which the consumption of tea has recently been reported to be preventive (Ascherio et al., 2001; Checkoway et al., 2002).

REFERENCES

Ascherio, A., Zhang, S.M., Hernan, M.A., Kawachi, I., Colditz, G.A., Speizer, F.E. and Willett, W.C. (2001) Prospective study of caffeine consumption and risk of Parkinson's disease in men and women. *Annals of Neurology*, 50, 56–63.

Aucamp, J., Gaspar, A., Hara, Y. and Apostolides, Z. (1997) Inhibition of xanthine oxidase by catechins from tea (*Camellia sinensis*). *Anticancer Research*, 17, 4381–4385.

Bu-Abbas, A., Clifford, M.N. and Walker, R. (1994) Marked antimutagenic potential of aqueous green tea extracts: mechanism of action. *Mutagenesis*, 9, 325–331.

Chan, M.M., Fong, D., Ho, C.T. and Huang, H.I. (1997) Inhibition of inducible nitric oxide synthase gene expression and enzyme activity by epigallocatechin gallate, a natural product from green tea. *Biochemical Pharmacology*, 54, 1281–1286.

Chan, P.H. (2001) Reactive oxygen radicals in signaling and damage in the ischemic brain. *Journal of Cerebral Blood Flow and Metabolism*, 21, 2–14.

Checkoway, H., Powers, K., Smith-Weller, T., Franklin, G.M., Longstreth, W.T. and Swanson, P.D. (2002) Parkinson's disease risks associated with cigarette smoking, alcohol consumption, and caffeine intake. *American Journal of Epidemiology*, 155, 732–738.

Dreosti, I.E., Wargovich, M.J. and Yang, C.S. (1997) Inhibition of carcinogenesis by tea: the evidence from experimental studies. *Critical Reviews of Food Science and Nutrition*, 37, 761–770.

Eliasson, M.J.L., Sampei, K., Mandir, A.S., Hurn, P.D., Traystman, R.J., Bao, J. et al. (1997) Poly(ADP-ribose) polymerase gene disruption renders mice resistant to cerebral ischemia. *Nature Medicine*, 3, 1089–1095.

Fujita, T. (1994) Osteoporosis in Japan: factors contributing to the low incidence of hip fracture. *Advances in Nutrition Research*, 9, 89–99.

Graham, H. (1992) Green tea composition, consumption, and polyphenol chemistry. *Preventive Medicine*, 21, 334–350.

Graham, S.H. and Chen, J. (2001) Programmed cell death in cerebral ischemia. *Journal of Cerebral Blood Flow and Metabolism*, 21, 99–109.

Hiramaoto, K., Ojima, N., Sako, K. and Kikugawa, K. (1996) Effect of plant phenolics on the formation of spin-adduct of hydroxyl radical and the DNA strand breaking by hydroxyl radical. *Biological and Pharmaceutical Bulletin*, 19, 558–563.

Hong, J.T., Ryu, S.R., Kim, H.J., Lee, J.K., Lee, S.H., Kim, D.B. et al. (2000) Neuroprotective effect of green tea extract in experimental ischemia-reperfusion brain injury. *Brain Research Bulletin*, 53, 743–749.

Hong, J.T., Ryu, S.R., Kim, H.J., Lee, J.K., Lee, S.H., Yun, Y.P. et al. (2001) Protective effect of green tea extract on ischemia/reperfusion-induced brain injury in Mongolian gerbils. *Brain Research*, 888, 11–18.

Horiba, N., Maekawa, Y., Ito, M., Matsumoto, T. and Nakamura H. (1991) A pilot study of Japanese green tea as a medicament: antibacterial and bactericidal effects. *Journal of Endodology*, 17, 122–124.

Hou, S.T. and MacManus, J.P. (2002) Molecular mechanisms of cerebral ischemia-induced neuronal death. *International Review of Cytology*, 221, 93–148.

Imai, K. and Nakachi, K. (1995) Cross sectional study of effects of drinking green tea on cardiovascular and liver diseases. *British Medical Journal of Clinical Research*, 310, 693–696.

Islekel, H., Islekel, S., Guner, G. and Ozdamar, N. (1999) Evaluation of lipid peroxidation, cathepsin L and acid phosphatase activities in experimental brain ischemia-reperfusion. *Brain Research*, 843, 558–563.

Kang, W.S., Lim, I.H. and Yuk, D.Y. (1999) Antithrombotic activities of green tea catechins and (−)-epigallocatechin gallate. *Thrombosis Research*, 96, 229–237.

Kang, W.S., Chung, K.H., Chung, J.H., Lee, J.Y., Park, J.B., Zhang, Y.H. et al. (2001) Antiplatelet activity of green tea catechins is mediated by inhibition of cytoplasmic calcium increase. *Journal of Cardiovascular Pharmacology*, 38, 875–884.

Katiyar, S.K., Elmets, C.A., Agarwal, R. and Mukhtar, H. (1995a) Protection against ultraviolet-B radiation-induced local and systemic suppression of contact hypersensitivity and edema responses in C3H/HeN mice by green tea polyphenols. *Photochemistry and Photobiology*, 62, 855–861.

Katiyar, S.K. and Mukhtar, H. (1996) Tea in chemoprotection of cancer: epidemiologic and experimental studies. *International Journal of Oncology*, 8, 221–238.

Katiyar, S.K., Rupp, C.O., Korman, N.J., Agarwal, R. and Mukhtar, H. (1995b) Inhibition of 12-O-tetradecanoylphorbol-13-acetate and other skin tumor-promoter-caused induction of epidermal interleukin-1 alpha mRNA and protein expression in SENCAR mice by green tea polyphenols. *Journal of Investigative Dermatology*, 105, 394–398.

Kawase, M., Murakami, K., Fujimura, M., Morita-Fujimura, Y., Gasche, Y., Kondo, T. et al. (1999) Exacerbation of delayed cell injury after transient global ischemia in mutant mice with CuZn superoxide dismutase deficiency. *Stroke*, 30, 1962–1968.

Khan, S.G., Katiyar, S.K., Agarwal, R. and Mukhtar, H. (1992) Enhancement of antioxidant and phase II enzymes by oral feeding of green tea polyphenols in drinking water to SKH-1 hairless mice: possible role in cancer chemoprevention. *Cancer Research*, 52, 4050–4052.

Kohlmeier, L., Weterings, K.G.C., Steck, S. and Koh, F.J. (1997) Tea and cancer prevention: an evaluation of the epidemiologic literature. *Nutrition and Cancer*, 27, 1–13.

Kondo, K., Kurihara, M., Miyata, N., Suzuki, T. and Toyoda, M. (1999) Scavenging mechanisms of (−)-epigallocatechin gallate and (−)-epicatechin gallate on peroxyl radicals and formation of superoxide during the inhibitory action. *Free Radicals in Biology and Medicine*, 27, 855–863.

Kono, S., Ikeda, M., Tokudome, S. and Kuratsune, M. (1988) A case-control study of gastric cancer and diet in Northern Kyusyu, Japan. *Japanese Journal of Cancer Research*, 79, 1067–1074.

Langley-Evans, S.C. (2000) Consumption of black tea elicits an increase in plasma antioxidant potential in humans. *International Journal of Food Science and Nutrition*, 51, 309–315.

Lee, S.F., Liang, Y.C. and Lin, J.K. (1995) Inhibition of 1,2,4-benzenetriol-generated active oxygen species and induction of phase II enzymes by green tea polyphenols. *Chemical and Biological Interactions*, 98, 283–301.

Lee, S.R., Suh, S.I. and Kim, S.P. (2000) Protective effects of the green tea polyphenol (−)-epigallocatechin gallate against hippocampal neuronal damage after transient global ischemia in gerbils. *Neuroscience Letters*, 287, 191–194.

Levites, Y., Amit, T., Youdim, M.B.H. and Mandel, S. (2002a) Involvement of protein kinase C activation and cell survival/cell cycle genes in green tea polyphenol (–)-epigallocatechin-3-gallate neuroprotective action. *The Journal of Biological Chemistry*, 277, 30574–30580.

Levites, Y., Youdim, M.B.H., Maor, G. and Mandel, S. (2002b) Attenuation of 6-hydroxydopamine (6-OHDA)-induced nuclear factor-kappaB (NF-kappaB) activation and cell death by tea extracts in neuronal cultures. *Biochemical Pharmacology*, 63, 21–29.

Levites, Y., Weinreb, O., Maor, G., Youdim, M.B.H. and Mandel, S. (2001) Green tea polyphenol (–)-epigallocatechin-3-gallate prevents N-methyl-4-phenyl-1,2,3,6-tertrahydropyridine-induced dopaminergic neurodegeneration. *Journal of Neurochemistry*, 78, 1073–1082.

Lin, Y.L. and Lin, J.K. (1997) (–)-Epigallocatechin-3-gallate blocks the induction of nitric oxide synthase by down-regulating lipopolysaccharide-induced activity of transcription factor nuclear factor-κB. *Molecular Pharmacology*, 52, 465–472.

Matsuo, Y., Kihara, T., Ikeda, M., Ninomiya, M., Onodera, H. and Kogure, K. (1995) Role of neutrophils in radical production during ischemia and reperfusion of the rat brain: effect of neutrophil depletion on extracellular ascorbyl radical formation. *Journal of Cerebral Blood Flow and Metabolism*, 15, 941–947.

Matsuo, Y., Kihara, T., Ikeda, M., Ninomiya, M., Onodera, H. and Kogure, K. (1996) Role of platelet-activating factor and thromboxane A2 in radical production during ischemia and reperfusion of the rat brain. *Brain Research*, 709, 296–302.

Matsuoka, Y., Hasegawa, H., Okuda, S., Muraki, T., Uruno, T. and Kubota, K. (1995) Ameliorative effects of tea catechins on active oxygen-related nerve cell injuries. *Journal of Pharmacology and Experimental Therapeutics*, 274, 602–608.

Mukhtar, H. and Ahmad, N. (2000) Tea polyphenols: prevention of cancer and optimizing health. *American Journal of Clinical Nutrition*, 7 (Suppl.), 1698S-1702S.

Nagai, K., Jiang, M.H., Hada, J., Nagata, T., Yajima, Y., Yamamoto, S. et al. (2002) (–)-Epigallocatechin-3-gallate protects against NO stress-induced neuronal damage after ischemia by acting as an anti-oxidant. *Brain Research*, 956, 319–322.

Nagakawa, K. and Miyazawa, T. (1997) Absorption and distribution of tea catechin, (–)-epigallocatechin-3-gallate, in the rat. *Journal of Nutrition Science and Vitaminology*, 43, 679–684.

Nagakawa, K., Okuda, S. and Miyazawa, T. (1997) Dose-dependent incorporation of tea catechins, epigallocatechin-3-gallate and (–)-epigallocatechin, in human plasma. *Bioscience, Biotechnology and Biochemistry*, 59, 2134–2136.

Nakayama, M., Toda, M., Okubo, S. and Shimamura, T. (1990) Inhibition of influenza virus infection by tea. *Letters of Applied Microbiology*, 11, 38–40.

Natella, F., Nardini, M., Giannetti, I., Dattilo, C. and Scaccini, C. (2002) Coffee drinking influences plasma antioxidant capacity in humans. *Journal of Agricultural and Food Chemistry*, 50, 6211–6216.

Nie, G., Cao, Y. and Zhao, B. (2002) Protective effects of green tea polyphenols and their major component, (–)-epigallocatechin-3-gallate (EGCG), on 6-hydroxydopamine-induced apoptosis in PC12 cells. *Redox Reports*, 7, 171–177.

Prior, R.L. and Cao, G. (1999) Antioxidant capacity and polyphenolic components of teas: implications for altering *in vivo* antioxidant status. *Proceedings of the Society of Experimental Biology and Medicine*, 220, 255–261.

Serafini, M., Ghiselli, A. and Ferro Luzzi, A. (1996) In vivo antioxidant effects of green tea and black tea in man. *European Journal of Clinical Nutrition*, 50, 28–32.

Siesjö, B.K. (1993) A new perspective on ischemic brain damage? *Progress in Brain Research*, 96, 1–9.

Stensvold, I., Tverdal, A., Solvoll, K. and Foss, O.P. (1992) Tea consumption: relationship to cholesterol, blood pressure, and coronary and total mortality. *Preventive Medicine*, 21, 546–553.

Suganuma, M., Okabe, S., Oniyama, M., Tada, Y., Ito, H. and Fujiki, H. (1998) Wide distribution of [^{3}H](–)-epigallocatechin gallate, a cancer preventive tea polyphenol, in mouse tissue. *Carcinogenesis*, 19, 1771–1776.

Sung, H., Nah, J., Chun, S., Park, H., Yang, S.E. and Min, W.K. (2000) In vivo antioxidant effect of green tea. *European Journal of Clinical Nutrition*, 54, 527–529.

Szabo, C. and Dawson, V.L. (1998) Role of poly(ADP-ribose) synthetase in inflammation and ischemia-reperfusion. *Trends in Pharmacological Sciences*, 19, 287–298.

Tijburg, L.B.M., Wiseman, S.A., Meijer, G.W. and Westrate, J.A. (1997) Effects of green tea, black tea and dietary lipophilic antioxidants on LDL oxidizability and atherosclerosis in hypercholesterolaemic rabbits. *Atherosclerosis*, 135, 37–48.

Toschi, T.G., Bordoni, A., Hrelia, S., Bendini, A., Lercker, G. and Biagi, P.L. (2000) The protective role of different green tea extracts after oxidative damage is related to their catechin composition. *Journal of Agriculture and Food Chemistry*, 48, 3973–3978.

Uchida, S., Ozaki, M., Akashi, T., Yamashita, K., Niwa, M. and Taniyama, K. (1995) Effects of (–)-epigallocatechin-3-O-gallate (green tea tannin) in the life span of stroke-prone spontaneously hypertensive rats. *Clinical and Experimental Pharmacology and Physiology*, 22 (Suppl.), S302–S303.

Uchida, S., Ozaki, M., Suzuki, K. and Shikita, K. (1992) Radioprotective effects of (–)-epigallocatechin-3-O-gallate (green-tea tannin) in mice. *Life Sciences*, 50, 147–152.

Watanabe, T., Yuki, S., Egawa, M. and Nishi, H. (1994) Protective effects of MCI-186 on cerebral ischemia: possible involvement of free radical scavenging and antioxidant actions. *Journal of Pharmacology and Experimental Therapeutics*, 268, 1597–1604.

Wei, H., Zhang, X., Zhao, J.F., Wang, Z.Y., Bickers, D. and Lebwohl, M. (1999) Scavenging of hydrogen peroxide and inhibition of ultraviolet light-induced oxidative DNA damage by aqueous extracts from green and black teas. *Free radicals in Biology and Medicine*, 26, 1427–1435.

Weisburger, J.H. (1996) Tea antioxidants and health, in *Handbook of Antioxidants*, Cadenas, E. and Packer, L., Eds., Marcel Dekker, New York, pp. 469–486.

Weisburger, J.H., Rivenson, A., Garr, K. and Aliaga, C. (1997) Tea, or tea and milk, inhibit mammary gland and colon carcinogenesis in rats. *Cancer Letters*, 114, 323–327.

Xu, M., Bailey, A.C., Hernaez, J.F., Taoka, C.R., Schut, H.A. and Dashwood, R.H. (1996) Protection by green tea, black tea, and indole-3-carbinol against 2-amino-3-methylimidazol[4,5-f]quinole-induced DNA adducts and colonic aberrant crypts in the F344 rat. *Carcinogenesis*, 17, 1429–1434.

Yamaguchi, Y., Hayashi, M., Yamazoe, H. and Kunimoto, M. (1991) Preventive effects of green tea extract on lipid abnormalities in serum, liver and aorta of mice fed an atherogenic diet. *Nippon Yakurzigaku Zasshi*, 97, 329–337.

Yang, J.A., Choi, J.H. and Rhee, S.J. (1999) Effect of green tea catechin on phospholipase A2 activity and antithrombus in streptozotocin diabetic rats. *Journal of Nutrition Science and Vitaminology*, 45, 337–346.

15 The Biology and Psychology of Chocolate Craving

David Benton

CONTENTS

Abstract 205
Chocolate Craving..206
    Craving..206
    Guilt 207
Chocolate Craving throughout the Menstrual Cycle...208
Biological Mechanisms Underlying the Action of Chocolate 209
    Phenylethylamine...209
    Methylxanthines...209
    Caffeine 210
    Theobromine 210
    Carbohydrate Intake and Serotonin Synthesis 210
    Endorphins 212
    Other Possible Biological Mechanisms..213
A Physiological or Psychological Reaction? 213
Discussion 214
References 216

ABSTRACT

A parsimonious explanation is that the attractiveness of chocolate reflects its taste and mouth-feel, leading to endorphin release. There is no convincing evidence that there are substances in chocolate that act directly on the brain in a pharmacological manner.

Chocolate is by far the most common food item that people report that they crave. Those who crave chocolate tend to do so when they are emotionally distressed, although a separate dimension, whether one feels guilt, is also important. There are two major explanations of chocolate craving. First, it is said to result from a pleasant taste. Alternatively, it has been suggested to reflect physiological mechanisms, including increased serotonin production; the release of endorphins; the actions of methylxanthines, phenylethylamine, and anandamides; and the supply of magnesium.

A meal containing almost exclusively carbohydrates will increase the level of tryptophan in the blood that, when taken into the brain, will increase the synthesis of serotonin. The phenomenon is, however, a laboratory artifact; most foods, for example chocolate, contain sufficient protein to ensure that the mechanism is not stimulated. Chocolate contains the methylxanthines theobromine and caffeine; however, the levels are so small that it is unlikely that they influence mood. Similarly, the levels of phenylethylamine and anandamides are too low to have a psychopharmacological action.

The administration of the pharmacological constituents of chocolate, together with white chocolate, is unable to satisfy chocolate craving. It seems, therefore, that the major factor that underlies chocolate craving is the hedonic experience. The attraction of chocolate lies in its taste. The combination of sweetness and fat approaches the ideal hedonic combination.

Food cravings are common, although highly selective. A Canadian survey reported that 68% of men and 97% of women experienced powerful food cravings where 85% said that they often gave in to them (Weingarten and Elston, 1991). Food cravings are selective; chocolate was by far the most commonly and intensely craved food. Chocolate is a special food; for many it has a uniquely attractive taste. It has a cultural importance and is used frequently as a gift on special occasions or to say "thank you" or to apologize. Such is its attraction that sections of the population will admit to craving chocolate. Some will even describe themselves as "chocoholics," although from a scientific perspective the analogy with addiction may not be justified. This chapter explores the origin of the popularity of chocolate. Two types of explanation are considered: (1) that chocolate contains "druglike" substances that influence the brain's chemistry and (2) that psychological mechanisms are predominantly important; that is, the attractive taste is a major reason for the eating of chocolate and/or it reflects an attempt to improve mood.

CHOCOLATE CRAVING

In the limited literature dealing with food craving, the term has tended to rely on the lay definition: it is a strong desire or urge for a particular food. Arguably, research has been inhibited by the way that cravings have been conceived and measured. In the majority of cases subjects have been asked simply to rate their desire to eat a particular food. A single-item scale is unreliable and it involves an implicit assumption that craving can be explained using a single dimension. To answer these concerns, Benton et al. (1998) asked 330 people to respond to 80 statements concerning chocolate and statistically established the underlying dimensions.

CRAVING

The first dimension was labeled *craving*. Those high on this dimension had a considerable preoccupation with chocolate and even acted compulsively. The questions that defined this first dimension fell into two groups. First, chocolate was a source of some distraction. It was said to be "overpowering," it "preyed on the mind," and respondents were unable to "take it or leave it." Liking the taste and mouth-feel of chocolate was associated with scoring heavily on this factor, but this first dimension was also associated with a second type of question. Those who craved chocolate displayed a weakness for chocolate when under emotional stress. It was eaten "when I am bored," "to cheer me up," "when I am upset," and "when I am down." Thus, a link between negative mood and an intense desire to consume chocolate was described statistically; there was evidence of "comfort" or "emotional" eating.

An unusual aspect of the study by Benton et al. (1998) was that the relationship between negative mood and chocolate craving was found in a sample chosen to be representative of the general population, rather than in those with a history of psychiatric complaints. Previously, groups of patients had been reported to "self-medicate" with chocolate, for example, those with hysteroid dysphoria (Schuman et al., 1987), that is, depression associated with feeling rejected, or those who were anxious or had a dysphoric mood (Hill et al., 1991). Thus, there is consistent evidence that chocolate craving is associated with depression and other disturbances of mood.

There is only one study that has experimentally manipulated mood while monitoring the intake of chocolate. Willner et al. (1998) rewarded subjects by giving them chocolate buttons when they pressed the space bar on a computer. The number of presses required to earn a chocolate button increased after each reinforcement according to a fixed ratio, 2, 4, 8, 16, 32, or 64 presses, and so on. The measure was the number of presses made to obtain chocolate, which reflected the motivation to obtain chocolate. Those with a higher craving score were prepared to press the space bar more often to receive chocolate buttons. When mild depression was induced, by playing miserable rather than happy music, the space bar was pressed more often to receive additional chocolate.

Guilt

The second dimension found by Benton et al. (1998) was labeled *guilt*. Again, it included two types of questions. First, there were comments associating chocolate with negative emotions; respondents felt "unattractive," "sick," "guilty," "depressed," or "unhealthy" after eating chocolate, and they often wished they had not done so. The second type of question that weighted on the guilt dimension related to weight and body image. Those who felt guilty often dieted and looked at the caloric value of a chocolate snack. If they ate less chocolate, they thought that they would have a better figure.

Although those sampled were not chosen for having eating disorders, those with high guilt scores were more likely to say that they forced themselves to be sick after eating chocolate. High craving scores were not related to this question. However, both high craving and high guilt scores were associated with the comment that they did not eat chocolate for several days and then ate a large amount at one time. Similarly, both dimensions were associated with saying that they continued eating chocolate when they did not really want it. When four groups were distinguished to reflect the four combinations of high and low levels of craving and guilt, it was the craving dimension that greatly influenced the weekly intake of chocolate. Guilt had a nonsignificant influence.

A third factor that accounted for less of the variance reflected a pragmatic approach to chocolate. It was eaten when it served a useful purpose: "to keep my energy levels up when doing physical exercise," "in the winter when it is colder," "only when I am hungry," and "as a reward when everything is going well." Unlike the guilt and craving factors, this third factor was not associated with mood.

The finding of orthogonal guilt and craving factors has added to our understanding of the reaction to chocolate. Because they are independent factors, high chocolate cravings are not necessarily associated with guilt. There was no reason to believe that any of the Benton et al. (1998) sample suffered with an eating disorder, so it was surprising that there was a small but significant tendency to report behaviors that may reflect a predisposition to eating disorders. A high guilt score was associated with a tendency to report that one was likely to make oneself vomit, to eat large amounts in one sitting, and to continue eating chocolate when it was not really wanted.

Those who craved chocolate tended to do so in response to low mood and scored highly on the external-eating dimension of the Dutch Eating Questionnaire. Those who experienced guilt when eating chocolate had low self-esteem and body dissatisfaction and were restrained eaters (Benton, 2001). There is a need to distinguish these two reactions. The association between the eating of highly palatable chocolate and low mood was distinct from the experience of guilt after its consumption. Such an observation demands a consideration of the impact of palatable foods in individuals distinguished in terms of the tendency to display craving/external eating and at the same time distinguished in terms of guilt/restrained eating. Virtually all the research in this area has treated mood as a single dimension. For example, the failure to distinguish these two emotional dimensions may well have been the cause of confusion. The failure to find that eating improved the mood of restrained eaters (Polivy et al., 1994) is likely, in part at least, to reflect the guilt felt by this group. The possibility should be considered that it is those not prone to guilt after eating attractive foods who respond to a palatable food with improved mood.

There is an assumption in those suffering with eating disorders that food cravings result from, or at least accompany, dieting or restrained eating (Wardle, 1987). Mitchell et al. (1985) found that 70% of bulimic women attributed the onset of binging to food cravings, most commonly for sweet items. It may be that a factor other than craving should be considered. Benton et al. (1998) found that a tendency to binge and vomit was associated more with the guilt rather than the craving factor.

A prediction that should be considered is that a predisposition to eating disorders is associated with the experience of both high cravings and high guilt (Benton, 1999).

CHOCOLATE CRAVING THROUGHOUT THE MENSTRUAL CYCLE

It has been commonly suggested that chocolate craving increases in the premenstrual stage. For example, a survey found that 61% of women reported an increased desire for sweet foods at the premenstrual stage (Vlitos and Davies, 1996). The consistency of this type of finding has led to the view that food craving is a symptom of the premenstrual syndrome.

There is a considerable body of evidence that energy intake increases during the luteal phase of the cycle. Vlitos and Davies (1996), when they reviewed the topic, listed 13 studies that reported this finding. The increase in energy intake between the first and second half of the cycle ranged from 87 to 674 kcal a day, that is, a 4 to 35% rise from the follicular phase. Over the menstrual cycle, in both animals and humans, there is also substantial evidence for changes in the basal metabolic rate (Buffenstein et al., 1995). For example, Webb (1986) found that 80% of women showed a rise of between 8 and 16% in basal metabolic rate between the follicular and luteal phases of the cycle. Thus, the premenstrual appetite for sweet food items, energy intake, and metabolic rate increase in parallel.

A related question is whether the anecdotal reports that there is an increased craving for sugary items results in changes in the pattern of macronutrient consumption. When Vlitos and Davies (1996) reviewed the topic they could find no clear evidence that the intake of carbohydrate increased during the premenstrual stage. If there is no general increase in carbohydrate intake, is there a specific increase in the craving for sweet foods? The intake of sucrose has been reported to be higher in the luteal phase (Gong et al., 1989). Also, the intake of dietary fiber was lower towards the end of the cycle, reflecting at menstruation an increased preference for chocolate foods, rather than similar nonchocolate alternatives (Tomelleri and Grunewald, 1987). Fong and Kretsch (1993) found that carbohydrate intake was higher when bleeding, rather than at ovulation, to a large extent a reflection of increased chocolate consumption. Thus, there is little evidence that carbohydrate as such is craved at the premenstrual stage, although there is an increased preference for pleasant-tasting, high-fat/high-carbohydrate foods.

It is possible that the common view that carbohydrate craving is associated with the premenstrual stage reflects a generally increased appetite, rather than a specific increase in carbohydrate intake. There is, however, some evidence that the premenstrual and menstrual periods are associated with a higher sugar intake. In particular, the consumption of chocolate increases, although not exclusively.

A possibility that has not been systematically considered is that the attraction of chocolate in the premenstrual stage reflects an attempt to increase the intake of magnesium. The levels of both plasma (Posaci et al., 1994) and erythrocyte magnesium (Sherwood et al., 1986) are lower in those suffering with the premenstrual syndrome. In a double-blind trial, Fachinetti et al. (1991) studied the impact of magnesium supplementation in women with premenstrual problems. Magnesium decreased the symptoms, but only in the second menstrual cycle.

Given the high levels of magnesium that are found in chocolate, there have been suggestions that the craving for chocolate is an attempt to self-medicate. The explanation is not convincing. There are only 26 mg of magnesium in a 50-g bar of milk chocolate and 50 mg in plain chocolate. Those who responded to the supplementation of magnesium took 360 mg of magnesium, three times a day, for the second half of the menstrual cycle. It is improbable that many would on a daily basis consume the amount of chocolate required to increase magnesium intake to a sufficient extent. The much smaller amount of magnesium offered by chocolate, and the time scale involved, makes it impossible that the eating of chocolate in the premenstrual phase is an attempt to increase the intake of magnesium.

In summary the picture that has emerged is that chocolate is by far the most common food item that is craved, particularly in females. Although there is considerable evidence that chocolate craving and poor mood are related, the origin of this relationship is unclear. Potentially, the mood-enhancing properties of chocolate could reflect either psychological or biological mechanisms.

BIOLOGICAL MECHANISMS UNDERLYING THE ACTION OF CHOCOLATE

Initially, a series of potential biological mechanisms will be considered. The suggestion has been made repeatedly that chocolate contains "druglike" substances that act on the central nervous system, adding to the attractiveness of chocolate.

Phenylethylamine

A 50-g bar of chocolate, at the most, contains about one third of 1 mg of phenylethylamine. Phenylethylamine occurs naturally in low concentrations in the brain, where it has distinct binding sites, although it acts as a neuromodulator rather than as a neurotransmitter. It is present in low concentrations (< 10 ng/g) and has a rapid turnover (half-life of 5 to 10 min). It is similar to amphetamine in that when injected into the brain of animals it causes stereotyped behavior. Although phenylethylamine does not bind directly to dopamine sites, its effects can be blocked by dopamine antagonists. Phenylethylamine is therefore assumed to release dopamine (Webster and Jordan, 1989). Monoamine oxidase B preferentially oxidizes phenylethylamine, although this enzyme also oxidizes dopamine.

Based on their treatment of "love-addicted" women, two New York psychoanalysts associated passionate love and chocolate because their patients produced a large amount of phenylethylamine. The production of phenylethylamine decreased when the women's infatuation stopped (Weil, 1990). The suggestion was made that chocolate can be a substitute for love.

Given the role played in the neuromodulation of dopamine by phenylethylamine, it was obvious to suggest that it may be responsible for the attraction of chocolate. In humans, addiction to drugs of abuse involves brain dopamine (see Chapter 9 of this book).

As with all drugs of abuse, phenylethylamine will increase the rate of lever pressing to receive stimulation of "reward centers" in the brain (Greenshaw, 1984). It is a plausible suggestion that if chocolate supplies sufficient phenylethylamine, then addiction would occur. The question is whether the level of the compound in chocolate is sufficient to cause addiction and thus account for craving.

Although the level of phenylethylamine in chocolate is higher than in most foods (Hirst et al., 1982), some cheeses and sausage contain higher levels, yet they are rarely craved. The behavior of rats trained to press a lever to obtain a pleasurable electrical stimulation of the hypothalamus was influenced by doses of 25 and 50 mg/kg of phenylethylamine (Greenshaw et al., 1985). Goudie and Buckland (1982) reported that 20 to 60 mg/kg influenced food-rewarded behavior. If the doses of phenylethylamine are effective in humans at the same dose as in rats, humans would need to consume 2 or 3 g. Consistent with this suggestion, a dose of 2 to 6 g/d was reported to enhance the mood of depressed patients (Sabelli and Javaid, 1995). At one third of 1 mg per bar of chocolate, the most extensive chocolate binge could not offer anything approaching the effective dose. In fact, the rapid rate at which phenylethylamine is broken down by monoamine oxidase makes it largely ineffective in animals, unless they are treated with a drug that inhibits this enzyme. In conclusion, although phenylethylamine can influence mood, the levels in chocolate are far too low to influence central nervous system activity.

Methylxanthines

Another suggestion is that the methylxanthines in chocolate have a psychotrophic action. Although the stimulant action of caffeine (1,3,7-trimethylxanthine) is well established, theobromine (3,7-dimethylxanthine) has been considered rarely. The levels differ from chocolate bar to chocolate bar; however, Shively and Tarka (1984) calculated that milk chocolate products average about 2 mg/g of theobromine and 0.2 mg/g of caffeine (about 88 mg of theobromine and 9 mg of caffeine per 44-g bar).

Are these small levels of caffeine high enough to influence neural functioning? Brewed coffee has 85 mg of caffeine/150 ml, whereas tea offers 50 mg/150 ml, yet neither of these drinks is commonly craved. With theobromine there is the additional problem that there is considerable doubt as to whether it influences psychological functioning at all.

CAFFEINE

The conclusion that caffeine influences cognitive functioning is to a large extent based on doses of at least 200 mg, at which level there is clear evidence of stimulant activity. There is, for example, consistent evidence that 200 mg of caffeine improves reaction times, although the findings with lower doses are less consistent (Lieberman, 1992). The data with lower doses are more limited, although 32 mg of caffeine has been reported to improve reaction times (Lieberman et al., 1987). However, Kuznicki and Turner (1986) were unable to find a significant finding with 20 mg. A typical chocolate bar would offer less than half the latter dose. Doses over 100 mg similarly improve the ability to sustain attention, a finding also reported with a dose as low as 32 mg (Lieberman et al., 1987).

Mood is the most commonly examined psychological parameter when studying the influence of caffeine. One hundred milligrams of caffeine increases measures of alertness and vigor and decreases boredom and fatigue (Fagan et al., 1988). However, there have been very few studies of low doses. Lieberman et al. (1987) found that 64 mg increased feelings of alertness and vigor. The finding that low doses of caffeine improved mood is supported by the reports that low doses of caffeine are reinforcing. When, under double-blind conditions, subjects were given the choice of two coffees, one without caffeine and the other with varying doses, there was evidence that some individuals choose a dose of 25 mg but were unable to distinguish 12.5 mg from a placebo (Oliveto et al., 1991). Again, this finding needs to be put into the context of the dose likely to be consumed in chocolate. It is unlikely that the eating of a typical chocolate bar would offer a dose of caffeine sufficient to influence functioning.

Most reliable psychological responses to caffeine have been observed with doses in excess of 100 mg. Where lower doses have been found to be active, in no case has this been 9 mg or less. A chocolate product offering caffeine towards the top of the dose range (31 mg) could influence some psychological measures in a weak manner. Similarly, a binge could offer a pharmacologically active dose. It is, however, not possible for such mechanisms to account for the widespread reporting of craving throughout the population.

THEOBROMINE

The behavioral response to theobromine is less than that to caffeine, to the extent that, based on animal studies, some have concluded that it is inert (Snyder et al., 1981). However, others have reported a modest impact on animal behavior (Katims et al., 1983), although reviews have concluded that theobromine has no behavioral influence in humans (Stavric, 1988).

Among the most sensitive methods for establishing subtle drug effects are studies of drug discrimination. In rats able to discriminate 32 mg/kg caffeine from saline, up to 75 mg/kg of theobromine did not create a caffeine-like response (Carney et al., 1985). In conclusion, there is no basis to suggest that theobromine accounts for craving.

CARBOHYDRATE INTAKE AND SEROTONIN SYNTHESIS

Another suggested mechanism is that the high level of carbohydrate in chocolate increases the level of brain serotonin. Increased serotonergic activity changes aggressiveness, mood, and pain sensitivity. There are, however, serious problems with the view that the high carbohydrate content of chocolate leads to enhanced serotonin synthesis (Benton, 2002).

Wurtman and Wurtman (1989) suggested that a meal high in carbohydrate increases in the blood the ratio of tryptophan to "large neutral amino acids" (tyrosine, phenylalanine, leucine, isoleucine, and valine). Based on the study of rats a sequence of events was proposed. After a meal an increase of blood glucose causes the release of insulin from the pancreas. In turn, insulin stimulates the uptake of most amino acids, but not tryptophan, by peripheral tissues such as muscle. Tryptophan binds to blood albumin, an action increased by insulin. Therefore, after a meal that is almost totally carbohydrate, the ratio of tryptophan to the other amino acids in the blood increases. Tryptophan and the other large neutral amino acids compete with each other for a transporter molecule that allows entry into the brain. Thus, when the ratio of tryptophan to other large amino acids increases, relatively more tryptophan is transported into the brain. In the brain, tryptophan is metabolized into the neurotransmitter serotonin. It was therefore proposed that a high-carbohydrate meal increased the level of tryptophan in the blood that, when transported into the brain, increased the synthesis of serotonin (Wurtman et al., 1981).

Benton and Donohoe (1999) summarized the results of 30 human studies that had looked at the influence of meals that differed in the percentage of calories that came from protein rather than carbohydrate. There was clear support for the theory of Wurtman in that the ratio of carbohydrate to protein in a meal influenced the ratio between tryptophan and long-chain neutral amino acids. Their data did not, however, give support for other than the first step of the theory.

When protein offered less than 2% of the calories, the level of tryptophan rather than that of other long-chain neutral amino acids was greater. However, as little as 5% of the calories in the form of protein ensured that the level of tryptophan was not increased. These data cause serious problems for the hypothesis that a high-carbohydrate meal increases the synthesis of brain serotonin. It is difficult to find meals that contain so little protein that the uptake of tryptophan by the brain will increase. In potatoes, 10% of the calories come in the form of protein, in bread 15%, and in milk chocolate 13%. With these foods that are often described as high in carbohydrate no increase in the availability of tryptophan occurs.

The time scale of the reaction causes additional problems for the theory. The response to mood after eating chocolate takes place in a few minutes, rather than after the hour or longer it takes to release amino acids into the blood. Much of the digestion of protein takes place in the intestine, so the release of amino acids into the blood is not immediate.

Wurtman and Wurtman (1989) proposed that a large carbohydrate intake reflected an attempt at self-medication; they believed that carbohydrate intake enhanced serotonin synthesis. They hypothesized that a subset of obese patients, whose weight problems were associated with depression and uncontrolled carbohydrate intake, consumed carbohydrate for its psychopharmacological effects. For example, they reported that when those suffering with carbohydrate-craving obesity were offered snacks, they consumed almost entirely high-carbohydrate foods (Wurtman and Wurtman, 1989). When asked why they snacked, the response was more likely to be that it made them calm or clear-headed rather than that they were hungry. Those who were carbohydrate cravers felt less depressed and more alert. Those who did not crave carbohydrate felt sleepy and fatigued. Benton (2002), however, noted that the snacks consumed contained a level of protein high enough to make it unreasonable to suggest that the level of blood tryptophan, the precursor of serotonin, would have increased.

Even the existence of "carbohydrate-craving" patients has been questioned. Toornvliet et al. (1997) gave three types of snack to "carbohydrate-craving" and "non-carbohydrate-craving" obese patients and found that although the tryptophan:large neutral amino acid ratio increased significantly after high-carbohydrate meals, mood was similar irrespective of the composition of the snack. This Dutch study found that the ingestion of a carbohydrate-rich snack did not improve mood. They concluded that "from a therapeutic point of view it was useless to maintain the concept of carbohydrate craving … the existence of carbohydrate craving patients has never been established …."

Those suffering with seasonal affective disorder (SAD) typically eat more in the winter and put on weight. From the Wurtmans' perspective, this is an attempt to decrease depressive symptoms by eating carbohydrate-rich foods. It has been reported that the eating of a carbohydrate-rich/protein-poor meal was associated with improved mood in those suffering with SAD (Rosenthal et al., 1989). Because the meal contained very little protein, it significantly increased the level of tryptophan in the blood. However, the unusual nature of the meal means that the findings cannot be generalized to any normal meal. It would be instructive to offer those suffering with SAD snacks that were equally palatable but differing in their carbohydrate content. It is possible that there would be a differential response to foods depending on palatability rather than carbohydrate content. Is it important that palatability is often associated with sweetness? These alternative hypotheses would also predict an increased intake of carbohydrate by those suffering with SAD, but not exclusively carbohydrate.

In summary, there is no reason to suggest that the attraction of chocolate results from an increased availability of tryptophan in the blood. Rather than being attracted to foods high in carbohydrate, it may be that patients with a poor mood are attracted to palatable foods that are high in both fat and carbohydrate.

ENDORPHINS

The endorphins are a family of peptides that act in the brain at the same site as morphine, and there is increasing evidence that the response to high-fat/sweet foods is endorphin-mediated. In rats the consumption of chocolate has been associated with an increased release of beta-endorphin (Dum et al., 1983). The number of beta-endorphin-occupying receptors in the rat hypothalamus has been reported to increase when chocolate milk and candy were eaten (Dum et al., 1983). In animals, the preference for and intake of sweet solutions was increased by an opiate agonist and decreased by an opiate antagonist such as naloxone or naltrexone (Reid, 1985). The palatability of food is important; naloxone in the rat decreased the consumption of chocolate-chip cookies more than the intake of standard rat food (Giraudo et al., 1993). Thus, there is increasing evidence that in rodents endogenous opiates regulate food intake by modulating the extent to which pleasure is induced by palatable foods.

Similarly, in humans both spontaneous eating (Davis et al., 1983) and the consumption of glucose (Getto et al., 1984) have been associated with an increased release of beta-endorphin. Opioid antagonists have been found to decrease feelings of hunger, thinking about food (Wolkowitz et al., 1988), and food intake (Trenchard and Silverstone, 1982). Naltrexone reduced the preference for sucrose (Fantino et al., 1986).

Mandenoff et al. (1982) proposed that when a monotonous diet was eaten in a predictable environment the endogenous opiate system is not necessary for the control of eating. However, when stressed, fasting, or after the consumption of highly palatable foods, the opioid mechanisms play a role. In the rat a stressor, such as pinching the tail, will induce a naloxone-reversible (opioid antagonist) increase in eating (Koch and Bodnar, 1993). Mandenoff et al. (1982) suggested that if a stress-induced release of endorphin is not enough to protect the animal, it was adaptive to eat and in that manner increase the levels of blood glucose levels. In this way further endorphin release can be stimulated. There are parallels between the stress-induced increase in rodents' eating and the stressed human who snacks on palatable foods. As was discussed initially, many a negative mood induces chocolate craving (Benton, 1999).

The suggestion that opiate mechanisms modulate the pleasure associated with palatable food was made when the impact of nalmefene, a long-lasting opioid antagonist, was considered (Yeomans et al., 1990). Treatment with nalmefene decreased caloric intake by 22%, without altering the subjective ratings of hunger; the intake of fat and protein, but not of carbohydrate, decreased. Nalmefene selectively influenced the intake of palatable foods, for example, high-fat cheese such as brie. The choice was between savory food items; chocolate and sweet foods were not an offer.

Drewnowski et al. (1992) similarly reported that the opioid antagonist naloxone selectively decreased the intake of palatable high-fat/high-sugar foods.

Because bingeing is typically associated with food cravings, Drewnowski et al. (1995) examined the hypothesis that the influence of opiate antagonists on taste preferences and food consumption would be greater in those with bulimia nervosa. Naloxone significantly reduced total energy intake, but most markedly the intake of high-sugar/high-fat foods, including chocolate, declined.

In summary, there are increasing data suggesting that the intake of highly palatable foods and that the pleasure associated with eating such foods is modulated via opioid mechanisms. A major theory is that the eating of palatable foods, such as chocolate, is associated with the release of endorphins.

OTHER POSSIBLE BIOLOGICAL MECHANISMS

Chocolate is chemically complex, containing many potentially pharmacologically active compounds, albeit in low concentrations. For example, it contains histamine, tryptophan, serotonin, and octopamine but these are found in higher levels in other food items without the appeal of chocolate, so it is improbable that they play a role it the attractiveness of chocolate.

Chocolate is a good source of iron: a 50-g bar of plain chocolate offers 1.2 mg of iron and milk chocolate 0.8 mg. These figures should be compared with the U.S. Recommended Daily Amount of 15 mg/d for an adult female and 10 mg/d for an adult male. Fordy and Benton (1994) found in young British adults that 52% of females and 11% of males had levels of ferritin, the storage protein for iron, below the recommended level. Both in industrialized and developing countries, any source of iron is likely to be valuable, given widespread iron-deficient anemia. There is, however, no reason to believe that in the short-term an enhanced intake of iron will improve your mood. The replacement of red blood cells takes many months. However, iron-deficiency anemia is associated with feelings of lethargy and lowered mood, and over time chocolate as an iron-containing food could offer a useful source of the mineral.

Anandamide is a brain lipid that binds to cannabinoid receptors similarly to the active ingredients of cannabis (DiMarzo et al., 1994). Because anandamide is released from neurons, it may act as an endogenous cannabinoid neurotransmitter or, alternatively, as a neuromodulator. It is interesting that anandamide has been found as a constituent of chocolate (DiTomaso et al., 1996), leading to speculation that the endogenous cannabinoid system may be responsible for the subjective feelings associated with eating chocolate and with chocolate craving. These findings must be treated as very preliminary because they are based on *in vitro* studies. It remains to be shown that the anandamide in chocolate is present in high enough levels to be active *in vivo*. It also remains to be shown that it can survive digestion and absorption, and that it crosses the blood-brain barrier in sufficient amounts to influence the activity of brain areas with cannabinoid receptors. It is highly improbable that chocolate consumed in normal amounts will be able to supply sufficient anandamide to influence neuronal activity.

A PHYSIOLOGICAL OR PSYCHOLOGICAL REACTION?

The above review considers the possible impact of some of the constituents of chocolate and argues that they are unlikely to be present in sufficiently high amounts to generate a physiological effect. Totally convincing evidence, however, only comes from studies of the active ingredients at a level that would be typically consumed. There is only one study that has attempted to compare the relative contributions of the psychological and physiological mechanisms that underlie chocolate craving.

Cocoa butter is the fat that, when removed from chocolate liquor, leaves cocoa powder. The known pharmacological ingredients are all in the cocoa powder. Therefore, if one eats white chocolate, made from the cocoa butter, one has the fat and sugar intake of chocolate but not the pharmacological constituents. If one consumes cocoa powder, one takes the pharmacological ingredients but not the fat and sugar. Michener and Rozin (1994) studied subjects who reported cravings for chocolate and assessed the ability of the various constituents of chocolate to satisfy these cravings.

If it is the sensory experience that is important, then chocolate itself, and to a lesser extent white chocolate, should be satisfying. If it is the increase in blood glucose that is important, then brown and white chocolate should help cravings. If the pharmacological ingredients are important, then both cocoa powder and chocolate should satisfy cravings. Only chocolate itself, and to a lesser extent white chocolate, had the ability to satisfy chocolate craving. Capsules containing the possible pharmacological ingredients had an effect similar to taking nothing. The adding of cocoa capsules to white chocolate did not increase the less than optimal effect of white chocolate. The obvious conclusion was that it was the sensory experience associated with eating chocolate, rather than pharmacological constituents, that was important.

DISCUSSION

There is little, if any, evidence that chocolate craving reflects a druglike biological response to a constituent of chocolate. The conclusion is that rather than a biological mechanism, the attraction of chocolate reflects its taste and various psychological reactions.

Chocolate contains a range of compounds that in appropriate doses would have psychotrophic properties. These include caffeine, phenylethylamine, magnesium, and anandamide. A common reason why they are unlikely to have any significant impact is that with any likely consumption of chocolate they are certain to be provided in a dose that is inactive. For example, to consume the minimal active dose of 1 g of phenylethylamine one would need to rapidly eat 15 kg of chocolate. In addition, to prevent its breakdown by the liver the taking of a monoamine oxidase inhibitor is to be recommended. Rogers and Smit (2000) calculated that one would need to consume 25 kg of chocolate to obtain a psychoactive dose of anandamide. These calculations illustrate the impossibility that the reaction to chocolate is "druglike."

In contrast, it seems to be particularly important that chocolate tastes good; we particularly prefer foods that are both sweet and high in fat. When the palatability of combinations of fat and sugar were compared, the optimal combination was found to be 7.6% sugar with cream containing 24.7% fat (Drewnowski and Greenwood, 1983). The fat content of chocolate is close to this ideal figure, although the sugar content of chocolate is greater. An explanation is that more sugar is needed to counteract the bitterness of chocolate. Chocolate, by chance, appears to reflect an optimal combination of sweetness and fat, giving it a uniquely attractive taste. The melting of chocolate just below body temperature, with the resulting mouth-feel, adds to the hedonic experience. The critical biological mechanism seems to be that when we eat something that tastes pleasant endorphin mechanisms are stimulated. Drewnowski (1992) suggested that the craving for foods high in fat and/or sweet carbohydrate results in activity of the endogenous opioid system. As discussed above, opioid antagonists such as naloxone influence the eating of pleasant-tasting food such as chocolate in both animals and humans.

It is believed that addiction to a range of drugs involves mechanisms in the brain that have as their normal function the control of rewarding activities such as eating or drinking (Di Chiara and North, 1992). Opioids play an important role in the initiation and maintenance of drug dependence; for example, alcohol craving is reduced after taking naltrexone (Van Ree, 1996). Interestingly, heroin (Shufman et al., 1997), alcohol (Kampov-Polevoy et al., 1997), and nicotine addictions (Kos et al., 1997) are all associated with the perception of a sweet taste as more pleasant.

There is a widespread understanding that drug craving and relapse can be triggered by environmental cues. Learning the environmental cues associated with the availability of a biological reward has an obvious evolutionary advantage. Schroeder et al. (2001) have suggested that drug abuse reflects the inappropriate recruitment of these neural learning mechanisms. In rats, they demonstrated that the environmental cues associated with both nicotine and chocolate consumption similarly stimulated the activation of Fos, a marker for neuronal activation, in the prefrontal cortex and limbic regions of the brain. They commented that it was reasonable to predict that the environmental cues associated with chocolate would activate the mesocorticolimbic dopamine system, which would affect areas associated with reward expectancy, such as the prefrontal cortex. Tuomisto et al. (1999) used a classic conditioning paradigm to predict that chocolate acts as an unconditioned stimulus that results in greater arousal and the eating of and craving for chocolate. Those who described themselves as chocolate addicts, when faced with chocolate cues, experienced more negative affect, experienced more cravings, and ate more chocolate.

Wise (1988) suggested that addiction to drugs of abuse could reflect the stimulation of one of two mechanisms; neural systems have been described by which drugs of abuse either give pleasure or alternatively decrease distress. It is possible to see chocolate craving as a reflection of both types of neural mechanism. High chocolate craving is associated with a positive view of the physical characteristics of chocolate, such as taste and mouth-feel (Benton et al., 1998). It is known that the pleasant taste of chocolate plays a large part in making it attractive; as discussed above, other factors such as "druglike" constituents are unimportant. Drugs that are positively reinforcing tend to be craved. It is easy to suggest that the uniquely attractive combination of fattiness and sweetness, and the flavor and mouth-feel, of chocolate make it more positively reinforcing than other foods, resulting in craving. Other drugs that become craved reduce distress. The association between chocolate craving and the eating of chocolate in emotionally distressing situations (Benton et al., 1998) suggests that cravings may also reflect the neural mechanisms important in reducing distress. Thus, it is possible that the high frequency of chocolate craving reflects its ability to tap both of the types of neural mechanisms that underlie craving.

The question arises as to whether the term *addiction* is appropriate when considering chocolate. It is relatively easy to obtain a sample of subjects to take part in a study if you advertise in a newspaper for "chocoholics." Large sections of the population will readily admit to craving chocolate; some will even claim to be addicted (Hetherington and MacDiarmid, 1993). Although used in the vernacular, the question that arises is whether it is scientifically appropriate to apply the term *addiction* to chocolate consumption.

Definitions of drug addiction emphasize compulsion, loss of control, discomfort after drug withdrawal, and a positive psychological response when it is taken. Dependence implies that the drug is needed to function within normal limits. There is no evidence that eating chocolate leads to physical dependence. With drugs of abuse tolerance typically occurs, that is, with repeated use there is a need for a higher dose to obtain the same effect. Although there is evidence that chocolate is often consumed to improve mood, this does not necessarily imply that it is addictive. It may simply be pleasant to eat. Most people eat chocolate on a regular basis without any signs of its getting out of control, without signs of tolerance or dependence.

Rogers and Smit (2000) argued that the term *addiction* should not be applied to chocolate; rather, they suggested that cognitive influences are particularly important. Eating can be stimulated by external cues such as the time of day and the place. In fact, neutral stimuli associated with food consumption can later stimulate eating in the absence of deprivation. The attitude to chocolate maybe critical; it is not viewed as a staple food, but rather as a treat or an indulgence. Culturally we are instructed that it is something that should be eaten in moderation; however, if we inhibit the eating of something that tastes good this increases our desire to eat it. Given the perception of chocolate as nonessential, if not adverse in nutritional terms, the increased desire to eat chocolate is not labeled hunger, but rather craving. We crave because we resist consumption. The inability to resist chocolate leads to the self-explanation that it is addictive — a psychological attempt at understanding rather than an accurate description.

There are similarities between the way that Tiffany (1990) views drug craving and the craving for chocolate. He sees drugs as being craved only when their use is resisted. Drug use is largely controlled by a series of cognitive tasks that, with repeated use, become automatic and effortless when triggered by external cues. In this view, the processes that control drug intake are separate from those that stimulate craving. There are parallels between the craving for both chocolate and drugs; both are stimulated by environmental cues.

Rogers and Smit (2000) argued that at least part of the problem is that there is a tendency to treat addiction as if it is an all-or-nothing phenomenon. The problem may be one of definition. There is no doubt that chocolate and drug consumption differ in many respects, and thus the analogy may be unhelpful. There are, however, similarities in terms of the ability to stimulate underlying biological mechanisms, albeit in a more restricted fashion and without the widespread development of adverse consequences.

REFERENCES

Benton, D. (1999) Chocolate craving: biological or psychological phenomenon? in *Chocolate and Cocoa: Health and Nutrition,* Knight, I., Ed., Blackwell Science, Oxford, pp. 256–278.

Benton, D. (2001) Psychological and pharmacological explanations of chocolate craving, in *Food Cravings and Addiction,* Hetherington, M., Ed., Leatherhead International, Leatherhead, U.K., pp. 265–293.

Benton, D. (2002) Carbohydrate ingestion, blood glucose and mood. *Neuroscience and Biobehavioral Reviews,* 26, 293–308.

Benton, D. and Donohoe, R.T. (1999) The effects of nutrients on mood. *Public Health Nutrition,* 2, 403–409.

Benton, D., Greenfield, K. and Morgan, M. (1998) The development of the attitudes to chocolate questionnaire. *Personality and Individual Difference,* 24, 513–520.

Buffenstein, R., Poppitt, S.D., McDevitt, R.M. and Prentice, A.M. (1995) Food intake and the menstrual cycle: a retrospective analysis with implications for appetite research. *Physiology and Behavior,* 58, 1067–1077.

Carney, J.M., Holloway, F.A. and Modrow, H.E. (1985) Discriminative stimulus properties of methylxanthines and their metabolites in rats. *Life Sciences,* 36, 913–920.

Davis, J.M., Lowy, M.T., Yim, G.K.W., Lam, D.R. and Malven, P.V. (1983) Relationships between plasma concentrations of immuno-reactive beta-endorphin and food intake in rats. *Peptides,* 4, 79–83.

Di Chiara, G. and North, R.A. (1992) Neurobiology of opiate abuse. *Trends in Pharmacological Sciences,* 131, 185–193.

DiMarzo, F.A., Caadas, H., Schinelli, S., Cimino, G., Schwaartz, J.C. and Piomelli, D. (1994) Formation and inactivation of endogenous cannabinoid anandamide in central neurons. *Nature,* 372, 686–691.

DiTomaso, E., Beltramo, M. and Piomelli, D. (1996) Brain cannabinoids in chocolate. *Nature,* 382, 677–678.

Drewnowski, A. (1992). Food preferences and the opioid peptide system. *Trends in Food Science and Technology,* 3, 97–99.

Drewnowski, A. and Greenwood, M.R.C. (1983) Cream and sugar: human preferences for high-fat foods. *Physiology and Behavior,* 30, 629–633.

Drewnowski, A., Krahn, D.D., Demitrack, M.A., Nairn, K. and Gosnell. B.A. (1992) Taste responses and preferences for sweet high-fat foods: evidence for opioid involvement. *Physiology and Behavior,* 51, 371–379.

Drewnowski, A., Krahn, D.D., Demitrack, M.A., Nairn, K. and Gosnell, B.A. (1995) Naloxone and opiate blocker reduces the consumption of sweet high-fat foods in obese and lean female binge eaters. *American Journal of Clinical Nutrition,* 61, 1206–1212.

Dum, J., Gramsch, C.H. and Herz, A. (1983) Activation of hypothalamic beta-endorphin pools by reward induced by highly palatable food. *Pharmacology, Biochemistry and Behavior,* 18, 443–447.

Fachinetti, F., Borella, P., Sances, G., Fioroni, L., Nappi, R.E. and Genazani, A.R. (1991) Oral magnesium successfully relieves premenstrual mood changes. *Obstetrics and Gynecology,* 78, 177–181.

Fagan, D., Swift, C.G. and Tiplady, B. (1988) Effects of caffeine on vigilance and other performance tests in normal subjects. *Journal of Psychopharmacology,* 2, 19–25.

Fantino, M., Hosotte, J. and Apfelbaum, M. (1986) An opioid antagonist naltrexone reduces preference for sucrose in humans. *American Journal of Physiology,* 251, R91–R96.

Fong, A.K.H. and Kretsch, M. J. (1993) Changes in dietary intake, urinary nitrogen and urinary volume across the menstrual cycle. *American Journal of Clinical Nutrition,* 234, E243–E247.

Fordy, J. and Benton, D. (1994) Does low iron status influence psychological functioning? *Journal of Human Nutrition and Dietetics,* 7, 127–133.

Getto, C.J., Fullerton, D.T. and Carlson, I.H. (1984) Plasma immunoreactive beta-endorphin response to glucose ingestion. *Appetite,* 5, 329–335.

Giraudo, S.Q., Grace, M.K., Welch, C.C., Billington, C.J. and Levine, A.S. (1993) Naloxone's anoretic effect is dependent upon the relative palatability of food. *Pharmacology, Biochemistry and Behavior,* 46, 917–921.

Gong, E.J., Garrel, D. and Calloway, D.H. (1989) Menstrual cycle and voluntary food intake. *American Journal of Clinical Nutrition,* 49, 252–258.

Goudie, A.J. and Buckland, C. (1982) Serotonin receptor blockade potentiates the behavioural effects of beta-phenylethylamine. *Neuropharmacology,* 21, 1267–1272.

Greenshaw, A.J. (1984) β-Phenylethylamine and reinforcement. *Progress in Neuro-psychopharmacology and Biological Psychiatry,* 8, 615–620.

Greenshaw, A.J., Sanger, D.J. and Blackman, D.E. (1985) Effects of d-amphetamine and beta-phenylethylamine on fixed interval responding maintained by self-regulated lateral hypothalamus stimulation in rats. *Pharmacology, Biochemistry and Behavior,* 23, 519–523.

Hetherington, M.M. and MacDiarmid, J.I. (1993) Chocolate addiction: a preliminary study of its description and its relationship to problem eating. *Appetite,* 21, 233–246.

Hill, A.J., Weaver, C.F.L. and Blundell, J.E. (1991) Food craving, dietary restraint and mood. *Appetite,* 17, 187–197.

Hirst, W.J., Martin, R.A., Zoumas, B.L. and Tarka, S.M. (1982) Biogenic amines in chocolate: a review. *Nutrition Reports International,* 26, 1081–1087.

Kampov-Polevoy, A., Garbutt, J.C. and Janowsky, D. (1997) Evidence of preference for a high-concentration sucrose solution in alcoholic men. *American Journal of Psychiatry,* 154, 269–270.

Katims, J.J., Annau, Z. and Snyder, S.H. (1983) Interactions in the behavioural effects of methylxanthines and adenosine derivatives. *Journal of Pharmacology and Experimental Therapeutics,* 227, 167–173.

Koch, J.E. and Bodnar, R.J. (1993) Involvement of mu1 and mu2 opioid receptor subtypes in tail-pinch feeding in rats. *Physiology and Behavior,* 53, 603–605.

Kos, J., Hassenfratz, M. and Battig, K. (1997) Effects of a 2-day abstinence from smoking on dietary cognitive subjective and physiologic parameters among younger and older female smokers. *Physiology and Behavior,* 61, 671–678.

Kuznicki, J.T. and Turner, L.S. (1986) The effects of caffeine on caffeine users and non-users. *Physiology and Behavior,* 37, 397–408.

Lieberman, H.R. (1992) Caffeine, in *Handbook of Human Performance,* Vol. 2, Smith, A.P. and Jones, D.M., Eds., Academic Press, London, pp 49–72.

Lieberman, H.R., Wurtman, R.J., Embe, G.G. and Coviella, I.L.G. (1987) The effects of caffeine and aspirin on mood and performance. *Journal of Clinical Psychopharmacology,* 7, 315–320.

Lieberman, H.R., Wurtman, R.J., Embe, G.G., Roberts, C. and Coviella, I.L. (1987) The effects of low doses of caffeine on human performance and mood. *Psychopharmacology,* 92, 308–312.

Mandenoff, A.F., Fumerton, M., Apfelbaum, M. and Margules, D.L. (1982) Endogenous opiates and energy balance. *Science,* 215, 1536–1537.

Michener, W. and Rozin, P. (1994) Pharmacolgical versus sensory factors in the satiation of chocolate craving. *Physiology and Behavior,* 56, 419–422.

Mitchell, J.E., Hatsukami, D., Eckert, E.D. and Pyle, R.L. (1985) Characteristics of 275 patients with bulimia. *American Journal of Psychiatry,* 142, 482–485.

Oliveto, A.H., Hughes, J.R., Pepper, S.L., Bickel, W.K. and Higgins S.T. (1991) Low doses of caffeine can serve as reinforcers in humans. *NIDA Research Monographs,* 105, 442.

Polivy, J., Zeitlin, S.B., Herman, C.P. and Beal, A.L. (1994) Food restriction and binge eating: a study of former prisoners of war. *Journal of Abnormal Psychology,* 103, 409–411.

Posaci, C., Erten, O., Uren, A. and Acar, B. (1994) Plasma copper, zinc and magnesium levels in patients with premenstrual tension syndrome. *Acta Obstetrica et Gynecologica Scandinavica,* 73, 452–455.

Reid, L.D. (1985) Endogenous opioid peptides and regulation of drinking and feeding. *American Journal of Clinical Nutrition,* 42, 1099–1132.

Rogers, P.J. and Smit, H.J. (2000) Food craving and food "addiction": a critical review of the evidence from a biopsychosocial perspective. *Pharmacology, Biochemistry and Behavior,* 66, 3–14.

Rosenthal, N., Genhart, M., Caballero, B., Jacobsen, F.M., Skwerer, R.G., Coursey, R.D. et al. (1989) Psychological effects of carbohydrate- and protein-rich meals in patients with seasonal affective disorder. *Biological Psychiatry,* 25, 1029–1040.

Sabelli, H.C. and Javaid, J.I. (1995) Phenylethylamine modulation of affect: therapeutic and diagnostic implications. *Journal of Neuropsychiatry and Clinical Neuroscience*, 7, 6–14.

Schroeder, B.E., Binzak, J.M. and Kelley, A.E. (2001) A common profile of prefrontal cortical activation following exposure to nicotine- or chocolate-associated contextual cues. *Neuroscience*, 105, 535–545.

Schuman, M., Gitlin, M.J. and Fairbanks, L. (1987) Sweets, chocolate and atypical depressive traits. *Journal of Nervous and Mental Disease*, 175, 491–495.

Sherwood, R.A., Rocks, B.F., Stewart, A. and Saxton, R.S. (1986) Magnesium and the premenstrual syndrome. *Annals of Clinical Biochemisty*, 23, 667–670.

Shively, C.A. and Tarka, S.M. (1984) Methylxanthine composition and consumption patterns of cocoa and chocolate products, in *The Methylxanthine Beverages and Foods: Chemistry, Consumption and Health Effects*, Spiller, G., Ed., Alan R. Liss, New York, pp 149–178.

Shufman, P.E., Vas, A., Luger, S. and Steiner, J.E. (1997) Taste and odor reactivity in heroin addicts. *Israel Journal of Psychiatry and Related Science*, 34, 290–299.

Snyder, S.H., Katims, J.J., Annau, Z., Bruns, R.F. and Daly, J.W. (1981) Adenosine receptors and behavioural actions of methylxanthines. *Proceedings of the National Academy of Sciences of the U.S.A.*, 78, 3260–3264.

Stavric, B. (1988) Methylxanthines: toxicity to humans 3. Theobromines, paraxanthine and the combined effects of methylxanthines. *Food Chemistry and Toxicology*, 26, 725–733.

Tiffany, S.T. (1990). A cognitive model of drug urges and drug-use behavior: the role of autonomic and non-automatic processes. *Psychological Reviews*, 97, 147–168.

Tomelleri, R. and Grunewald, K.K. (1987) Menstrual cycle and food cravings in young college women. *Journal of the American Dietetics Association*, 87, 311–315.

Toornvliet, A.C., Pijl, H., Tuinenburg, J.C., Elte-de Wever, B.M., Pieters Frolich, M., Onkenhout, W. and Meinders, A.E. (1997) Psychological and metabolic responses of carbohydrate craving patients to carbohydrate, fat and protein-rich meals. *International Journal of Obesity and Related Metabolic Disorders*, 21, 860–864.

Trenchard, E. and Silverstone, T. (1982) Naloxone reduces the food intake of normal human volunteers. *Appetite*, 4, 249–257.

Tuomisto, T., Hetherington, M.M., Morris, M.-F., Tuomisto, M.T., Turjanmaa, V. and Lappalainen, R. (1999) Psychological and physiological characteristics of sweet food "addiction." *International Journal of Eating Disorders*, 25, 169–175.

Van Ree, J.M. (1996) Endorphins and experimental addiction. *Alcohol*, 13, 25–30.

Vlitos, A.L.P. and Davies, G.J. (1996) Bowel function, food intake and the menstrual cycle. *Nutrition Research Reviews*, 9, 111–134.

Wardle, J. (1987) Compulsive eating and dietary restraint. *British Journal of Clinical Psychology*, 26, 47–55.

Webb, P. (1986) Twenty-four hour energy expenditure and the menstrual cycle. *American Journal of Clinical Nutrition*, 44, 614–619.

Webster, R.A. and Jordan, C.C. (1989) *Neurotransmitters, Drug and Disease*. Blackwell Scientific Publications, Oxford.

Weil, A. (1990) *Natural Health, Natural Medicine*. Houghton Mifflin, Boston, MA.

Weingarten, H.P. and Elston, D. (1991) Food cravings in a college population. *Appetite*, 17, 167–175.

Willner, P., Benton, D., Brown, E., Cheeta, S., Davies, G., Morgan, J. and Morgan, M. (1998) Depression increases craving for sweet rewards in animal and human models of depression and craving. *Psychopharmacology*, 136, 272–283.

Wise, R.A. (1988) The neurobiology of craving: implications for the understanding and treatment of addiction. *Journal of Abnormal Psychology*, 97, 118–132.

Wolkowitz, O.M., Doran, M.R., Cohen, R.M., Cohen, T.N., Wise, T.N. and Picckar, D. (1988) Single-dose naloxone acutely reduced eating in obese humans: behavioral and biochemical effects. *Biological Psychiatry*, 24, 483–487.

Wurtman, R.J. and Wurtman, J.J. (1989) Carbohydrates and depression. *Scientific American*, 260, 50–57.

Wurtman, R.J., Hefti, F. and Melamed, E. (1981) Precursor control of neurotransmitter synthesis. *Pharmacological Reviews*, 32, 315–335.

Yeomans, M.R., Wright, P., Macleod, H.A. and Critchley, J.A.J.H. (1990) Effects of nalmefene on feeding in humans. *Psychopharmacology*, 100, 426–432.

16 Is There a Relationship between Chocolate Consumption and Headache?

Lidia Savi

CONTENTS

Introduction 219
Mechanism of Action and Biochemical Aspects 220
Double-Blind, Placebo-Controlled Trials 221
Conclusions 223
References 225

INTRODUCTION

The role and the importance of dietary constituents as triggers of headache and migraine attacks is still a matter of intensive debate. According to the International Headache Society (IHS) Classification (Headache Classification Committee, 1988), a group of headaches associated with substance use or withdrawal are coded in Chapter 12. Headaches associated with the use of specific food components or food additives are included. These particular forms are sometimes described as dietary headaches. Unfortunately, it is not easy to know whether there is only a casual relationship between a particular headache and the acute use of, or exposure to, a specific substance or combination of substances. To prove it, double-blinded, placebo-controlled experiments are necessary. The first step to establish whether a substance really induces a particular type of headache is to determine whether it fulfills the diagnostic criteria proposed by the IHS for substance-induced headache: (1) headache occurs within a specified time after substance intake; (2) a certain required minimum dose should be indicated; (3) headache has occurred in at least 50% of exposures and at least three times; and (4) headache disappears when the substance is eliminated or within a specified time thereafter (Headache Classification Committee, 1988).

It is generally thought that certain foods can provoke typical migraine attacks in migraine sufferers, and this form is generally called dietary migraine (Dalessio, 1972). One of the foods most frequently reported both by patients and doctors is chocolate. Foods have been considered trigger factors of headache attacks since ancient times. Probably, the presence of nausea and vomiting during the migraine attack induced early authors to associate headache with gastric troubles due to particular foods eaten. John Fothergill (1712–1780), a physician who described his own migraine attacks, felt that dietary constituents were the most important trigger factors. He was probably the first to incriminate chocolate as a precipitant of attacks, as Pearce (1971b) notes. In 1925, Curtis-Brown proposed a protein theory of migraine, suggesting that all nitrogenous food, either animal or vegetable, contains a potential poison. He also cited chocolate, among many others foodstuffs. At the 1982 Annual Meeting of the American Association for the Study of Headache

(AASH), the relationship between foods and headache was discussed. Many different opinions emerged. Blau and Diamond (1985) decided to try to clarify this problem. Five hundred and fifty questionnaires were sent to members of the AASH as well as to British physicians with a known interest in migraine, asking their opinion about different aspects connected with foodstuffs and headache. Replies were received from 321 respondents, 74% of whom indicated that the frequency of migraine attacks induced by foods ranged from 0 to 20%. Chocolate was the most frequent migraine trigger cited, it was indicated by 72% of the AASH and 87% of U.K. respondents.

Many studies confirm that even patients believe that foods can provoke their migraine attacks, and chocolate is one of the foodstuffs they more frequently mention. Hannington et al. (1970), after examining 500 migraine patients who mentioned foods as possible trigger factors of their migraine attacks, found that chocolate was the most commonly cited constituent (75%). Dalton (1975) found chocolate responsible in 33% of cases of dietary migraine. He also noted that in women food sensitivity differed according to the stage of the menstrual cycle. At the onset of the menstrual cycle 30% were affected; 7% were affected at midcycle, and 13% were affected before menstruation began. Littlewood et al. (1982) interviewed 1310 patients referred to Princess Margaret Migraine Clinic at Charing Cross Hospital, London, and found that about one quarter thought their attacks could be induced by dietary constituents.

Peatfield et al. (1984) found that of 490 patients with migraine, 19% reported that their headaches could be precipitated by chocolate. Over 10 years later, Peatfield (1995) found that of 429 patients with migraine, 16.5% reported that their headache could be precipitated either by cheese or by chocolate, and nearly always by both, while none of the 40 patients with tension-type headache reported sensitivity to foods. He concluded that foods have mechanisms that are in some way more closely related to migraine than to tension-type headache. In a study of 112 patients under current treatment for migraine, Ciervo et al. (1996) observed that more than 70% believed that diet contributed to the occurrence of headache. Chocolate was the most frequent provocative factor cited. Studying a group of 390 subjects referred to the Headache Center of the University of Turin suffering from migraine, tension-type headache, and combined migraine and tension-type headache, it was observed that 35.8% migraine patients, 25.7% of tension-type headache patients, and 40.3% of combined migraine and tension-type headache patients ascribed the onset of their headache attacks to foods (Savi et al., 2002). Forty-four different substances were identified as headache triggers by these food-sensitive patients. Chocolate was the first after alcoholic drinks. It was indicated by 30% of migraine and 27.7% of tension-type headache patients.

The supposition that diet plays a role in triggering headache has been both supported (Hannington, 1967; Pearce, 1971a; Savi et al., 1998) and challenged (Hannington and Harper, 1968; Ryan, 1974; Medina and Diamond, 1978). Some studies on dietary restriction have reported a decrease in headache occurrence rate after participation in an elimination diet (Egger et al., 1983; Mansfield et al., 1985). In contrast, other studies have found that dietary restrictions do not significantly decrease headache occurrence rate (Kohlenberg, 1981), or that placebo ingestion is as likely to induce a headache attack as challenge food ingestion (Moffett et al., 1974). It has been suggested that subject selection plays a large role in the outcome of these studies (Littlewood et al., 1982). So, the concept of "dietary migraine" as a clinical entity is still not widely accepted.

MECHANISM OF ACTION AND BIOCHEMICAL ASPECTS

Vasoactive amines related to serotonin (5HT) and norepinephrine (NE) contained in the foods implicated as possible triggers for migraine are believed to play a role in inducing headache (Marcus, 1993; Moskowitz and Macfarlane, 1993), either directly by affecting blood vessels (Meyer et al., 1986; Olsen, 1990) or by causing the release of epinephrine (EP) and NE (Hannington, 1983), thus indirectly affecting blood vessels. A variety of amines have been implicated in the development of headache, most commonly tyramine (TYR), histamine (HIS), and beta-phenylethylamine (PEA).

Chocolate is especially rich in a variety of vasoactive amines, including PEA (Tarka, 1982; Craig and Nguyen, 1984), that can cross the blood-brain barrier and affect cerebral blood flow (Hannington, 1967; Glover et al., 1984). PEA is metabolized by monoamine oxidase (MAO), and headache may be related to a deficient metabolism. Sandler et al. (1974) reported reduced oxidative capacity of MAO for PEA in migraineurs and identified a headache occurrence rate of 50% in migraine headache patients exposed to PEA, compared to 6% in those receiving a placebo. Glover et al. (1977) also observed a reduction in MAO activity during migraine attacks. Unfortunately, neither dietary migraine patients nor nondietary migraine patients showed any difference in platelet MAO activity. As noted before, only a small percentage of headache sufferers identify foods as triggers, and it is unclear why only a limited proportion of headache patients is affected. Moreover, researchers have suggested that the percentage of headache sufferers who actually "have" some foods as triggers is significantly less than the percentage of those who "identify" some foods as triggers (Hannington and Harper, 1968). For those who do identify foods as triggers, some vasoactive amine-rich foods may be identified as consistent headache triggers, whereas others are not. This suggests that if food acts as a headache trigger in some individuals, vasoactive amines may not play as strong a causative role as it was once thought. Contrary to previous views, chocolate has a low concentration of both TYR and PEA. So, PEA may not be the strongest headache trigger present in chocolate, as previously thought.

Littlewood et al. (1982) found that migraine patients who believe that dietary factors can induce their attacks have significantly lower mean platelet phenolsulphotransferase P activity than controls, and this observation was later confirmed by Soliman et al. (1987). This fact could indicate that this enzyme might play a role in diet-sensitive migraine. Chocolate is known to contain phenolic flavonoids, which are known to inhibit phenolsulphotransferase (Gibb et al., 1991). However, other pharmacologically active compounds may be found in relatively large amounts in chocolate, in particular theobromine (Tarka, 1982), a methylxanthine that is similar in chemical structure to caffeine. So, the next step will be to identify the relevant chemical agent or agents responsible for initiating a migraine attack.

It has been suggested that migraine patients susceptible to dietary provoking agents might suffer from a food allergy. However, while several workers claim to have demonstrated an abnormal allergic response to food in these patients (Monro et al., 1980; Egger et al., 1983), others have not found enough evidence to link allergy with dietary migraine (Merrett et al., 1983; Bentley et al., 1984; Nattero et al., 1994). Therefore, the fact that supersensitivity of certain migraine patients to certain foods is mediated by the immune system remains to be established.

Using the differential sugar absorption test with mannitol and lactulose, Nattero et al. (1994) found an abnormal bowel permeability in patients who recognized some foods as triggers of their migraine attacks. This alteration could modify the absorption of some foods constituents and could explain why some migraine patients are sensitive to some foods while others are not.

Another possibility is the presence of a genetic factor that predisposes the patient to food sensitivity. Peatfield et al. (1985), studying genetic tendencies in migraine, found that patients with dietary migraine were more likely to have mothers with dietary migraine compared with patients with nondietary migraine. A double-blind, placebo-controlled study (Walton et al., 1993) showed that patients with unipolar depression, a disease related to a strong genetic predisposition, were more sensitive to aspartame side effects, including headache, than healthy controls. So, it is possible that genetic factors may predispose the patients to headache attacks induced by specific chemical constituents of ingested foods.

DOUBLE-BLIND, PLACEBO-CONTROLLED TRIALS

In spite of the fact that many patients, and many doctors too, believe that eating chocolate may induce migraine attacks, the results of the studies are very controversial. Up to now only three double-blind, placebo-controlled studies have been performed to examine the possible role of chocolate in migraine.

In the first study, Moffett et al. (1974) selected a group of 25 subjects (23 females and 2 males; age range 22–62 years; mean age 49 years) suffering from migraine from 332 subjects who answered a questionnaire published in the *Journal of the British Migraine Association* requesting information about dietary precipitants of migraine. All these subjects had found that even small amounts of chocolate precipitated their migraine. Two separated double-blind, fully balanced studies were performed, using two different type of chocolate and matching placebos, one for each study, prepared by two different factories. Both placebos consisted of a synthetic fat whose physical quality approximated cocoa butter but that was made of vegetable oils not containing cocoa. Sugar, coloring, and flavoring were added to this. The texture and taste of the real chocolate were disguised by additives.

In the first of these two studies, each subject was sent two separate samples, the second 2 weeks after the first. One sample consisted of chocolate and the second of its matching placebo. The samples were of similar weight and were identically wrapped in silver foil. The subjects were only told that they would be asked to eat two different sorts of chocolate. They were asked to respond by questionnaire 48 h after eating the chocolate, indicating whether they had experienced a headache or not, and in case of a positive answer, whether it was similar to their usual migraine. They were requested to return any uneaten portions of chocolate with their questionnaire so that there could be no direct comparison between the two samples; the second sample was only sent after this had been done. There was not direct contact between subjects and investigators at any time. In the second study, 15 of the subjects who had taken part in the first study took two additional samples of chocolate. The design of this study was exactly the same as that of the first one.

In the first study, 15 headaches occurred in 50 sessions; 8 of these occurred after eating chocolate only and 5 after eating placebo only. One subject had headache after eating both chocolate and placebo, and 11 had no headache after either sample. In the second study, 10 headaches occurred in 30 sessions, 5 of these after eating chocolate only and 3 after eating placebo only; 1 subject had headache after eating both.

Of the 15 subjects who took all four samples, only 5 responded consistently in both studies, 2 responded to chocolate alone on both occasions, and 3 had no headache after any of the four samples. All the other subjects behaved differently in the two studies.

In conclusion, in these two studies 25 headaches were reported in 80 subject sessions but only 13 of these occurred after chocolate alone. Only two subjects responded to chocolate alone in both studies.

Gibb et al. (1991) carried out a double-blind, placebo-controlled trial to test the hypothesis that chocolate is able to initiate a migraine attack in some patients who believe themselves to be sensitive to it. They selected 20 patients (17 females and 3 males; age range 23 to 64 years; mean age 39.5 years) attending Princess Margaret Migraine Clinic at Charing Cross Hospital, London, on the basis of the belief that their migraine attacks could be provoked by eating chocolate. They were suffering from migraine diagnosed according to the criteria of Valquist. These patients participated in a double-blind, parallel-group study and were challenged, under supervision, with either a bar of chocolate (12 patients) or a matching placebo (8 patients) containing no cocoa products. Patients were observed for the first 3 h following the challenge and were contacted by telephone 32 h later. The two groups were roughly matched for age and sex. Chocolate and matching placebo bars were identical in appearance and coded to allow the experiment to be carried out as a double-blind study. "Chocolate" made from carob and cocoa bean is quite different in character, but this difference was successfully disguised by the addition of carob to the cocoa product as well as to the placebo, and by the use of a peppermint masking flavor. To test the quality of blindness to the taste of the chocolate and placebo, a group of 26 control subjects was asked to choose at random a sample of either and to state whether or not what they had eaten was chocolate; 13 randomly selected chocolate and 13 placebo. Among those who ate chocolate, five identified it correctly. Among those who selected placebo, seven thought it was chocolate. The differences were not significant. In the parallel-group study, a migrainous headache developed in 5 out of 12 patients challenged with chocolate (41.6%), but none of the 8 patients who were given placebo developed a migraine attack ($p = .051$, one-tailed). The median time interval between chocolate consumption and the onset of symptoms was 22 h (range 3.5 to 27 h).

Finally, Marcus et al. (1997) carried out another double-blind, placebo-controlled provocative study to evaluate whether chocolate provokes headache in a large sample of patients with migraine, tension-type, or combined migraine and tension-type headache. Through posters placed across the main branch of the University of Pittsburgh campus, they recruited 81 women suffering from migraine, tension-type headache, or combined migraine and tension-type headache according to the IHS criteria. In order to ensure the blinded condition of the subjects, chocolate and placebo (carob) bars were identical in appearance and wrapping. Chocolate and placebo recipe formulas were also identical to the ones used in the study by Gibb et al. (1991). The only difference was in the weight of the bars (40 g for Gibb, 60 g in this case). A group of 21 adults were recruited for a double-blind taste test of the chocolate and carob products to determine whether they were able to identify which sample contained the actual chocolate product. Subjects ate both a chocolate and a carob sample on two different days; most subjects ate the samples 1 d apart. At each testing, subjects were asked to record whether they believed they were eating chocolate or not. Subjects guessed that they were eating chocolate 66.7% of the time, and a kappa statistic was not significant, demonstrating that subjects could not accurately determine what they were eating. Thus, the samples appeared to be adequate for use in subsequent trials.

Sixty-one women (age range 18 to 64 years; mean age ± SD 28.3 ± 10.7 years; 50% suffering from migraine, 37.5% from tension-type headache, and 12.5% from combined migraine and tension-type headache) completed the study. Eleven subjects (17.5%) reported that chocolate was a trigger for their headaches (seven were migraine patients, two were tension-type headache sufferers, and two were combined headache sufferers). Subjects were placed on a restricted diet adapted from Theisler (1990). This diet restricts vasoactive amine-rich foods. After completing a 2-week washout period on the diet, subjects began the series of four provocative trials with chocolate or the carob placebo. They remained on the diet for the duration of the trial period. The trials were double-blind. Subjects were randomly selected to receive any of the six possible presentation orders of the two carob trials and the two chocolate trials. Food trials were scheduled during a nonmenstrual week, with at least 3 d between each trial. At the time of the food trials, subjects were asked to make their best guess as to whether they were or were not eating chocolate. A total of 260 taste questionnaires were completed. A kappa statistic was calculated comparing subjects' guesses to what they actually ate; it was not significant.

At the end of the study, 245 food trials were analyzed. The onset of a headache attack within 12 h after the ingestion of the sample was observed in 11 (17.2%) cases with chocolate and in 26 (40.6%) cases with placebo. Thirty-two women (51% of the sample) did not report a headache on any occasion after eating either sample. Three women reported a headache after both chocolate samples and not after either of the placebo samples, and three women also reported a headache after both placebo samples and not after either of the chocolate samples. Six women consistently reported a headache after all four samples. Among the 11 women who believed chocolate was a trigger for their headache, a headache attack was observed in two (18.21%) trials after chocolate samples and in four (36.4%) after placebo samples.

CONCLUSIONS

Although chocolate is frequently identified as a migraine trigger by medical texts, doctors, and patients, the scientific data actually available are very controversial.

In the double-blind, placebo-controlled studies previously described, the results are quite different and conflicting. Moffet et al. (1974) and Marcus et al. (1997) found no relationship between eating chocolate and the onset of a migraine attack, while Gibb et al. (1991) showed the existence of such a relationship. It should also be considered that the three studies present some important differences and several limitations that restrict the interpretation of the results.

First of all, the headache diagnosis was made according to different criteria (not specified — probably Ad Hoc Committee on Headache Classification criteria — for the study by Moffett et al. (1974), Valquist's criteria for the study by Gibb et al. (1991), IHS criteria for the study by Marcus et al. (1997)). The study of Moffett et al. (1974) was conducted entirely by mail, without direct contact with patients either for establishing a diagnosis or for administering the challenge. In the studies by Moffett et al. (1974) and Gibb et al. (1991), the patients were volunteers who represented a headache-suffering population in general, not a clinical population of treatment-seeking headache sufferers. Furthermore, in these two studies the patients continued to take their habitual antimigraine treatment and to follow their usual diet, and all of them believed that chocolate could provoke their migraine attacks. It is possible that ingestion of other foods also containing vasoactive amine acts as a "primer" for other headache triggers. In other words, although chocolate alone may be inadequate to trigger headaches, when it is combined with other headache-triggering foods there may be synergy, the combination providing an adequate trigger.

The subjects included in the study of Marcus et al. (1997) were all females, and this limits the generalizability of the findings to one gender. The sample of women who participated tended to be younger than the typical headache patient and reported significantly less pain and life interference associated with headache in comparison to treatment seekers. There was also a significant amount of noncompliance with the restrictive diet. The authors noted that although they found no difference in headache incidence between subjects who fully complied with the diet and subjects who did not, they cannot truly report that chocolate was ingested on all occasions in isolation from other vasoactive amines. In this study, it was considered a headache attack if it occurred within 12 h after the ingestion of the sample, while Moffett et al. (1974) considered a headache attack to occur within 48 h after the ingestion of the sample and Gibb et al. (1991) within 32 h. According to the latter author, the median time interval between chocolate consumption and the onset of symptoms following chocolate was 22 h.

In all three studies, it is not stated whether the patients were fasting when eating chocolate or placebo bars. This fact could influence their time of absorption, so the subsequent appearance of migraine, and the time interval before it, could be different. Both Moffett et al. (1974) and Gibb et al. (1991) utilized only migraine sufferers who specifically identified chocolate as a trigger of their headaches and who had even decided to eliminate it from their diets because of this. Therefore, a very select group of headache sufferers was involved in these studies. On the contrary, the Marcus study was designed to investigate the incidence of chocolate-triggered headache in a general sample of headache sufferers, and they did not utilize the same screening criterion. It is possible that their findings were negative simply because they used a general headache-suffering group, not individuals who had identified chocolate as a trigger.

On the other hand, we have to consider that patients' experiences with chocolate cravings may lead to an erroneous assumption of a causal relationship between chocolate and headache. Sweet craving has been identified as a prodrome to the onset of headache (Blau, 1992). Fulfilling this craving with chocolate could then lead to the belief that chocolate caused the headache. In addition, in about 60% of women, headache is related to menses. Premenstrual sweet craving may cause patients to associate chocolate with the menstrual headache. Finally, stress has been identified as a headache trigger for the majority of chronic headache sufferers. Stress has also been linked to sweet cravings, permitting the sweet, rather than the original stress, to be identified as the headache trigger.

In conclusion, we must say that in spite of what many doctors and many patients believe, the relationship between chocolate and migraine is not clear at all. Further studies are needed that should also consider other important aspects that have not yet been analyzed up to now. In particular, it is necessary to determine the time after the ingestion of chocolate in which headache can occur, the amount of chocolate that can provoke it, the association with other foods, the type of headache that is more frequent, and further characteristics of subjects' sensitivity to this substance. It should also be important to identify which component of chocolate may induce headache and by which mechanism. To clarify this last aspect, however, it would probably be necessary to know more about the pathophysiology of migraine in general.

REFERENCES

Ad Hoc Committee on Classification of Headache (1962) Classification of headache. *Journal of the American Medical Association*, 179, 717–718.
Bentley, D., Katchburian, A. and Brostoff, J. (1984) Abdominal migraine and food sensitivity in children. *Clinical Allergy*, 14, 499–500.
Blau, J.N. (1992) Migraine triggers: practice and theory. *Pathology and Biology*, 40, 367–372.
Blau, J.N. and Diamond, S. (1985) Dietary factors in migraine precipitation: the physician's view. *Headache*, 25, 184–187.
Ciervo, C.A., Gallagher, M., Muller, L. and Perrino, D. (1996) The role of diet in treated migraine patients. *Headache Quarterly*, 7, 319–323.
Craig, W.J. and Nguyen, A. (1984) Caffeine and theobromine levels in cocoa and carob products. *Journal of Food Sciences*, 49, 302–305.
Curtis-Brown, R.A. (1925) Protein poison theory: its application to the treatment of headache and especially migraine. *British Medical Journal*, i, 155–156.
Dalessio, D.J. (1972) Dietary migraine. *American Familial Physician*, 6, 60–64.
Dalton, K. (1975) Food intake prior to a migraine attack: study of 2,313 spontaneous attacks. *Headache*, 15, 188–193.
Egger, J., Carter, C.M., Wilson, J., Turner, M.W. and Soothill, J.F. (1983) Is migraine food allergy?: a double-blind controlled trial of oligoantigenic diet treatment. *Lancet*, ii, 865–869.
Gibb, C.M., Davies, P.T.G., Glover, V., Steiner, C.J., Rose, C.F. and Sandler, M. (1991) Chocolate is a migraine-provoking agent. *Cephalalgia*, 11, 93–95.
Gibb, C.M., Glover, V. and Sandler, M. (1987) In vitro inhibition of phenolsulphontransferase by food and drink constituents. *Biochemical Pharmacology*, 36, 2325–2329.
Glover, V., Littlewood, J., Sandler, M., Peatfield, R., Petty, R. and Rose, F.C. (1984) Dietary migraine: looking beyond tyramine, in *Progress in Migraine Research 2*, Rose, F.C., Ed., Pitmann, London, pp. 34–39.
Glover, V., Sandler, M., Grant, E., Rose, F.C., Orton, D., Wilkinson, M. and Stevens, D. (1977) Transitory decrease in platelet monoamine oxidase activity during migraine attacks. *Lancet*, 1, 391–393.
Hannington, E. (1967) Preliminary report on tyramine headache. *British Medical Journal*, 1, 550–551.
Hannington, E. (1983) Migraine, in *Clinical Reactions to Food*, Lessof, M.H., Ed., John Wiley, New York, pp. 142–148.
Hannington, E. and Harper, A.M. (1968) The role of tyramine in the aetiology of migraine and related studies on the cerebral and extracerebral circulations. *Headache*, 8, 84–97.
Hannington, E., Hain, M. and Wilkinson, M. (1970) Further observations of the effects of tyramine, in *Background to Migraine: 3rd Migraine Symposium*, Cochrane, A.L., Ed., Heinemann, London, pp. 113–119.
Headache Classification Committee of the International Headache Society (1988) Classification and diagnostic criteria for headache disorders, cranial neuralgia and facial pain. *Cephalalgia*, 8 (Suppl. 7), 1–96.
Kohlenberg, R.J. (1981) Tyramine sensitivity in dietary migraine: a double-blind study. *Headache*, 22, 30–34.
Littlewood, J., Glover, V., Sandler, M., Petty, R., Peatfield, R. and Rose, F.C. (1982) Platelet phenolsulphotransferase deficiency in dietary migraine. *Lancet*, 1, 983–986.
Mansfield, L.E., Vaughn, T.R. and Waller, S.F. (1985) Food allergy and adult migraine, double-blind and mediator confirmation of an allergic etiology. *Annals of Allergy*, 55, 126–129.
Marcus, D.A. (1993) Serotonin and its role in headache pathogenesis and treatment. *Clinical Journal of Pain*, 9, 159–167.
Marcus, D.A., Scharff, L., Turk, D. and Gourley, L.M. (1997) A double-blind provocative study of chocolate as a trigger of headache. *Cephalalgia*, 17, 855–862.
McCulloch, J. and Harper, A.M. (1977) Phenylethylamine and cerebral blood flow: a possible involvement of phenylethylamine in migraine. *Nature*, 27, 817–821.

McCulloch, J. and Harper, A.M. (1979) Factors influencing the response of the cerebral circulation to phenylethylamine. *Neurology*, 29, 201–207.

McQueen, J., Loblay, R.H., Savain, A.R., Anthony, M. and Lance, J.W. (1989) A controlled trial of dietary modification in migraine, in *New Advances in Headache Research*, Rose, F.C., Ed., Smith-Gordon, London, pp. 235–242.

Medina, J.L. and Diamond, S. (1978) The role of diet in migraine. *Headache*, 18, 31–34.

Merrett, J., Peatfield, R., Rose, F.C. and Merrett, T.G. (1983) Food related antibodies in headache patients. *Journal of Neurology, Neurosurgery and Psychiatry*, 46, 738–742.

Meyer, J.S., Zetuskv, W., Jonsdottir, M. and Mortel, K. (1986) Cephalic hyperemia during migraine headaches: a prospective study. *Headache*, 26, 388–397.

Moffett, A.M., Swash, M. and Scott, D.F. (1974) Effect of chocolate in migraine: a double-blind study. *Journal of Neurology, Neurosurgery and Psychiatry*, 37, 445–448.

Monro, J., Brostoff, J., Carini, C. and Zilkha, K. (1980) Food allergy in migraine. *Lancet*, ii, 1–4.

Moskowitz, M.A. and Macfarlane, R. (1993) Neurovascular and molecular mechanisms in migraine headaches. *Cerebrovascular and Brain Metabolism Reviews*, 5, 159–177.

Nattero, G., Savi, L., Di Pentima, M., Inconis, T., Cadario, G. and Paradisi, L. (1994) Gastrointestinal permeability and dietary migraine, in *New Advances in Headache Research 3*, Rose, F.C., Ed., Smith-Gordon, London, pp. 157–162.

Olsen, T.S. (1990) Migraine with and without aura: the same disease due to cerebral vasospasm of different intensity. A hypothesis based on CBF studies during migraine. *Headache*, 30, 269–272.

Pearce, J.M.S. (1971a) General review, some etiological factors in migraine, in *Background to Migraine*, Cumings, J.N., Ed., Heinemann, London, pp. 1–17.

Pearce, J.M.S. (1971b) Insulin-induced hypoglycaemia in migraine. *Journal of Neurology, Neurosurgery and Psychiatry*, 34, 154–156.

Peatfield, R.C. (1995) Relationship between food, wine, and beer-precipitated migrainous headaches. *Headache*, 35, 355–357.

Peatfield, R.C., Glover, V., Littlewood, J.T., Sandler, M. and Rose, C.F. (1984) The prevalence of diet-induced migraine. *Cephalalgia*, 4, 179–183.

Peatfield, R.C., Sandler, M. and Rose F.C. (1985) The prevalence and inheritance of dietary migraine, in *Migraine, Chemical and Therapeutic Advances*, Rose, F.C., Ed., Karger, Basel, pp. 218–224.

Ryan, R.E. (1974) A clinical study of tyramine as an etiological factor in migraine. *Headache*, 14, 43–48.

Sandler, M., Youdim, M.H. and Hanington, E. (1974) A phenylethylamine oxidizing defect in migraine. *Nature*, 250, 33–37.

Savi, L., Iosca, L., Rainero, I. and Pinessi, L. (1998) Study of dietary risk factors in primary headaches. *Neuroepidemiology*, 17, 46–47.

Savi, L., Rainero, I., Iosca, L., Sabbatini, F. and Pinessi, L. (1998) Study of dietary risk factors in primary headaches, in *Management of Pain: A World Perspective III*, De Vera, J.A., Parris, W. and Erdine, S., Eds., Monduzzi Editore S.p.A., Bologna, pp. 307–310.

Savi, L., Rainero, I. and Pinessi, L. (1998) Dietary migraine or dietary headache? *Functional Neurology*, XIII, 165.

Savi, L., Rainero, I., Valfrè, W., Gentile, S., Lo Giudice, R. and Pinessi, L. (2002) Food and headache attacks: a comparison of patients with migraine and tension-type headache. *Panminerva Medical*, 44, 27–31.

Scharff, L., Turck, D.C. and Marcus, D.A. (1995) Triggers of headache episodes and coping responses of headache diagnostic groups. *Headache*, 35, 397–403.

Soliman, H., Pradalier, A., Launay, J.-M., Dry, J. and Dreux, C. (1987) Decreased phenol and tyramine conjugation by platelets in dietary migraine, in *Advances in Headache Research*, Rose, F.C., Ed., John Libbey, London, pp. 117–121.

Tarka, S.M., Jr. (1982) The toxicology of cocoa and methylxanthines: a review of the literature. *CRC Critical Reviews of Toxicology*, 9, 275–312.

Theisler, C.W. (1990) *Migraine Headache Disease: Diagnostic and Management Strategies*, Aspen Publishers Inc., Austintown, OH, pp. 111–112.

Van den Bergh, V., Amery, W.K. and Waelkens, J. (1987) Trigger factors in migraine: a study conducted in the Belgian migraine society. *Headache*, 27, 191–196.

Vaughan, W.T. (1927) Allergic migraine. *Journal of the American Medical Association*, 88, 1383–1386.

Walton, R.G., Hudak, K. and Geen-Waite, R.J. (1993) Adverse reactions to aspartame: double-blind challenge in patients from a vulnerable population. *Biological Psychiatry*, 34, 13–17.

Index

[³H]naloxone, 189
1-methyl-4-pheyl-1,2,3,6-tetrahydropyridine. *See* MPTP
24-hour endogenous arousal cycle, 14
3,4-dicaffeoyl-1,5-quinide. *See* DICAQ
4-caffeoyl-1,5-quinide. *See* 4-CAQ
4-CAQ, 189
4-chloro-*m*-cresol, 5
8-(3-chlorostyryl)caffeine. *See* CSC

A

A_1 receptors, 3, 153
    antiepileptic effect of, 167
    catecholamine and serotonin level increase due to blocking of, 76
    chronic caffeine exposure and susceptibility to seizures, 168
    competitive blockade of in peripheral vasculature, 118
    effect on excitatory neurotransmission, 165
A_2 receptors, effect on ischemic neuronal damage, 166
A_{2A} receptors
    acute treatment for neuroprotection, 166
    blocking of by caffeine, 35
    contribution of to death of striatal medium spiny neurons, 155
    down-regulation of in patients with major depression, 76
    genetic inhibition of by caffeine, 36
    mechanisms of protection of antagonists in Parkinson's disease, 156
    neuroprotection by specific antagonists in Parkinson's disease, 153
    psychostimulant properties of caffeine due to blockade of, 139
    role of in epilepsy, 167
    tonic adenosine activation of, 3
Abuse, caffeine, 40, 133
Acetaminophen, efficacy of in headaches when combined with caffeine, 176
Acetylcholine
    implication of in the neurobiology of depression, 75
    release of by adenosine A_{2A} receptors, 168
Acoustic startle reflex, delayed habituation of due to caffeine, 56
ACTH, 114
    combined effect of caffeine and behavioral stress on, 119
Activation-Deactivation Adjective Check List, 39
Addictive disorders, 215

beneficial effect of coffee consumption on, 189
Adenosine
    action in the brain, 36
    action in the heart, 118
    anticonvulsant possibilities of, 167
    inhibition of transporter by roasted coffee, 190
    monophosphate, 76
    pharmacology, 154
    receptors (*See also* specific receptors)
        altered development due to caffeine in maternal milk, 103
        antinociceptive actions of caffeine on, 176
        blockade of by caffeine, 2, 114, 117
        differences in young *vs.* mature animals, 166
        influence on processes in development of postischemic damage, 165
    sleep-inducing factors of, 14
Adenylyl cyclase, removal of G_i input by adenosine receptor blockade by caffeine, 3
Adjunctive analgesic properties of caffeine, 176
Adolescents
    beneficial effects of caffeine on cognitive function in, 37
    caffeine dependence of, 41
ADP. *See* cyclic ADP-ribose sensitive calcium release channels
Adrenal system
    effect of caffeine on functions of, 114
    release of catecholamines from chromaffin cells, 6
    release of cortisol by cortex, 114
Adrenaline levels, elevation of by caffeine, 18
    tolerance to, 137
Adrenocortical functions, effect of caffeine on, 115, 119
Adrenocorticotropic hormone. *See* ACTH
Affective disorders, 74
    euthymic effect of caffeine and, 76
Age
    -associated cognitive decline, 86
        restoration of, 91
    effects of caffeine depending upon, 37, 85
    optimal doses of caffeine related to, 92
    trends in caffeine intake, 87
Alcohol
    caffeine consumption in patients with dependence, 77, 138
    effect of use on sleep, 17
    withdrawal seizures, 168
Alertness
    cycles of, 15
    effect of caffeine on, 39
Alkylxanthines, adenosine receptor blockade in effects of, 4

Alzheimer's disease, 36
    Apo-E4 role in, 86
    lower risk of in conjunction with higher caffeine intake, 88
    reduced risk of due to coffee consumption, 192
American Association for the Study of Headache (AASH), 220
AMP. *See* cAMP
Amphetamines, 139
Analgesics
    as a risk factor for restless leg syndrome, 44
    caffeine from, 25
    caffeine-containing, 44
    inhibition of morphine with coffee extract, 189
    methylxanthines as adjuncts in, 1
Anandamides, in chocolate, 205, 213
Angiotensin II type 2 receptor gene expression, change in due to maternal caffeine consumption, 107
Animals
    caffeine withdrawal in, 135
    metabolic body weight comparisons with humans, 99
    reinforcing effects of caffeine in, 137
    susceptibility to caffeine of, 97
    tolerance to caffeine in, 136
Anorectic, caffeine and ephedrine as an, 1
Anticraving properties of roasted coffee, 187
Antidepressant agents, 75
    caffeine and, 77
Antinociceptive properties of caffeine, 176
Antioxidant activity of caffeic acid, protection from free radical oxidation of brain cells, 187
Antipsychotics, effects of caffeine on, 44
Anxiety. *See also* specific disorders
    disorders exacerbated by caffeine, 36, 42
    due to caffeine withdrawal, 27
    due to high doses of caffeine, 133
    production of by high doses of caffeine, 3, 40
    tolerance to caffeine-induced, 137
Anxiety dreams, 16
Apnea of infants, methylxanthines as treatment for, 1, 103, 167
Apo-E4, increase in risk of early cognitive decline due to, 86
Apolipoprotein E. *See* Apo-E4
Arginine vasopressin (AVP), secretion of, 114
Arousal
    decrease in baseline in elderly adults, 92, 94
    interaction of caffeine and state of, 65
    inverted U-shaped function of, 38
    relationship to cognitive performance, 91
    traits and states in dual-interaction theory, 36
Aspirin, efficacy of in headaches when combined with caffeine, 176
Asthma, methylxanthines as treatment for, 1
Athletes, caffeine abuse of, 41
ATP receptors, response to theophylline of, 2
Attention
    effect of aging on, 86
    effect of caffeine on, 62
Attention deficit hyperactivity disorder (ADHD), effects of caffeine on, 36, 43

Auditory Evoked Potential, effect of caffeine on, 56
Auditory information, effects of caffeine on processing of, 55
Auditory Vigilance Task (AVT), 63
Automatic target detection, 63
Autonomic responses
    caffeine and, 113
    failure, 117

B

Bakan task, 63
Basal forebrain, 36
Basic information processing speed, 86
Beck Depression Inventory (BDI), correlation of caffeine use with, 78
Behavioral stimulation, 1
    caffeine and, 113
    due to blockade of adenosine receptors, 4
    properties of caffeine, 138
Benzodiazepine agonists, 5
Benzodiazepines, interaction with caffeine, 44
Beta-endorphin, 114
    release of following chocolate consumption, 212
Beta-phenylethylamine, as a trigger for headache, 220
Bingeing, 213
Bipolar disorder, 74
    caffeine consumption in patients with, 77
    effects of caffeine on, 43
Black tea, 197
Blood pressure
    effect of caffeine on, 117
        tolerance to, 137
    expectancy and placebo effects on, 26
    high, cognitive aging and, 86
    increase in due to caffeine, 113
    increase in due to caffeine and stress, 120
Brain
    adenosine receptors in, 3
    attenuation of ischemic injury by treatment of caffeine and ethanol, 155
    caffeine activation of circuits underlying dependence to drugs, 139
    decrease in weight due to caffeine in maternal milk, 103
    effect of caffeine on fetal brain development, 98
    P3b component of potentials, 61
Brain-function impairment, due to maternal caffeine intake, 102
Brainstem Auditory Evoked Potential (BAEP), effect of caffeine on, 55
Bright light, combined with caffeine to counteract fatigue, 21, 28
Bronchial asthma, methylxanthines as treatment for, 1, 167
Bupropion, 75

C

c-fos expression, reduction in after seizures in animals treated with caffeine, 169

Index

Caffeic acid, antioxidant properties of, 187, 191
Caffeine, 1
    abuse and dependence, 40, 133
    acute *vs.* long term use of, 165
    aids to, 21
    antinociceptive properties of, 176
    as an additive in foods, 98
    behavioral effects of, 133
    biobehavioral effects of, 35
    blockade of adenosine receptors by, 2
    bradycardiac effects of, 119
    cerebral vasoconstrictive properties of, 180
    cognitive enhancement with, 86, 91
    concentrations of, 2
    consumption, 73
        by shift workers, 20
        diurnal trends of, 15
        motives for, 16
    dose-related reinforcement, 138
    effect on ACTH and cortisol during rest, 115
    effect on the demise of nigrostriatal dopaminergic neurons, 152
    effects on gestation and lactation, 97
    epilepsy and, 167
    euthymic effect of, 76
    half-life, 25
        in pregnancy and neonates, 98
    hormonal differences in effects of, 152
    IC_{50} values for, 5
    in chocolate, 205, 210
    in over-the-counter analgesics, 25, 44, 169
    individual differences in effect of, 122
    influence on sleep of, 14
    inhibition of iron absorption due to, 104
    instruments used to evaluate effects on mood, 76
    interactions with psychopharmacological treatment, 44
    intoxication (*See* caffeinism)
    ischemic brain damage and, 165
    long-term treatment of ischemia and epilepsy with, 168
    longitudinal effects of habitual intake on cognitive performance, 91
    lower risk of Parkinson's disease due to, 150
    measuring intake, 24
    metabolization of, 85
    neural inhibitory action of on adenosine receptor substrates, 36
    peripheral effects of, 6
    phosphodiesterase inhibition by, 4
    protective effect of habitual use on brain integrity, 88
    protective effect on dopaminergic neurons, 147
    psychostimulant properties and dopamine transmission, 139
    reinforcement properties of, 137
    sensitivity differences, 29, 40
    sleep and, 14
    slow-release, 21
    sources of, 25
    stress reactions to, 115
    teratogenic effects on fetuses due to, 102
    tolerance to the effects of, 136
    withdrawal effects, 16, 135
    workplace studies of blood pressure elevation due to, 121
    worldwide consumption of, 134
Caffeinism, 2, 36, 42
Caffeoyl, antioxidant activity of, 191
Cake, caffeine from, 25
Calcium
    effects of caffeine on channels, transporters and modulatory sites, 4
    release channels, effect of caffeine on, 2
Calcium carbonate, as an element in treatment of migraines, 179
Calcium release channels, inhibition of by adenosine A_1 receptors, 3
Camellia sinensis, 197
CAMP, 4
    -dependent protein kinase (PKA), 101
    interference of caffeine with, 115
    reduction in dopaminergic neurons in Parkinson's disease, 158
Cancellation, improved task performance due to caffeine, 57
Cancer
    antioxidant properties of tea and, 197
    gastric, reduction of risk in green tea drinkers, 198
    pain management with caffeine, 44
    therapeutic target for caffeine, 1
Candy, caffeine from, 25, 134
Carbohydrate craving, menstrual cycle and, 208
Cardiac output measurements, 118
Cardiac stimulants, methylxanthines as, 1
Cardiovascular effects of caffeine
    adenosine and, 114
    during rest, 117
    during stress, 119
Case-control studies, 148
Catecholamines
    effect of caffeine on release, 116
    increase in levels due to cortisol release, 115
    levels in habitual coffee drinkers, 18
Caudate nucleus, increase of dopamine in due to caffeine, 139
Central nervous system
    A_{2A} receptor modulation of generalized processes in, 158
    caffeine activation of, 113
    CFFT as an index state of, 55
    effects of cortisol on HPA axis in, 115
    excitability due to caffeine in maternal milk, 103
    long term effects due to maternal caffeine consumption, 107
    mechanisms of caffeine on, 2
Cerebral blood flow
    increase in during caffeine withdrawal, 136
    relationship to headache of changes in, 180
Cerebral energy metabolism, lack of caffeine tolerance, 137
Cerebral glucose utilization, increases in due to drugs of abuse, 139
Cheese
    as a trigger for headaches, 220
    influence of nalmefene on intake of, 212

Children
    caffeine intake of, 27
    caffeine withdrawal of, 136
    effects of caffeine on cognitive function in, 37
Chloride flux, inhibition of by caffeine in synaptoneurosomes, 5
Chlorogenic acids, in coffee beans, 188
Chocolate
    as a trigger for headaches, 220
    biological mechanisms of, 209
    craving, 206
    craving throughout the menstrual cycle, 208
    guilt associated with consumption of, 207
    hot, caffeine from, 25
    psychological basis for cravings, 214
    taste and mouth-feel, 205
    vasoactive amines in, 221
Choice reaction time
    performance relation to caffeine, 90
    use of in caffeine studies, 56
Cholesterol synthesis, effect of caffeine on, 103
Chronic pain, effect of caffeine in patients with, 44
Circadian rhythm, effect of caffeine on, 18
Clozapine, increase in toxicity of due to caffeine, 44, 79
Cocaine, caffeine administration in animals and, 138
Cocoa
    beverages (*See also* chocolate)
        caffeine in, 134
    powder, 213
Coffee. *See also* caffeine
    age trends in intake of, 87
    antiaddiction effects of, 188
    antidepressant effects of, 191
    antioxidant effects of, 191
    association of lower risk of Parkinson's disease, 148
    attitudes linked to quantity consumed, 133
    consumption of in Western society, 85, 134
    diurnal trends of consumption, 15
    effect on mood of, 77
    estrogenic effects of, 192
    inhibition of morphine-induced analgesia using extract, 189
    methylxanthine in, 73
    noncaffeine constituents in, 187
Cognitive aging, 86
Cognitive functions
    effect of caffeine on, 35, 37, 54, 56
        age-dependency of, 85
        mediating factors, 65
    protective effect of habitual caffeine use on, 88
    restoration of in aging, 91
Cognitive stressors, 120
Cohort studies, 148
Cola beverages
    age trends in intake of, 87
    methylxanthine in, 73
Cold pressor challenge, 121
Cold remedies, caffeine from, 25
Color discrimination, effects of caffeine on, 55
Competitive blockade of adenosine receptors, 114. *See also* adenosine

Concentration, effect of caffeine on performance test for, 58
Concept shifting test, 90
Conviviality, as a motive for coffee consumption, 16
Coronary heart disease, antioxidant properties of tea and, 197
Corticosterone, increase of due to caffeine, 6
Corticotropin-releasing factor (CRF)
    implication of in the neurobiology of depression, 75
    stimulation of by cAMP, 115
Corticotropin-releasing hormone (CRH), 114
Cortisol
    combined effect of caffeine and behavioral stress on, 119
    levels in habitual coffee drinkers, 18
    release of by adrenal cortex, 114
Craving, chocolate, 206
Cresol, 5
Critical flicker fusion threshold (CFFT), effects of caffeine on, 55
CSC, neuroprotective effect of, 154
Cyclic ADP-ribose sensitive calcium release channels, effect on calcium sensitivity of caffeine, 4
Cyclic AMP. *See* cAMP
Cyclic nucleotides, 2
Cyclooxygenase, 200
Cytochrome P450 CYP1A2, 36
    inhibition of due to fluvoxamine, 78
    metabolism of caffeine and estrogen by, 152

D

DARPP-32, caffeine involvement in phosphorylation of, 169
Decaffeinated coffee, 16
Decisional processes, effect of caffeine on, 56
Deep sleep, 14
Delayed free recall, effect of caffeine on, 60
Delayed recall, enhancement of performance due to caffeine, 38
Dependence, caffeine, 41
Depressant effects, of xanthines due to phophodiesterase inhibitors, 4
Depression, 74
    antidepressant effects of coffee constituents, 187, 191
    chocolate craving and, 206
    effects of caffeine on, 36, 42
    methylxanthine and, 76
    neurobiology of, 75
    pharmacological treatment for, 75
    sleep and, 17
    unipolar, food sensitivity in, 221
Dexamethasone suppression test, caffeine ingestion and false positive results, 115
Diabetes
    cognitive aging and, 86
    therapeutic target for caffeine, 1
Diagnostic and Statistical Manuals (DSM), 74
Diazepam, interference of caffeine with treatment in infants, 104
DICAQ, 189

Dicinnamoylquinides, inhibition of human adenosine transporter by, 190
Dietary headaches, 219
DIFEQ, 189
Digit span task, assessment of caffeine on working memory using, 62
Dihydrocaffeic acid (DCA), antioxidant effects of, 192
Dihydroxyphenylacetic acid. See DOPAC
Discriminatory processes, effect of caffeine on, 56
Disease
  effect on cognitive aging, 86
  fetal origins of adult, 98, 106
Display load, effect of caffeine on, 61
Diuretics, methylxanthines as, 1
Divided attention
  effect of caffeine on, 63
  improved performance of tasks due to caffeine, 38
DNA, effect of caffeine on fetal brain, 100
  in malnourished women, 105
DOPAC, attenuation of toxin-induced loss of with caffeine, 153
Dopamine
  abnormalities found in depressive states, 75
  attenuation of MPTP-induced loss of transporter, 154
  content in adult brain of rats exposed to caffeine in utero, 101
  D_2 receptors, expression of A_{2A} receptors in, 3
  neuromodulation of by phenylethylamine, 209
  release in prefrontal cortex due to caffeine, 76
  role in drug dependence of, 139
Dose-effect trends, for caffeine in cognitive function studies, 90, 92
DPCPX, 166
Driving
  study on effect of caffeine and naps on, 22
  study on effect of caffeine on, 18
Drug dependence, 214
  caffeine reinforcement in persons with, 138
  criteria for, 134
Drugs of abuse, brain circuits targeted by, 139
Dual-interaction theory, 36
Dutch Eating Questionnaire, 207
Dysphoria, self-medication with chocolate for, 206
Dysthymia, 74
  early-age use of caffeine in patients, 78

E

Eating disorders
  chocolate craving and, 207
  effects of caffeine on, 36, 43
Eicosanoids, 200
Elderly people
  acute caffeine effects in, 93
  effects of caffeine in, 37
  effects of caffeine on blood pressure of, 124
  memory effects of caffeine in, 38
Electroconvulsive therapy
  improved efficacy of with caffeine, 167
  methylxanthine use in, 1

Encoding efficiency, enhancement of performance due to caffeine, 38
Endocrine responses, caffeine and, 113
  adenosine effects, 114
Endogenous arousal cycle, 14
Endoplasmic reticulum, release of intracellular calcium due to caffeine, 4
Endorphin release, due to chocolate, 205, 212
Energy drinks, age trends in intake of, 87
Enkephalin receptors, expression of A_{2A} receptors in, 3
Epicatechins, in green tea, 197
Epilepsy, caffeine and, 167
Epinephrine
  effect of caffeine on secretion of, 117
  elevated plasma levels of due to caffeine, 6
Ergotamine, methylxanthine in combination with for migraines, 1
Estrogen
  in coffee constituents, 192
  inhibition of caffeine metabolism by, 152
Eudistomins, 5
Event related potentials (ERPs), effects on of caffeine across age groups, 93
Excitotoxic neuronal injury, protection against by adenosine receptor blockade, 153
Exercise, effects of caffeine on blood pressure following, 120
Expectancy, influence on reporting of caffeine-related symptoms, 26
Explicit memories, 59

F

Fatigue, 13
  decrease in due to slow-release caffeine, 22
Fertilization, effects of caffeine prior to, 101
Feruloylquinide, 189
Fetal cerebral weight, effect of caffeine on, 100
Filter tasks, 62
Flavonoids, 199
Fluoxetine, caffeine consumption and, 77
Focused attention, effect of caffeine on, 62
Food
  allergy, 221
  caffeine from, 25
  cravings, 206
    bingeing and, 213
Free fatty acids, increase of due to caffeine, 6
Free recall, effect of caffeine on, 60

G

G-aminobutryic acid. See GABA
GABA, 2
  -ergic neurons, effect of A_{2A} receptors on, 3
  A_{2A} receptor stimulation of GABAergic striatopallidal neurons and increase in, 156
  effects of caffeine on, 5
  implication of in the neurobiology of depression, 75

increase in receptor complex due to coffee consumption, 191
methylxanthine-induced seizures, 167
Gastric lesions, development of due to maternal caffeine consumption, 106
G_b, g, effect of adenosine receptor blockage on, 3
Generalized anxiety disorder, 42
Genetic mechanisms, influence on behavioral effects of caffeine, 36
Gestation, effect of caffeine during, 97
G_i input, removal of to adenylyl cyclase by caffeine blockade of A_1 receptors, 3
Glial cells, 36
    adenosinergic modulation of, 158
    effect of *in vitro* caffeine on number of, 103
Glucose
    liberation of by the liver as effect of cortisol, 114
    release of insulin due to increased levels in blood, 211
Glutamate release
    facilitation of by A_{2A} receptor stimulation, 158
    seizures and, 167
Glutamatergic projections, 156
Glycine, 2
    -elicited chloride current inhibition by caffeine, 5
    effects of caffeine on receptors, 5
Green tea, 197
    as an iNOS inhibitor, 201
    pharmacological properties of catechins in, 198
Growth retardation, of fetuses due to caffeine exposure, 101
Guilt, association of chocolate with, 207

H

Headaches
    contribution of caffeine to relief of, 175
    due to caffeine withdrawal, 27, 136, 179
    role of dietary constituents as triggers, 219
    treatment of with caffeine, 44
    vascular changes during, 179
Health and Lifestyle Survey, 90
Health Professionals Follow-up Study, 149
Heart. *See also* cardiovascular effects of caffeine
    effects of caffeine on, 118
Heterocyclic ring systems, as selective antagonists, 3
Hippocampal cortex
    seizure activity in, 167
    use of A_{2A} receptors to prevent excitotoxic death of neurons in, 155
Hippocampal neurons, inhibition of GABA receptor-elicited chloride currents by caffeine, 5
Histamine
    as a trigger for headache, 220
    in chocolate, 213
Histologic defects, due to maternal caffeine intake, 102
Histone deacetylase activity, induction of by theophylline, 6
Honolulu Heart Program, 149
Hormonal differences in effects of caffeine, 152
Hot chocolate, caffeine from, 25
HPA axis
    caffeine and stress responses in, 119
    individual differences in, 122
    effect of caffeine on during rest, 114
Huntington's disease, role of A_{2A} receptors in, 155
Hyaluronan secretion, effect of *in vitro* caffeine on, 103
Hyperactive behavior, due to maternal ingestion of caffeine, 104
Hypertension, sensitivity to caffeine of people with, 122
Hypnic "alarm clock" headache syndrome, 27
Hypothalamic paraventricular nucleus. *See* PVN
Hypothalamic-pituitary-adrenocortical axis. *See* HPA axis
Hysteroid dysphoria, self-medication with chocolate for, 206

I

IB-MECA, neuroprotective properties against seizures, 168
Ibuprofen, efficacy of in headaches when combined with caffeine, 177
Implicit memories, 59
Impulsivity
    caffeine and cognitive task performance in subjects with, 58
    caffeine and learning task performance on subjects with, 59
Incidental learning, effect of caffeine on, 59
Incidental verbal memory, performance relation to caffeine, 90
Indirect pathway, A_{2A} receptors in neurons of, 3
Infants
    caffeine intake of Guatemalan, 27
    caffeine withdrawal in, 136
    half-life of caffeine in, 99
    treatment of seizures in, 104
Inflammation, antioxidant properties of tea and, 197
Inosine, A_3-adenosine receptor sensitivity to, 2
Inpatients, caffeine abuse of, 41
Insomnia
    due to high doses of caffeine, 133
    exacerbation of in depressed patients due to caffeine, 43
Instant coffee. *See also* coffee
    feruloylquinides in, 189
Insulin sensitivity, due to elevated plasma levels of epinephrine, 6
Intentional learning, effect of caffeine on, 58
International Headache Society, 219
Intracellular calcium release channels, effect of caffeine on, 2
Inverted U-shaped arousal function, 38
Ion channels
    action on by caffeine, 2
    effect of caffeine on calcium, 4
IP_3, -induced release of calcium by caffeine, 4
Iron
    deficiency anemia, due to maternal coffee intake, 104
    from chocolate, 213
Irregular sleep, 17
Irritation, due to caffeine withdrawal, 27

Index

Ischemic brain injury
    attenuation of by treatment of caffeine and ethanol, 155
    caffeine and, 165
    long-term caffeine use following, 168
    tea phenols and protection from, 199
Ischemic neuronal injury, protection against by adenosine receptor blockade, 153

J

Jenkins sleep questionnaire, 16
Jet-lag, effect of caffeine on, 19

K

Kaldi, 37
Knock-out mice, A_{2A} adenosine receptor, 3

L

L-type calcium channels, inhibition of by caffeine, 5
Lactation, effect of caffeine during, 97, 102
Lateral amygdala, 36
Laterodorsal core, 139
Learning, effects of caffeine on, 38, 58
Leisure World Cohort, 149
Lethargy, due to caffeine withdrawal, 27
Lifestyle choices, effect on cognitive aging, 86
Light sleep, 14
Lipoxygenase, 200
Lithium, increased excretion of due to caffeine, 44, 79
Locomotor activity, changes in due to prenatal exposure to caffeine, 102
Logical reasoning, improvement of due to caffeine, 57
Long-term memory, 59
Low back pain, influence of caffeine use on, 16

M

Maastricht Aging Study, 90
Magnesium, in chocolate, 205
    effect on premenstrual syndrome, 208
Maintenance of wakefulness test (MWT), 24
Major depressive disorder, 42. *See also* depression
Malignant hyperthermia, caffeine as a diagnostic tool for, 1
Malnutrition, and caffeine consumption in pregnant women, 105
Malondialdehyde, 200
Manic symptoms, increase in due to caffeine, 78
Manic-depressive illness. *See* bipolar disorder
Mate, methylxanthine in, 73
Maternal caffeine consumption
    effect on developing fetus, 97
    effect on placental blood supply in final trimester, 101
    effects on brain development, 100
Medioventral shell, 139
Medulla, 36
Melatonin levels, caffeine and, 19
Memory
    -search paradigm, 61
    effects of caffeine on, 35, 38, 59
    load, 61
    span, 62
Menstrual cycle
    chocolate craving throughout, 208
    migraines during, 224
Mental fatigue, 64
    attenuation of with caffeine, 92, 94
Mental illness, global burden of, 74
Mental performance. *See also* cognitive functions
    effect of caffeine on, 65
Mental stressors, 120
Mesocorticolimbic dopamine transmission system, deficiency in related to depressive disorders, 75
Mesolimbic dopamine system, role in drug dependence of, 139
Mesopontine area, 36
Metabolic body weight, 99
Metabolism, capacity-limited of caffeine, 19
Methylxanthines, 1
    consumption, 73
    in chocolate, 205, 209
    proconvulsant effects of, 167
    tolerance in animals, 136
    vasodilation due to, 180
Migraines
    caffeine in analgesics used for, 176
    cerebral blood flow changes in, 179
    chocolate as a trigger for, 220
        studies, 222
    efficacy of caffeine in the treatment of, 178
    methylxanthine in combination with ergotamine for, 1, 178
    role of dietary constituents as triggers, 219
    treatment of with caffeine, 44
Mild cognitive impairment (MCI), Apo-E4 role in, 86
Mirtazapine, 75
Monoamine oxidase, 221
Monoamine oxidase inhibitors (MAOIs), 75
Monoaminergic systems, abnormalities found in depressive states, 75
Monotonous tasks, benefit of caffeine for performance of, 19
Mood
    disorders, 74
    effects of caffeine on, 17, 26, 35, 39, 76
    effects of slow-release caffeine on, 23
    influence of phenylethylamine on, 209
    instruments used to evaluate effect of caffeine on, 76
    link to chocolate craving, 206
Morphine-induced analgesia, inhibition of with coffee extract, 189
Motor processes
    effect of caffeine on, 56
    maternal caffeine consumption and development of newborns, 105

MPTP
    model of Parkinson's disease, 152
    toxicity and A_{2A} antagonists, 154, 158
MSLT, assessment of sleepiness with, 19
MTX. See methylxanthine
Mu opioid receptor antagonists, 189
Muscarinic cholinergic receptors, increase in due to noncaffeine constituents of coffee, 187
Muscimol, 5
Myelin formation, effect of caffeine on, 103

N

N-methyl-D-aspartate. See NMDA receptors
NADH, Parkinson's disease and, 149
Nalmefene, 188
    decrease in caloric intake due to, 212
Naloxone, decrease in cravings for sweets due to, 212
Naltrexone, 188
    reduction in sucrose preference due to, 212
Naps
    combined with caffeine to counteract fatigue, 21
    slow-release caffeine and, 22
Neonatal apnea, methylxanthines as treatment for, 1, 103, 167
Nestlé Visual Analog Mood Scale (NVAMS), 76
Neural-tube closure, delay in due to maternal caffeine consumption, 101
Neurocognitive measures, caffeine dose-related trends, 90
Neurodevelopment, effects of caffeine on, 98
Neuroendocrine stress responses, 18
Neuroepithelium evagination, acceleration of into telencephalic vesicles in fetal mice, 101
Neurogenesis, delay due to maternal caffeine consumption and malnutrition, 106
Neuroinflammation, A_{2A} receptor activation and reduction of, 192
Neuronal cells
    aggravation of damage by caffeine or theophylline, 165
    release of intracellular calcium due to caffeine, 4
Neuropeptide Y (NPY), implication of in the neurobiology of depression, 75
Neurophysiological expressions, influence on behavioral effects of caffeine, 36
Neurotransmitter receptor hypothesis, 75
Niacin
    as an element in treatment of migraines, 179
    inverse association of with the risk of Parkinson's disease, 149
Nicotinamide adenine dinucleotide. See NADH
Nicotine use
    dopamine release in the shell of nucleus accumbens due to, 139
    in patients with schizophrenia, 78
    rarity of Parkinson's disease in smokers, 148
Night shift-work, benefit of caffeine for performance of, 19
Nigrostriatal dopaminergic neurons, effect of caffeine on the demise of, 152
Nitric oxide, in ischemia/reperfusion-induced cell damage, 200

NMDA receptors
    decrease in activation due to activation of A_1 receptors, 166
    interactions with adenosine A_1 receptors, 3, 36
Nocturnal sleep, shortage of, 13
Non-REM sleep. See NREM sleep
Noradrenaline, effect of caffeine on, 19
    tolerance to, 137
Norepinephrine
    -related vasoactive amines as a trigger for migraine, 220
    abnormalities found in depressive states, 75
    and dopamine reuptake inhibitor (bupropion), 75
    in patients with autonomic failure, 117
NREM sleep, 14
Nucleus accumbens
    A_{2A} receptors in, 3
    drugs of abuse and, 139
Nucleus acubens, 36
Nucleus caudatus, A_{2A} receptors in, 3
Nurses' Health Study, 149
Nutritional
    products, methylxanthine in, 73
    status of pregnant women and caffeine effects, 105

O

Obsessive-compulsive disorder, 42
    caffeine consumption in patients with, 77
Octopamine, in chocolate, 213
Oddball paradigm, 38
    tasks, 63
Oolong tea, 198
Opioid receptor antagonists
    food intake and, 212
    reduction of craving by, 188
Oral contraceptives, delayed caffeine clearance due to, 92, 152
Over-the-counter medications, caffeine from, 25

P

P1 receptors. See adenosine receptors
P2 receptors. See ATP
P300 waves, association with attention, 38
P3b brain potential, latency of after caffeine, 61
Pain, effect of caffeine on, 44
Paired-associate learning, effect of caffeine on, 58
Palatable foods, impact of, 207
Pancreatic b-cells, inhibition of L-type calcium channels in by caffeine, 5
Panic disorder, 42
Parasomnias, exacerbation of in depressed patients due to caffeine, 43
Paraxanthine, 1
    accumulation of in the fetal brain, 102
    attenuation of MPTP toxicity with, 153
    inhibition of adenosine receptors by, 3
    ratio of to caffeine in increased doses, 19

Index

Parkinsonism
    case-control studies for risk factors, 148
    cohort studies for risk factors, 149
    dopaminergic neuronal degeneration in, 147
    MPTP model of, 152
    neuroprotective effects of caffeine in, 36
    reduced risk of due to coffee consumption, 192
    therapeutic target for caffeine, 1
Perception, effects of caffeine on, 55
Perinatologic risks, due to maternal caffeine consumption, 101
Periodic limb movement disorder (PLMD), caffeine intake and, 17
Pharmaceutical products, methylxanthine in, 73
Phenosulphotransferase P activity, in migraine sufferers, 221
Phenothiazines, effect of caffeine on oral doses of, 79
Phenylethylamine, in chocolate, 205, 209
Phosphodiesterases
    blockade of by caffeine, 2, 115
    inhibition of by caffeine, 4, 76
Pituitary gland, stimulation of POMC in, 114
Placebos, influence on caffeine-related symptoms, 26
Placenta, effect of caffeine on weight of, 100
Plasma levels
    assessment of in studies of caffeine intake, 54
    elevation of epinephrine due to caffeine, 6
    increase in volume due to cortisol release, 114
Polyphenols in tea, 198
POMC, cleavage of stimulated by CRH, 114
POMS, 76
    mood testing, 19, 39
Pontine inhibition of the brain stem, maternal caffeine intake and increase in, 98
Positive affect, increase in due to caffeine, 39
Positive Affect Negative Affect Schedule, 39
Postdural puncture headache, analgesic potency of caffeine in treatment of, 177
Postmenopausal hormones, interaction with caffeine, 152
Posttraumatic stress disorder, 16, 42
Potassium channels, activation of by adenosine A_1 receptors, 3
Potent antagonists, development of, 3
Prednisolone, caffeine and administration of, 115
Prefrontal cortex, 139
Pregnancy
    delayed caffeine clearance due to, 92
    effect of caffeine consumption during, 97, 100
Preoptic nucleus of the hypothalamus, 36
Pro-opiomelanocortin. *See* POMC
Processing speed, increase in due to caffeine, 38
Protein content, effect of caffeine on neonatal, 100
Protein malnutrition, 105
Psychological functioning, caffeine effects on, 35
Psychomotor functions, effect of caffeine on, 35, 37
Psychomotor stress, effect of caffeine combined with, 120
Psychotogenic effects of caffeine, in schizophrenics, 43
Psychotropic medication, caffeine and, 78
Pulse rate, expectancy and placebo effects on, 26
Putamen, A_{2A} receptors in, 3
PVN, initiation of HPA cascade at, 114

Q

Quinides
    as antagonists of the mu opioid receptor, 189
    in roasted coffee, 188
Quinolinic acid, 166

R

Rapid eye movement stage of sleep. *See* REM stage of sleep
Rapid Information Processing (RIP) task, 63
Rats, half-life of caffeine in, 99
Reaction times, effect of caffeine on, 35, 37, 56
Recall, effects of caffeine on, 60
Recognition memory
    delayed, effect of caffeine on, 60
    enhancement of performance due to caffeine, 38
Regular sleep, 17
Reinforcing effects of caffeine, 138
Relaxation, as a motive for coffee consumption, 16
REM stage of sleep, 14
Renin, increase of due to caffeine, 6
    tolerance to, 137
Reperfusion-induced brain injury, tea phenols and, 199
Response accuracy, effect of caffeine on, 56
Restless legs syndrome, 43
    caffeine intake and, 17, 36
Restlessness, due to high doses of caffeine, 133
Restrained eaters, 207
Reverse placebo effect, 28
Reversible monoamine oxidase inhibitors (RIMAs), 75
RO15-1788 benzodiazepine antagonist, 5
Roasted coffee. *See also* coffee
    antimorphine activity of, 188
Ryanodine-sensitive
    channel, 4 (*See also* cyclic ADP-ribose sensitive calcium release channels)
    receptor, 4

S

Saliva, levels of caffeine in, 25, 54
Sanders' Additional Factor Method (AFM), 55
Sarcoplasmic reticulum, release of intracellular calcium due to caffeine, 4
Schizophrenia
    caffeine consumption in patients with, 78
    effects of caffeine on, 36, 43
Seasonal affective disorder
    caffeine consumption in patients with, 77
    caffeine use and, 43
    carbohydrate-rich foods and, 212
Second messenger cyclic adenosine monophosphate, 76
Seizures
    chronic caffeine treatment and decrease in, 168
    prolonging of with caffeine, 167
Selective antagonists, development of, 3
Selective attention
    effect of caffeine on, 62

enhancement of due to caffeine, 38
Selective norepinephrine reuptake inhibitors (SNRI), 75
Selective serotonin reuptake inhibitors (SSRIs), 75
Self-focus, increased ability due to caffeine, 38
Self-medication theory, 77
Semantic memory, enhancement of performance due to caffeine, 38
Sensation, effects of caffeine on, 55
Serial learning, effect of caffeine on, 58
Serial position effect, 60
Serotonin
 -related vasoactive amines as a trigger for migraine, 220
 abnormalities found in depressive states, 75
 carbohydrate intake and synthesis of, 210
 increase in due to caffeine, 76
Serotonin and norepinephrine reuptake inhibitors (SNRIs), 75
Serotonin production, increase in due to chocolate, 205
Sexual functioning, improvement in veterans with PTSD due to reduction of caffeine, 42
Shift-work
 benefit of caffeine for night workers, 19
 consumption of caffeine by workers, 20
 slow-release caffeine and, 22
Signal transduction pathway dysfunction of monoamine receptors, in depressive disorders, 75
Simple reaction time
 performance relation to caffeine, 90
 use of in caffeine studies, 56
Sleep, 14
 adenosine and, 114
 correlation of coffee consumption with increased latency of, 17
 deprivation, 13
  effect on driving of, 18
  maintenance of cognitive alertness under conditions of, 20
  slow-release caffeine and, 22, 28
  studies, 24
 effect of caffeine on, 137
 objective and reported parameters, 26
 schedules, 17
 Ultradian cycles during, 15
Sleep onset latencies (SOL), 13
Sleep-impairing disorders, caffeine intake and, 17
Slow-release caffeine (SRC), 22, 37
 optimal dose of, 23
Slow-wave sleep (SWS), 14
Smoking. See nicotine use
Snoring, 16
Social phobia, 42
Social stressors, 120
Soft drinks, caffeine from, 25, 134
Somatostatin, implication of in the neurobiology of depression, 75
Speeded decision making, improvement of due to caffeine, 57
Spinal cord, adenosine receptors in, 3
Stanford Sleepiness Scale (SSS), 76
State-Trait Anxiety Inventory (STAI), correlation of caffeine use with, 78

Stimulation, as a motive for coffee consumption, 16
Strategy use, effect of aging on, 86
Stress
 combined effect with caffeine, 113
  individual differences in, 122
 mental and cognitive stressors, 120
 neuroendocrine responses, 18
 relief, as a motive for coffee consumption, 16
Striatal dopamine, attenuation of MPTP-induced loss, 154
Striatal dopaminergic reward system, 36
Striatopallidal neurons, blockade of postsynaptic A_{2A} receptors on, 156
Striatum, 36
Stroop Color and Word Test, 62, 90
 stress responses due to, 120
Students, caffeine intake of, 25
Studies, caffeine effects on mental performance, 54
Subject-controlled search, 63
Substance abuse
 definition of, 41
 early-age use of caffeine in patients with disorders related to, 78
Substitution, effect of caffeine on tasks requiring, 57
Subthalamic nucleus, enhanced release of glutamate from, 157
Suckling behavior, maternal caffeine consumption and emergence of, 105
Sudden infant death syndrome (SIDS), association of maternal caffeine consumption with, 98
Suicide, 74
 relationship of caffeine intake to risk of, 78
Surgery, risk of caffeine withdrawal headache surrounding, 179
Sustained attention, effect of caffeine on, 63
Sympathetic functions, effect of caffeine on, 114
Sympathetic nervous system
 cortisol and function of, 115
 dependence of caffeine-related catecholamine increase on state of, 117
Synaptoneurosomes, inhibition of chloride flux in by caffeine, 5

T

Tachycardia, due to high doses of caffeine, 133
Target detection, effect of caffeine on, 61
Task load, 61
Taurine, 77
Tea
 age trends in intake of, 87
 antioxidant properties of, 197
 caffeine from, 25, 134
 consumption of in Western society, 85
 effect on mood of, 77
 general protective properties of, 198
 methylxanthine in, 73
 neuroprotective effects of, 199
 origin of, 197
 polyphenols in, 198
  mechanisms of, 199

Telencephalic evagination, 101
Temperature, differing effects of caffeine depending upon, 37
Tension headaches
    caffeine in analgesics used for, 176
    efficacy of caffeine in treatment of, 177
Thea sinensis L., 197
Theobromine, 1
    accumulation of in the fetal brain, 102
    forms of, 73
    in chocolate, 205, 210, 221
Theophylline, 1, 180
    accumulation of in the fetal brain, 102
    adenosine receptor sensitivity to, 2
    adverse effects of, 167
    aggravation of neuronal damage by, 165
    attenuation of MPTP toxicity with, 153
    forms of, 73
    inhibition of adenosine receptors by, 3
    inhibition of phosphodiesterases by, 4
Therapeutic targets for caffeine, 1
Thromboxane A_2, 200
Time on task (TOT) effects, 63
Tobacco use
    delayed caffeine clearance due to, 92
    in schizophrenics, 78
Tonic activation, of adenosine receptors, 2
Total sleep time (TST), 15
Traffic accidents, due to fatigue, 13
Trazodone, 75
Tricyclic antidepressants, 75
    effects of caffeine on, 44
Tryptophan
    as an element of treatment for migraines, 179
    increase in due to carbohydrate intake, 211
    increase in serotonin synthesis due to, 205
Tuberculum olfactorium, A_{2A} receptors in, 3
Type A behavior, effect on sleep of, 17
Tyramine, as a trigger for headache, 220

U

Ultradian cycles of sleep, 15
Unipolar depression, food sensitivity in, 221

V

Vanderbilt Institute for Coffee Studies, 189
Vascular resistance, increase in due to caffeine, 118
Vasoactive amines in food, as triggers for migraine, 220
Verbal memory, enhancement of performance due to caffeine, 38
Vigilance performance, effects of caffeine on, 35, 37, 63
Visual Analog Mood Scale (VAMS), 76
Visual luminescence threshold, effect of caffeine on, 55
Visual search task, 63
Visuospatial reasoning, performance relation to caffeine, 90

W

Wakefulness, 13
White chocolate, 213
Withdrawal effects of caffeine, 16, 18, 27, 135
    on mood responses, 77
Working memory, 59
    effect of caffeine on in high-display load conditions, 61
    enhancement of performance due to caffeine, 38
World Health Organization's Global Burden of Disease Survey, 74

X

Xanthines, 166
    adenosine receptor sensitivity to, 2

Y

Yerkes-Dodson Law, 91

Z

Zeitgebers, 14
Zinc, decrease in fetal brain due to maternal caffeine consumption and malnutrition, 106